AF684483

PROTOZOAL ABORTION IN FARM RUMINANTS: GUIDELINES FOR DIAGNOSIS AND CONTROL

Protozoal Abortion in Farm Ruminants: Guidelines for Diagnosis and Control

Edited by

Luis M. Ortega-Mora

Universidad Complutense de Madrid, Departamento de Sonidad Animal, Ciudad Universitaria, Madrid, Spain

Bruno Gottstein

Institute of Parasitology, University of Bern, Bern, Switzerland

Franz J. Conraths

Friedrich-Loeffler-Institut, Federal Research Institute for Animal Health, Institute for Epidemiology, Wusterhausen, Germany

and

David Buxton

Moredun Research Institute, Division of Virology, Pentlands Science Park, Edinburgh, UK

www.cabi.org

CABI is a trading name of CAB International

CABI Head Office
Nosworthy Way
Wallingford
Oxfordshire OX10 8DE
UK

Tel: +44 (0)1491 832111
Fax: +44 (0)1491 833508
E-mail: cabi@cabi.org
Website: www.cabi.org

CABI North American Office
875 Massachusetts Avenue
7th Floor
Cambridge, MA 02139
USA

Tel: +1 617 395 4056
Fax: +1 617 354 6875
E-mail: cabi-nao@cabi.org

© CAB International 2007. All rights reserved. No part of this publication may be reproduced in any form or by any means, electronically, mechanically, by photocopying, recording or otherwise, without the prior permission of the copyright owners.

A catalogue record for this book is available from the British Library, London, UK.

A catalogue record for this book is available from the Library of Congress, Washington, DC.

ISBN-13: 978 1 84593 211 4

Typeset by Columns Design Ltd, Reading, UK
Printed and bound in the UK by Athenaeum Press, Gateshead

Contents

Contributors

Dr Gorka Aduriz, *Instituto Vasco de Investigación y Desarrollo Agrario (NEIKER), Departamento de Sanidad Animal, Berreaga 1, 48160 Derio, Vizcaya, Spain. Tel: +34 94 403 43 00; Fax: +34 94 403 43 10. gaduriz@neiker.net*

Dr Gema Alvarez-García, *Universidad Complutense de Madrid, Departamento de Sanidad Animal, Facultad de Veterinaria, Ciudad Universitaria, 28040 Madrid, Spain. Tel: +34 91 3943844; Fax: +34 91 3943908. gemaga@vet.ucm.es*

Dr Raquel Atxaerandio, *Instituto Vasco de Investigación y Desarrollo Agrario (NEIKER), Departamento de Sanidad Animal, Berreaga 1, 48160 Derio, Vizcaya, Spain. Tel: +34 94 403 43 00; Fax: +34 94 403 43 10. ratxaerandio@neiker.net*

Dr Chris Bartels, *Animal Health Service, PO Box 9, 7400 AA Deventer, The Netherlands. c.bartels@gdvdieren.nl; cjm.bartels@planet.nl*

Dr Camilla Björkman, *National Veterinary Institute and Swedish University of Agricultural Sciences, Department of Parasitology (SWEPAR), 751 89 Uppsala, Sweden. Tel: +46 18 67 1778; Fax: +46 18 67 3575. camilla.bjorkman@idmed.slu.se*

Dr David Buxton, *Moredun Research Institute, Division of Virology, Pentlands Science Park, Bush Loan, Edinburgh EH26 0PZ, UK. Tel: +44 (0)131 445 5111; Fax : +44 (0)131 445 6111. david.buxton@moredun.ac.uk*

Dr Carlos Campero, *Instituto Nacional de Tecnología Agropecuaria, C.C. 276-(7620), Balcarce, Argentina. Tel: 54–2266–439120; Fax: 54–2266–439101. ccampero@balcarce.inta.gov.ar*

Dr Franz J. Conraths, *Friedrich-Loeffler-Institut, Federal Research Institute for Animal Health, Institute for Epidemiology, Seestrasse 55, 16868*

Wusterhausen, Germany. Tel: +49 33 979 80 176; Fax: +49 33 979 80 200. franz.conraths@fli.bund.de

Dr Thomas Dijkstra, *Animal Health Service, PO Box 9, 7400 AA Deventer, The Netherlands. t.dijkstra@gdvdieren.nl*

Dr Ignacio Ferre Pérez, *Universidad Complutense de Madrid, Departamento de Sanidad Animal, Facultad de Veterinaria, Ciudad Universitaria, 28040 Madrid, Spain. Tel: +34 91 3943844; Fax: +34 91 3943908. iferrepe@vet.ucm.es*

Dr Ana García-Pérez, *Instituto Vasco de Investigación y Desarrollo Agrario (NEIKER), Departamento de Sanidad Animal, Berreaga 1, 48160 Derio, Vizcaya, Spain. Tel: +34 94 403 43 00; Fax: +34 94 403 43 10. agarcia@neiker.net*

Dr Bruno Gottstein, *Institute of Parasitology, University of Bern, Länggasss-Str. 122, 3001 Bern, Switzerland. Tel: +41 31 631 2418; Fax: +41 31 631 2622. bruno.gottstein@ipa.unibe.ch*

Dr Michael Haessig, *University of Zürich, Herd Health, Department of Farm Animals, Winterthurerstrasse 260, 8057 Zürich, Switzerland. Tel: +41 44 635 82 60; Fax: +41 44 635 89 04. mhaessig@vetclinics.unizh.ch*

Dr Anja R. Heckeroth, *Intervet Innovation GmbH, Zur Propstei, 55270 Schwabenheim, Germany. Anja.Heckeroth@intervet.com*

Dr Klaus Henning, *Friedrich-Loeffler-Institut, Federal Research Institute for Animal Health, Institute for Epidemiology, Seestrasse 55, 16868 Wusterhausen, Germany. Tel: +49 33 979 80 156; Fax: +49 33 979 80 222. Klaus.henning@fli.bund.de*

Dr Ana Hurtado, *Instituto Vasco de Investigación y Desarrollo Agrario (NEIKER), Departamento de Sanidad Animal, Berreaga 1, 48160 Derio, Vizcaya, Spain. Tel: +34 94 403 43 00; Fax: +34 94 403 43 10. ahurtado@neiker.net*

Dr Elisabeth A. Innes, *Moredun Research Institute, Division of Parasitology, Pentlands Science Park, Bush Loan, Edinburgh EH 26 0PZ. Tel: +44 (0)131 445 5111; Fax: +44 (0)131 445 6111. lee.innes@moredun.ac.uk*

Dr Fredi Janett, *University of Zürich, Department of Farm Animals, Winterthurerstrasse 260, 8057 Zürich, Switzerland. Tel: +41 44 63 58218. fjanett@vetclinics.unizh.ch*

Dr Laura Kramer, *Dipt. Di Produzione Animale, Facolta di Medicina Veterinaria, Via del Taglio 8, 43100 Parma, Italy. Tel: +31 0521 902 776; Fax: +39 0521 902 770. kramerlh@unipr.it*

Dr Bertrand Losson, *Faculty of Veterinary Medicine and Parasitology, University of Liège, BD. DE Colonster, 20 4000 Liège, Belgium. Tel: +32 4 366 4090; Fax: +32 4 366 4097. blosson@ulg.ac.be*

Dr Stephen Maley, *Moredun Research Institute, Division of Virology, Pentlands Science Park, Bush Loan, Edinburgh EH26 0PZ, UK. Tel: +44 131 445 5111; Fax: +44 131 445 6111. stephen.maley@moredun.ac.uk*

Dr Jens Mattsson, *National Veterinary Institute and Swedish University of Agricultural Sciences, Department of Parasitology (SWEPAR), 751 89 Uppsala, Sweden. jens.mattsson@sva.se*

Dr Norbert Müller, *Institute of Parasitology, Lànggass-strasse 122, University of Bern, 3012 Bern, Switzerland. Fax: +41 316312622. nmueller@ipa.unibe.ch*

Dr Luis M. Ortega-Mora, *Universidad Complutense de Madrid, Departamento de Sanidad Animal, Facultad de Veterinaria, Ciudad Universitaria, 28040 Madrid, Spain. Tel: +34 91 3944069; Fax: +34 91 3943908. luis.ortega@vet.ucm.es*

Dr Juana Pereira-Bueno, *Departamentto de Sanidad Animal, Facultad de Veterinaria, Universidad de León, 24071 León, Spain. dsajpb@ isidoro.unileon.es*

Dr Heinz Sager, *Institute of Parasitology, University of Bern, Länggasss-Str. 122, 3001 Bern, Switzerland. Tel: +41 31 631 24 75; Fax: +41 31 631 26 22. heinz.sager@ipa.unibe.ch*

Dr Gereon Schares, *Friedrich-Loeffler-Institut, Federal Research Institute for Animal Health, Institute for Epidemiology, Seestrasse 55, 16868 Wusterhausen, Germany. Tel: +49 33 979 80 193; Fax: +49 33 979 80 222. Gereon.schares@fli.bund.de*

Dr Smaragda Sotiraki, *Veterinary Research Institute, Ionia 57008, Thessaloniki, Greece. Tel: +30 2310 781701; Fax: +30 2310 781161. sotiraki@vri.gr; snso@kvl.dk*

Dr Astrid Tenter, *Institut für Parasitologie, Tierärztliche Hochschule Hannover, Bünteweg 17, 30559 Hanover, Germany. Tel: +49 511 953 87 17; Fax: +49 511 953 88 70. astrid.tenter@tiho-hannover.de*

Dr Rico Thun, *University of Zürich, Department of Farm Animals, Winterthurerstrasse 260, 8057 Zürich, Switzerland. Tel: +41 44 635 82 09; Fax: +41 44 635 89 03. rthun@vetclinics.unizh.ch*

Dr John Williams, *Conseiller Scientifique/Science Officer, Agriculture, Biotechnology and Food Science, COST-ESF, 147 Avenue Louise, 1050 Brussels, Belgium. jwilliams@cost.esf.org*

Dr Willem Wouda, *Animal Health Service, PO Box 9, 7400 AA Deventer, The Netherlands. w.wouda@gdvdieren.nl*

Dr Steve Wright, *Moredun Research Institute, Division of Parasitology, Pentlands Science Park, Bush Loan, Edinburgh EH26 0PZ, UK. Tel: +44 131 445 5111; Fax: +44 131 445 6111. steve.wright@moredun.ac.uk*

Preface

This book is one of the major results of COST Action 854. Formed in March 2002 with the objective of developing strategies to control reproductive diseases caused by protozoa in farm ruminants, it has successfully reunited a body of European experts to concentrate on this task. The guidelines for diagnosis contained in this book constitute an important step forward, and COST is pleased to be associated with its publication and hopes that veterinary practitioners in every part of Europe will use it in their routine work.

Benefit will accrue to farmers, as early diagnosis frequently leads to successful treatment at a lower cost and with lower or zero economic loss. The consumer also stands to gain, since some of the diseases are zoonotic. Early diagnosis and treatment reduces the risk of spreading the disease to man. Finally, while we may debate on the degree of suffering and discomfort that animals suffering from such diseases may experience, few would disagree that early diagnosis and treatment is one of the first and most important contributions that can be made to animal welfare overall.

A COST Action is essentially a network of scientists formed to facilitate collaboration between scientists of any of the COST member states. This network then links nationally funded research projects and permits scientific meetings, both large and small, to be organized as appropriate. This Action ended in June 2006 after 4 years' collaborative research that has involved 19 European countries. A fund of approximately €300,000 provided by the European Union will have been spent on this Action using the tools developed by COST to encourage cooperation. Besides scientific meetings, one of the most popular and effective tools is the short-term scientific exchange grants that enable scientists to work in a partner laboratory for up to 3 months. More often than not, the beneficiaries of these grants are doctoral or post-doctoral students gaining their first experience of working in another European country.

COST was the first attempt to link up European scientists in a common effort to work towards a common goal. It was created by a ministerial conference in 1971 and was, and still is, an intergovernmental initiative. It has grown and adapted itself over the years, but remains true to a few, simple rules that its founders shrewdly agreed on. First, there is no fixed programme: COST provides an open door for any idea or project devised by the scientists themselves representing a minimum of five member states. Secondly, there is an open invitation to the other member states to join in, but there is no obligation. A COST Action grows according to needs. And lastly, in the words of the late Hubert Curien, one of the founding fathers, 'COST works quickly, and that is very, very important'.

The concept of a European Research Area, or ERA, was formulated by Philippe Bousquin in 2001 when, as EC commissioner for research and development, he launched the sixth Framework Programme. The ERA includes the EU's Framework Programme, but also all of the nationally funded research programmes and projects carried out by the member states. This nationally funded research represents the major part of our research effort – estimates vary from 70 to 95% – but little of it benefits from any coordination.

This is precisely where COST fits into the ERA. By providing an open door, with flexible rules, very few constraints and no preconceived ideas, similar nationally funded projects can be networked and the result is synergistic. COST produces a multiplier effect, where the overall result is simply more than the simple sum of each national project. This book is a good example of COST synergy. Without COST, it is doubtful if it would have been published, and certain that it could not have been done so within a 4-year timeframe. Each contributor would have made individual and incremental advances in the diagnosis of protozoal diseases but this knowledge assembled here, and then disseminated across member states and worldwide, will have a far greater impact than the sum of each individual progress.

This book, like a COST Action, is a collective effort and many scientists have devoted hours of work to produce it. We appreciate their efforts, and if cows, sheep and goats could applaud, they would be on two feet clapping loudly! In conclusion, I should like to add a special word of thanks to the chair of this Action, Dr Franz J. Conraths, whose vision, perseverance and commitment were at the origin of this COST Action and whose careful, thoughtful and intelligent guidance have contributed in no small way to its success, epitomized by this book.

J. Williams
Brussels, June 2006

Introduction

The farming of cattle, sheep and goats forms the core of livestock agriculture in Europe and many other parts of the world. Each year, the farm ruminant industry suffers major economic losses from reproductive dysgenesis, an all-inclusive term indicating reproductive failure, regardless of cause and regardless of when these losses have occurred in the gestational period. Embryonic mortality and fetal death, two of the most important forms of reproductive dysgenesis, result not only in the loss of offspring and an increased calving (or lambing/kidding) interval, but in other other factors such as increased culling, reduced milk production and the reduced value of breeding stock, which also need to be taken into account.

Abortion, the expulsion before full term of a conceptus that is incapable of independent life, is a unique challenge for both clinical practitioners and veterinary laboratories. In most cases there are few, if any, clinical signs prior to abortion, macroscopic lesions are non-specific and identification of the potentially many causative agents requires special laboratory procedures. In a large percentage of cases the cause cannot be determined, but in the majority of diagnosed cases an infectious agent is identified and found to be responsible.

Abortion in ruminants may also involve a very considerable public health risk, as many of the pathogens responsible can pose a significant danger to humans. Thus rapid, accurate diagnosis is vital in order to be able to assess the degree of risk caused by potential ruminant abortifacients with zoonotic potential such as *Toxoplasma gondii*, *Chlamydophila abortus* (*Chlamydia psittaci*), *Coxiella burnetii*, *Listeria monocytogenes*, *Salmonella* spp., *Campylobacter* spp. and *Brucella* spp.

While *Neospora caninum*, *Tritrichomonas fetus* and *Sarcocystis* spp. are not currently considered to be zoonotic, rapid, accurate diagnostic methods for protozoal causes are essential in order to rule in or out more dangerous pathogens and to allow meaningful risk assessments.

The two principal agents causing protozoal abortion in ruminants are *N. caninum* in cattle and *T. gondii* in sheep and goats, and both are closely related. *Tritrichomonas fetus* is a serious cause of early pregnancy loss in cattle, and parasites of the genus *Sarcocystis* are widely distributed and may inflict infections affecting the reproductive tract of ruminants, which may compromise pregnancy. Protozoa of the genus *Hammondia* are not considered as being a cause of reproductive failure, but may play a role in the differential diagnosis due to their close relationship to *N. caninum* and *T. gondii*.

Neospora caninum is an important cause of infectious abortion and stillbirth in cattle worldwide. Infection is common, serological surveys suggesting that from 5–60% of cattle are seropositive, and the parasite is frequently passed from mother to calf (vertical transmission) with no signs of disease. Disease occurs when *N. caninum* multiplies in the developing calf and its placenta and causes sufficient damage to trigger abortion or stillbirth. Research suggests that if the mother passes infection to the fetus early in gestation, it is more likely to be fatal to the conceptus than if infection is passed later in gestation. However, it also appears that infection is more likely to be transmitted in late rather than early pregnancy.

Thus, the majority of vertically transmitted infections are not fatal, and in this way subclinical infections are maintained in a herd. Vertical transmission appears to be a very important method of spread of this parasite, but the ingestion of oocysts of *N. caninum*, produced by dogs and excreted in their faeces to contaminate food or water (horizontal transmission), is also of considerable significance.

Control of bovine neosporosis is difficult. Pharmaceutical preparations that will kill *N. caninum* are known, but their use in controlling infection and/or disease in cattle does not appear to be a practical option, and no effective vaccine is currently available. Control measures therefore rely on applying certain management strategies which, to date, are only partially satisfactory.

Control of toxoplasmosis also poses problems. *Toxoplasma gondii* is an important zoonotic infection, as well as being a major cause of abortion in sheep and goats. The majority of cases of human toxoplasmosis follow the consumption of uncooked or lightly cooked meat. There is also an added risk from drinking unpasteurized goats' milk, as well as from the ingestion of fruit and vegetables contaminated with soil containing *Toxoplasma* oocysts.

Control depends on preventing a primary infection in the pregnant sheep or goat. While certain management procedures may reduce the risk, elimination is not possible. A live commercial vaccine – for use in sheep – is sold in some EU member states. The availability of a vaccine that protects against abortion in sheep and goats would be considered very likely to reduce, if not prevent, the development of tissue cysts in muscle. This would make the meat (and milk) very much safer for human consumption.

Tritrichomonas fetus, a venereally transmitted bovine infection, is an important cause of pregnancy loss and abortion in naturally bred cattle throughout the world. Trichomonosis has been a List B disease by OIE classification for many years, and is therefore the subject of animal disease control in several countries. Therefore, bovine trichomonosis is one of the diseases included in the EU Directive that regulates the trade in bovine semen. Since the introduction of artificial insemination, the economic importance of this disease has decreased, but there is evidence of its re-emergence in extensive animal husbandry regimes in some European states. As *T. fetus* may also have spread unnoticed among other livestock and companion animals such as pigs and cats, a heightened awareness of the disease with state-of-the-art diagnostic procedures is required.

With a move to more extensive systems of agriculture, the risk of protozoal abortion in ruminants will remain and perhaps attain greater significance. This is the case with bovine trichomonosis, which continues to be an important cause of early pregnancy loss in extensive production systems where natural mating is the rule. In the case of toxoplasmosis in sheep and goats, infection is picked up from contaminated grass, hay and water, as well as from concentrated loose feed. It occurs just as commonly in extensive as intensive farming systems.

With bovine neosporosis, evidence is accumulating to indicate that the incidence of abortion is exacerbated by stress. In some situations, this may occur with more intensive farming methods (such as the feedlot systems encountered in California), while in other cases stress may occur in extensive systems of agriculture due to severe environmental conditions, as with extremes of weather.

Cases of fatal sarcocystiosis occur more frequently when extensively reared animals are moved to grassland nearer the farm or locations visited by people, due to contamination of the ground by dog and cat faeces. It should also be noted that *Sarcocystis hominis* has a life cycle in which humans and other primates are the definitive hosts and cattle the intermediate host. Thus, this parasite is also regarded as being a zoonotic agent.

The eventual control of these infectious conditions will be founded on knowledge of their incidence, which in turn relies on accurate diagnosis of both infection and disease. Accurate diagnosis, in turn, is dependent on all laboratories 'speaking the same diagnostic language', so that results between laboratories in all countries can be readily interpreted and understood.

It is with this goal in mind that this book has been created by the participating members of the EU COST action 854 – 'Protozoal reproduction losses in farm ruminants'. On the following pages, carefully selected methodologies are presented in a simple and practical form, written by experts who daily undertake these tests. It is the aim of this book to enable laboratories, both experienced with and new to these procedures, to be able to carry them out with precision. The ultimate goal is to improve the health and welfare of farm livestock, safeguard human health and reduce the financial risks of farming.

Luis M. Ortega-Mora, David Buxton, Bruno Gottstein
and Franz J. Conraths

Acknowledgements

The editors would like to thank Ms. Cristina Alonso Andicoberry (SALUVET-UCM) for her helpful secretarial work, Mr. Scott Hamilton (Moredun Research Institute, Edinburgh) for his drawing of the *Toxoplasma* and *Neospora* life cycles and Dr. Willem Wouda (AHS, Deventer) for providing histological images for the *Neospora* chapter. Thanks are due to all members of COST-Action 854 for their contributions and to the ESF-COST office and especially to Dr. John Williams for his support.

1

Background to Reproductive Problems in Cattle and Sheep

Rico Thun and Fredi Janett

1.1 Puberty

Puberty is defined in the female as the age at first expressed oestrus accompanied by spontaneous ovulation. The onset of puberty is a gradual process involving maturation of the hypothalamus–hypophyseal–ovarian axis (Kinder *et al.*, 1995). During prepuberty the hypothalamus of heifers and ewe lambs is hypersensitive to oestradiol, so even small amounts of oestrogen are sufficient to suppress gonadotropin-releasing hormone (GnRH) secretion.

When puberty approaches, a reduction in sensitivity (desensitization) of the hypothalamus to oestradiol occurs, resulting in increased frequency of GnRH pulses. This change in responsiveness to oestradiol and the development of positive feedback are known to be the major endocrine factors limiting pubertal onset. The development of larger follicles will produce enough oestradiol to induce behavioural oestrus, the LH surge and ovulation. During transition to sexual maturity, several types of incomplete cycles as short luteal phases, anovulatory oestrus or silent heat are common (Dodson *et al.*, 1988).

The age at which puberty is acquired varies among and within species and is influenced by genetic (gender and breed) and numerous environmental factors (nutritional status, climatic conditions, photoperiod, social interactions). In heifers, the average age of puberty is 10–12 months for dairy and 11–15 months for beef breeds. Ewes reach puberty at around 6–9 months of age. Growth rate with accumulation of a threshold amount of body fat seems to be most important in all female mammals before reproductive cycles can be initiated.

© CAB International 2007. *Protozoal Abortion in Farm Ruminants* (eds L.M. Ortega-Mora, B. Gottstein, F.J. Conraths and D. Buxton)

There is good evidence that some metabolic signals (leptin, glucose, fatty acids) affect the secretion of GnRH (Foster and Nagatani, 1999). Feeding animals a low nutritional plane will delay puberty. Photoperiod will also affect the age of puberty, particularly in ewes that are short-day breeders. Lambs born in the autumn require twice as much time to reach puberty than spring-born lambs, attaining pubertal age during the breeding season (Foster *et al.*, 1986). Introduction of a mature male during the prepubertal period will hasten the onset of puberty in heifers and lambs.

1.2 Oestrous Cycle

Length and Definition of Periods

After puberty, the female enters a period of reproductive cyclicity that continues throughout her life unless interrupted by pregnancy or abnormalities of genital organs. The cow is polyoestrous, having oestrous cycles of about 21 d which occur regularly throughout the entire year. Ewes are seasonally polyoestrous, exhibiting oestrous cycles of about 17 d only during a few months of the year (breeding season). Because they begin to cycle when day length decreases (autumn), they are known as short-day breeders.

The stages of the oestrous cycle are pro-oestrus, oestrus, metoestrus and dioestrus. Depending on the dominant structure present in the ovary, the cycle can be subdivided into follicular and luteal phases. The follicular phase (pro-oestrus and oestrus) is the period from the regression of the *corpus luteum* to ovulation and is dominated by oestradiol. During the luteal phase (metoestrus and dioestrus) the main ovarian structure is the *corpus luteum*, with progesterone as the primary hormone.

Oestrus is considered to be the period of time when the female is receptive to the male (standing heat) and last about 12–18 h in the cow. Behavioural signs of oestrus are mounting, restlessness, licking, butting and head-resting. Typical clinical signs are oedema and hyperaemia of the vulva, secretion of mucus from the cervix and vagina and increased uterine contractility. Oestrus in the ewe lasts about 24–36 h and is relatively inconspicuous in the absence of a ram. Vulvar swelling and mucus discharge may be present.

Metoestrus is the period immediately after ovulation when the *corpus luteum* forms and lasts about 3–4 d. Dioestrus is the period when the *corpus luteum* is fully functional, with a duration of about 14 d in cattle and 10 d in ewes. The period of pro-oestrus, lasting about 3–4 d in the cow and 2 or 3 d in sheep, begins with the regression of the *corpus luteum* and continues to the onset of oestrus.

Follicular Dynamics and Ovulation

Follicular dynamics (the process of growth and regression of ovarian follicles) occurs continuously throughout the entire oestrous cycle. Stimulation and differentiation of primordial follicles to becoming primary follicles capable of ovulation do not depend on gonadotropins, and the whole process may take as long as 60 d (Lussier *et al.*, 1987).

Development of antral follicles in the presence of gonadotropins has been shown to occur in regular patterns of two or three distinct follicular waves (Ginther *et al.*, 1997).

Each wave of follicle growth is characterized by recruitment of a cohort of small follicles. One follicle within the cohort is selected and becomes the largest (dominant) follicle by continuing to grow, while suppressing the remaining follicles (subordinate) and causing them to degenerate. During dominance, increased oestradiol and inhibin secretion reduce FSH and prevent recruitment of a new follicle wave. In the presence of a *corpus luteum* (high-progesterone environment), the selected dominant follicle undergoes atresia because LH secretion is not sufficient to maintain continuous growth.

Follicle waves emerge at about days 2 and 11 or 2, 9 and 16 for animals with two and three follicle waves, respectively (Sirois and Fortune, 1988).

Sheep do not have the distinct, wave-like patterns with strong follicular dominance mechanisms seen in cattle. Selection of the ovulatory follicle seems to be a passive process, allowing the development of multiple ovulatory follicles from the last, and also the second last, follicular wave of the cycle (Gibbons *et al.*, 1999; Driancourt, 2001).

With regression of the *corpus luteum*, any dominant follicle will continue its growth, reaching peak diameters of 16–20 mm in the cow and 6–8 mm in the ewe. High concentrations of oestradiol and low progesterone induce a pre-ovulatory surge of LH and subsequent ovulation. The preovulatory LH peak occurs early in oestrus and induces ovulation about 28 (cattle) and 26 (sheep) h later. Thus, cattle are unique among farm animals in that they ovulate about 10–12 h after termination of oestrus, whereas the ewe normally ovulates prior to or near the end of oestrus.

Formation and Regression of the Corpus Luteum

The pre-ovulatory LH surge causes theca and granulosa cells to differentiate into small and large luteal cells, respectively. Both cell types are steroidogenic, and luteinization is characterized by increased progesterone secretion. Small luteal cells possess most of the LH receptors while the large cells possess nearly all of the receptors for PGE2 and PGF2α (Braden *et al.*, 1988). In ruminants, large luteal cells also produce neurophysin and oxytocin during the cycle, and relaxin during pregnancy.

Under the influence of both luteotropic hormones LH and GH (growth hormone), the *corpus luteum* increases in mass until the middle of the cycle, when its size and function are maximal. During *corpus luteum* formation, local regulators such as growth factors (vascular endothelial and fibroblast growth factors), peptides, steroids and prostaglandins stimulate luteal cell proliferation and support the endocrine luteotropic action. In cattle, the *corpus luteum* can be palpated trans-rectally, but functional status is difficult to ascertain. Ultrasonography has proved more effective, as progesterone concentration in blood is correlated with the diameter of the *corpus luteum.*

Luteolysis is a process whereby the *corpus luteum* undergoes irreversible degeneration characterized by a dramatic drop in blood progesterone concentrations. In ruminants, it has been clearly demonstrated that PGF2α from the uterus acts as luteolysin that reaches the ipsilateral *corpus luteum* through a vascular countercurrent diffusion system between the uterine vein and the ovarian artery. This mechanism is particularly important because more than 90% of PGF2α is denaturated during one circulatory pass through the pulmonary system in the cow and the ewe. Uterine secretion of PGF2α late in dioestrus is dependent on the effects of progesterone, oestrogen and oxytocin on the endometrium (McCracken, 1998).

During the first half of the oestrous cycle (until day 12), progesterone prevents secretion of PGF2α by blocking the formation of oestrogen and oxytocin receptors in the uterus. In addition to its effects on these receptors, progesterone stimulates accumulation of lipid droplets and of cyclooxygenase, the enzyme that converts arachidonic acid to PGF2α (Charpigny *et al.*, 1997).

During the late luteal phase, however, the inhibitory influence of progesterone is lost (desensitization) and the endometrium regains responsiveness to oestrogen stimulation, which results in upregulation of oxytocin receptors. Oxytocin secreted from the *corpus luteum* and posterior pituitary can then interact with its receptor to stimulate the release of uterine PGF2α pulses, which again lead to a rapid release of ovarian oxytocin. Thus, oxytocin and PGF2α stimulate each other in a positive feedback manner until luteolysis is completed.

Besides intracellular mechanisms causing functional (blockade of LH action) and structural (apoptosis) luteolysis, additional factors as blood flow, vasoactive peptides, angiogenic growth factors – as well as inflammatory cytokines – participate in the luteolytic cascade (Schams and Berisha, 2004).

1.3 Pregnancy and Parturition

The Embryonic–Maternal Interaction

Freshly ovulated oocytes are picked up by the ciliated fimbriae of the infundibulum and directed into the ampulla of the oviduct, where fertilization takes place. The bovine oocyte remains viable for only 8–12 h after ovulation

(16–24 h in sheep), whereas spermatozoa retain their fertilizing capacity for about 30–50 h after insemination. After fertilization, the embryo remains near or at the ampullary-isthmic junction of the oviduct before entering the uterus at 4–5 d (3–4 d in sheep).

Migration of blastocytes between the uterine horns is rare in monovulatory cows and ewes. Elongation of the conceptus to the filamentous form, which usually extends into the contralateral uterine horn, represents the first step towards implantation. In order for embryogenesis to continue into an established pregnancy, the functionality of the *corpus luteum* must be maintained. Thus, maternal recognition of pregnancy must occur prior to luteolysis.

In ruminants, the antiluteolytic signal has been identified as interferon-tau (IFN-τ), produced by the trophectoderm as early as days 15–16 in the cow (Bartol *et al.*, 1985) and days 13–14 in sheep (Godkin *et al.*, 1982). IFN-τ acts directly on the endometrium to suppress pulsatile secretion of PGF2α by inhibiting the expression of endometrial oestradiol and oxytocin receptors and by inducing inhibitors of PGF2α synthesis (Thatcher, 1998).

In addition, cytokines produced by IFN-τ may prevent rejection of the conceptus by immunomudulatory effects (Demmers *et al.*, 2001). During early pregnancy, progesterone is obligatory for endometrial secretion of nutrients, growth factors, immunosuppressive agents, enzymes, ions and steroids that are essential for embryo development and formation of a placenta.

The Placenta

The placenta is a transient organ of metabolic interchange between the conceptus and the dam. Ruminants have a cotyledonary placenta, where chorionic villi are restricted to very distinct oval areas of the chorion (cotyledons) which overlie the uterine caruncles. Both structures form a unit known as the placentome. Firm attachment between chorionic villi protruding into caruncular crypts is well established by day 40 in cows and day 30 in ewes.

Microscopically, both cows and sheep have an epitheliochorial placenta with partial erosion of the endometrial epithelium (syndesmochorial). Replacement of the endometrial surface is achieved by binucleate giant cells which migrate from the trophoblast into the uterine epithelium. These form a syncytium between maternal and conceptus compartments, provide immunological protection of the conceptus and are known to secrete placental lactogen (somatomammotropin). As a major endocrine organ, the placenta produces various hormones that are important for the maintenance of pregnancy, for partitioning of nutrients from the mother to support fetal growth, for stimulating mammary function as well as for inducing parturition.

In the cow, progesterone is produced throughout pregnancy primarily by the *corpus luteum* and only temporarily by the placenta between days 180 and 250 of pregnancy. In the ewe, the *corpus luteum* is not needed for the entire gestational period as adequate production is taken over by the placenta after day 50 of pregnancy. The mean duration of gestation ranges between 280 and 290 d in cows and 145–150 d in sheep. Differences in gestation length are associated with breed, twinning, sex of the fetus and parity of the dam.

Parturition

Induction of parturition

Initiation of parturition is closely synchronized with the final stages of maturation of the fetus in its preparation for post-natal life. The key event for the induction of parturition resides in the activation of the fetal hypothalamic–pituitary–adrenal system (Thorburn *et al.*, 1977). About 20 d before parturition, the fetal adrenal cortex begins to grow and becomes increasingly sensitive to ACTH. The rise in fetal cortisol activates the appropriate enzymes (17-hydroxylase, 17-20 desmolase, aromatase) of the placentomes to convert progesterone to oestrogen. This change of placental steroid synthesis stimulates uterine production of PGF2α that leads to regression of the *corpus luteum* and an abrupt decline in maternal plasma progesterone concentrations about 24 h before parturition.

The drop in progesterone and the elevation of oestradiol increase myometrial sensitivity to oxytocin, enabling strong uterine contractions which push the fetus into the cervix and anterior vagina. Distension of the cervix activates stretch receptors which, by way of a nervous reflex (Ferguson reflex), stimulate the release of more oxytocin from the maternal pituitary gland. Oxytocin in turn causes further uterine contractions, escalating in a self-potentiating positive feed-forward system. The purpose of such a feed-forward system is to accelerate and intensify the whole process of birth which, once begun, must be achieved in a limited period of time. In addition to coordinated uterine contractions, abdominal muscles are also contracted at the same time and these combined efforts usually culminate in rapid delivery of the fetus.

Stages of parturition

During the first stage of parturition, several clinical signs indicate impending parturition. Under the influence of increasing amounts of oestrogens and peak peripartal relaxin levels (produced by the *corpus luteum* in the cow), softening of the collagen in the cervix, uterus, vagina and pelvic ligaments occurs, minimizing the resistance during expulsion. As calving nears, the cervix and vagina begin to produce a stringy mucus discharged by the vulva and the udder enlarges, with secretions taking on a yellowish-opaque appearance indicative of colostrum.

During the second stage of parturition, lasting 6–12 h, the cervix dilates and the fetus enters the birth canal. Signs of discomfort and restlessness, together with abdominal straining, appear and usually the animal lies down in a lateral position. When the uterine pressure reaches a certain level, fetal membranes rupture and the amniotic fluid provides lubrication of the birth canal.

During the third stage of parturition, expulsion of the fetus takes place, which normally lasts about 30–60 min. Difficulties in parturition (dystocia) usually occur during expulsion of the fetus and may be caused by excessive size or abnormal positioning of the fetus – as well as by multiple births. Stage four of parturition, the interval between fetal and placental expulsion, varies between 2–6 h in cows and 2–3 h in ewes. If placental delivery is delayed or is not completed by 12 h, a condition called retained placenta exists. This has serious consequences because the cervix cannot close down properly, tissue is left to autolyse within the uterine lumen and inflammation can occur.

1.4 The Post-partum Period

Uterine Involution

The post-partum period (puerperium) begins immediately after parturition and lasts until a functional anatomical and endocrinological utero-ovarain relationship is restored, allowing another pregnancy to occur. The puerperium is characterized by two major events, the involution of the uterus and resumption of normal ovarian activity, accompanied by visible oestrus symptoms. Uterine involution results from three overlapping processes: reduction in size, loss of tissue and tissue repair.

Immediately after parturition, the uterus is a large sac 1 m long and weighing about 10 kg. This weight should be reduced to 0.8–0.9 kg when involution is completed, at about 4–5 weeks post-partum (pp). The most profound reduction in both cell size and number occurs during the first few days post-partum (halved by day 5). Myometrial contractions facilitate discharge of fluids and tissue debris (lochia) from the uterus, compress the uterine vasculature and thus reduce the overall size of the uterus. Suckling or frequent milking enhances uterine activity, whereas stress and low prostaglandin levels reduce myometrial activity (Thun *et al.*, 1993).

Before surface epithelial repair can take place, the large mass of caruncular tissue must undergo necrosis and sloughing. Using histological criteria, epithelial repair of caruncles should be completed by 6–8 weeks pp. In sheep, the process of uterine involution is similar to that of cows and lasts about 3–4 weeks.

Endocrine Changes

The fall in both progesterone and oestrogen concentrations after calving results in the removal of the negative feedback and allows the anterior pituitary to regain its full sensitivity to GnRH-induced LH release. This period lasts about 10–15 d and is characterized by elevated plasma FSH concentrations, follicular growth and formation of a large dominant follicle within 2–3 d. Although there are follicles ready to ovulate around day 15 post-partum, ovulation will only occur when oestradiol production is sufficient to increase LH pulsatility, induce a positive feedback and an ovulatory LH surge.

When LH pulse frequencies are low, the dominant follicle either undergoes atresia, followed by a new follicular wave, or becomes cystic. Therefore, factors that suppress LH secretion in the post-partum period will delay first ovulation and prolong the anoestrous period. In general, healthy cows in good body condition should have resumed cycling activity by day 50 after parturition, with first ovulations occurring on average at about 30 (range 17–42) d (Lamming and Darwash, 1998; Opsomer *et al.*, 1998; Butler, 2003). The incidence of first post-partum ovulation without oestrous behaviour (silent heat) may be as high as 80%, and decreases in subsequent ovulations.

Thus, the first oestrus may not reflect the resumption of ovarian cyclicity. During periods of transition from a state of acyclicity to one of cyclicity (post-partum period and seasonal anoestrus in sheep), more than 50% of cows and ewes exhibit short first oestrous cycles with transient increases of progesterone of only 6–7 d duration. The uterus has most probably not regained normal function at the time of first ovulation (inadequate level of pre-ovulatory oestrogen production) and a premature release of uterine prostaglandin occurs resulting in luteolysis (Mann and Lamming, 2000).

Beef cows suckling calves have a longer post-partum acyclic period (no ovarian activity) than milked dairy cows, and usually do not resume cyclicity before 45–60 d pp. The prolongation of the anoestrous period is caused mainly by an interaction between social factors (cow–calf bond) and not by lactation itself (Williams *et al.*, 2001). Maternal offspring bonding – determined by visual and olfactory cues – inhibits pulsatile LH secretion and thus delays first ovulation. Resumption of ovarian cycles can be restored partially with temporary weaning of calves.

In ewes, post-partum intervals to first oestrus and ovulation are markedly affected by the season of parturition. If parturitions occur during anoestrus, ewes will not resume cyclical activity until the following season but, if they lamb during the breeding season, the first post-partum ovulations usually occur within 20 d and are usually associated with silent oestrus (Jainudeen and Hafez, 1993).

1.5 Fertility Problems

Over the last few decades, an intense genetic selection has significantly increased milk yields of dairy cattle and the production efficiency of the dairy industry. In order to maintain optimal reproductive performance in a dairy herd, cows should produce one calf each year. However, to obtain a calving interval of 12 months, breeding must begin before 60 d post-partum and conception should occur by 85 d post-partum.

Uterine and ovarian health – as well as successful conception and establishment of pregnancy – are time-limiting factors. The longer the puerperium, the longer the delay to a subsequent pregnancy and the less efficient the production process becomes. Today, enough evidence has been presented in Europe and the USA that this increase in milk yield has been accompanied by a decline in fertility of lactating cows (Lucy, 2001). For example, a remarkable decline in first-service conception rates from around 65% in the 1950s to well below 40% in 2001 has been reported by Butler (2003).

Uterine Health

Although energy balance is one of the most important factors in the prevention of early post-partal diseases, parturition problems such as stillbirth, twins, dystocia, caesarean section, prolapsed uterus or retained fetal membranes – as well as metabolic disturbances such as hypocalcaemia and ketosis – are well-known risk factors for uterine infections. Uterine bacterial infections are important because they disrupt the function of not only the uterus, but also the ovary and the higher control centres in the hypothalamus and the pituitary. The inflammatory and immune response to uterine bacterial infection delays uterine involution and reduces fertility.

During parturition, bacteria can enter the uterus through the open birth canal and, in fact, bacterial contamination can be detected in up to 90% of dairy cows during the first post-partum week. In most cases, bacteria are eliminated within 10 d post-partum without causing any clinical symptoms. A heavy bacterial load, combined with a weak intra-uterine defence, usually results in inflammation of the uterus.

Recognized uterine pathogens most often associated with clinical disease are *Arcanobacterium pyogenes*, *Escherichia coli* and *Fusobacterium necrophorum*. Persistance of these pathogens may cause acute metritis/endometritis, characterized by pyrexia up to 14 d post-partum with a fetid, purulent vulval discharge.

Subsequent to acute inflammation, a chronic endometritis may develop, which might persist beyond the puerperium. Early ovulation after parturition and formation of a *corpus luteum* before elimination of uterine bacteria predisposes to pyometra, because progesterone suppresses the immune response, making the uterus more susceptible to bacterial infection (Lewis, 2003).

Depending upon genetic and management factors in different herds and the method used for diagnosis, the incidence of endometritis is about 10–40% in dairy cattle (Sheldon and Dobson, 2004). In cows with endometritis, the conception rate is about 20% lower, the median calving to conception interval 30 d longer and there are 3% more animals culled for failure to conceive (LeBlanc *et al.*, 2002).

Ovarian Cysts

Ovarian follicular cysts (cystic ovaries, cystic ovarian degeneration) have been reported in many mammalian species, including cattle and sheep. Cysts are a major contributor to reduced reproductive efficiency of lactating dairy cows, causing considerable economic loss for dairy farmers. Ovarian cysts are more common in milking cows and the incidence has been reported to vary from about 6 to 20% (Garverick, 1997). Cows with cystic ovaries are often infertile as long as the condition persists, and have extended calving intervals from about 20 to 65 d (Peter, 2004). Ovarian cysts are anovulatory follicular structures, usually larger than 25 mm in diameter, which persist for at least 10 d.

The condition has a multifactorial aetiology, with a number of endogenous and exogenous factors responsible for the disease process. The pathophysiology appears to involve an neuroendocrinological dysfunction of the hypothalamus–pituitary–gonadal axis, resulting in failure of the pre-ovulatory LH surge to occur at the appropriate time in follicular maturation (Hamilton *et al.*, 1995). This malfunction appears to reside at the hypothalamic level and be associated with low progesterone concentrations blocking the LH surge and thus preventing ovulation (Hatler *et al.*, 2003).

If the mature follicle fails to ovulate at the appropriate time, it either regresses or persists as a follicular or luteal cyst, depending upon its steroidogenic capacity. Another contributing factor in the pathogenesis of persistent follicles in cattle and sheep is stress imposed by the environment in the post-partum period. The preovulatory LH surge is sensitive to the inhibitory effects of stress hormones – mainly endorphin, corticotrophin-releasing hormone, ACTH and cortisol.

Activation of the hypothalamus–pituitary–adrenal axis by various stressors (diseases, pain, high milk yield, hot climate, transport, fixation, isolation in sheep) causes an increase in plasma cortisol, reducing LH pulse frequency and follicular oestradiol secretion (Dobson *et al.*, 2000; Ribadu *et al.*, 2000). In cattle, it has been demonstrated that changes in cortisol, progesterone and LH after restraint stress are dependent on circulating oestradiol concentrations (Thun *et al.*, 1998; Hollenstein *et al.*, 2006).

Among the predisposing factors, heritability of cystic follicles is very low. However, it is possible that in breeds with genetic predisposition to ovarian cysts, selection for high milk production can increase the incidence of follicular cysts

(Hooijer *et al.*, 2001). The condition has been observed more frequently in older cows and when forages are fed containing high concentrations of oestrogenic compounds.

The roles of nutrition and milk production are interrelated. The metabolic demands of high milk production result in greater negative energy balance, causing ovarian dysfunctions such as ovarian cysts, persistent *corpus luteum* and non-functioning ovaries, with clinical symptoms of acyclicity (Opsomer *et al.*, 1998; Shresta *et al.*, 2004). In addition, post-partum uterine infection (Bosu and Peter, 1987) and bacterial endotoxins released in the uterus might stimulate PGF2α and cortisol secretion, predisposing cows to ovarian cysts.

Embryonic Mortality

Embryonic mortality has a major impact on reproductive efficiency in cattle. The embryonic period of gestation extends from conception to the end of the differentiation stage, at approximately 42 d of gestation, while the fetal period extends from day 42 (in sheep, day 35) of gestation to delivery. Two sources of pregnancy failure exist after breeding: fertilization failure and embryonic loss.

Fertilization failure may be due to the inability of sperm to reach the ovum and, in the case of artificial insemination, semen processing, storage and handling as well as insemination of non-oestrous cows may be at fault. Most embryos are lost before the critical time of maternal recognition at day 16 in cows (day 13 in ewes) and cycle length is not affected. When embryonic death precedes luteolysis, luteal regression is delayed by at least 3 d after the end of pregnancy. When the interval between insemination and return to oestrus is extended beyond day 24, it points to embryonic losses that had occurred following pregnancy recognition (Humblot, 2001).

The process of implantation starts around days 21–22 in cows and days 15–16 in ewes (Guillomot, 1995). Thus, in the cow, losses of pregnancy prior to day 24 indicate early embryonic losses and those between days 24 and 42 late embryonic losses. In cows, it is generally accepted that the fertilization rate is in the order of 90%, indicating that about 10% of the inseminations or matings are associated with lack of fertilization. Embryonic mortality has been estimated to be about 20–40% and fetal death about 5–10% (López-Gatius *et al.*, 1996). Pre-natal losses can be caused by infectious and by non-infectious factors, the latter accounting for 70% or more of the cases in cows. Non-infectious causes are often multifactorial and are difficult to diagnose.

Infectious causes

Embryonic mortality caused by systemic pathogens is usually related to fever during the infection. High fever can lead to early embryonic death as a result of denaturation of embryonic proteins or luteolysis by elevated prostaglandins. Infection of the embryonic environment can be caused by specific and non-specific uterine pathogens.

Specific pathogens (viruses, bacteria and protozoa) enter the uterus by the haematogenous route (e.g. *Toxoplasma gondii*, *Neospora caninum*) or via the vagina at natural service or at insemination with contaminated semen (e.g. *Campylobacter fetus* BVDV, BHV-1). BHV-1 and BVDV might already be present in follicular fluid or the granulosa cells of bovine oocytes (Brownlie *et al.*, 1997) and contaminate the embryo before hatching by adhering to the *zona pellucida.* An infection with BVDV during days 60–120 of pregnancy will not always cause abortion, but the surviving fetus will be born as persistently infected.

Some bacterial and protozoal infections, such as trichomoniasis and campylobacteriosis, are characterized by endometritis resulting in infertility and embryonic loss. Other infections, such as brucellosis, *Arcanobacter pyogenes* infection, candidiasis, leptospirosis, listeriosis and *Haemophilus somnus* infection, are more often associated with late embryonic losses, fetal death and abortion (Vanroose *et al.*, 2000).

Non-infectious causes

Among non-infectious causes, chromosomal aberrations and the influence of various stressors such as trauma, pain, diseases and high ambient temperature are major causes of early pregnancy failure (Dobson and Smith, 1995). In high-yielding dairy cows, it has only recently been reported that negative energy balance will affect oocyte quality through increased concentrations of non-esterified fatty acids in follicular fluid, leading to a lower fertilization rate and increased early embryonic death (Leroy *et al.*, 2005).

In addition, environmental toxicants, teratogenic compounds and mycotoxins may also affect the very early stages of conceptus (Christianson, 1992). Maternal factors causing embryonic death are hormonal imbalances and disturbances of the embryonic–maternal interactions. Both a late post-ovulatory progesterone rise and low luteal phase concentrations are associated with poor embryonic development and the production of insufficient interferon-τ to prevent luteal regression. The post-ovulatory rise of progesterone is of particular interest as this hormone maintains the synchrony between embryo and uterus (Mann *et al.*, 1998).

References

Bartol, F.F., Roberts, R.M., Bazer, F.W., Lewis, G.S., Godkin, J.D. and Thatcher, W.W. (1985) Characterization of proteins produced *in vitro* by periattachment bovine conceptuses. *Biology of Reproduction* 32, 681–693.

Bosu, W.T.K. and Peter, A.T. (1987) Evidence for a role of intrauterine infections in the pathogenesis of cystic ovaries in post-partum dairy cows. *Theriogenology* 28, 725–736.

Braden, T.D., Gamboni, F. and Niswender, G.D. (1988) Effects of prostaglandin F-induced luteolysis on the populations of cells in the ovine *corpus luteum*. *Biology of Reproduction* 39, 245–253.

Brownlie, J., Booth, P.J., Stevens, D.A. and Collins, M.E. (1997) Expression of non-cytopathogenic bovine viral diarrhoea virus (BVDV) in oocytes and follicles of persistently infected cattle. *Veterinary Record* 141, 335–337.

Butler, W.R. (2003) Energy balance relationships with follicular development, ovulation and fertility in post-partum dairy cows. *Livestock Production Science* 83, 211–218.

Charpigny, G., Reinaud, P., Tamby, J.P., Creminon, C., Martal, J., Maclouf, J. and Guillomot, M. (1997) Expression of cyclooxygenase-1 and -2 in ovine endometrium during the oestrous cycle and early pregnancy. *Endocrinology* 138, 2163–2171.

Christianson, W.T. (1992) Stillbirths, mummies, abortions, and early embryonic death. *Veterinary Clinics of North America, Food Animal Practice* 8, 623–639.

Demmers, K.J., Derecka, K. and Flint, A. (2001) Trophoblast interferon and pregnancy. *Reproduction* 121, 41–49.

Dobson, H. and Smith, R.F. (1995) Stress and reproduction in farm animals. *Journal of Reproduction and Fertility Suppl.* 49, 451–461.

Dobson, H., Ribadu, A.Y., Noble, K.M., Tebble, J.E. and Ward, W.R. (2000) Ultrasonography and hormone profiles of adrenocorticotrophic hormone (ACTH)-induced persistent ovarian follicles (cysts) in cattle. *Journal of Reproduction and Fertility* 120, 405–410.

Dodson, S.E., McLeod, B.J., Haresign, W., Peteres, A.R. and Lamming, G.E. (1988) Endocrine changes from birth to puberty in the heifer. *Journal of Reproduction and Fertility* 82, 527–538.

Driancourt, M.A. (2001) Regulation of ovarian follicular dynamics in farm animals. Implications for manipulation of reproduction. *Theriogenology* 55, 1211–1239.

Foster, D.L. and Nagatani, S. (1999) Physiological perspectives on leptin as a regulator of reproduction: role in timing puberty. *Biology of Reproduction* 60, 205–215.

Foster, D.L., Karsch, F.J., Olster, D.H., Ryan, K.D. and Yellon, S.M. (1986) Determinants of puberty in a seasonal breeder. *Recent Progress in Hormone Research* 42, 331–384.

Garverick, H.A. (1997) Ovarian follicular cysts in dairy cows. *Journal of Dairy Science* 80, 995–1004.

Gibbons, J.R., Kot, K., Thomas, D.L., Wiltbank, M.C. and Ginther, O.J. (1999) Follicular and FSH dynamics in ewes with a history of high and low ovulation rates. *Theriogenology* 52, 1005–1020.

Ginther, O.J., Kulick, L.J., Kot, K. and Wiltbank, M.C. (1997) Emergence and deviation of follicles during the development of follicular waves in cattle. *Theriogenology* 48, 75–87.

Godkin, J.D., Bazer, F.W., Moffatt, J., Sessions, F. and Roberts, R.M. (1982) Purification and properties of a major, low molecular weight protein released by the trophoblast of sheep blastocytes at Day 13–21. *Journal of Reproduction and Fertility* 65, 141–150.

Guillomot, M. (1995) Cellular interactions during implantation in domestic ruminants. *Journal of Reproduction and Fertility, Suppl.* 49, 39–51.

Hamilton, S.A., Garverick, H.A., Keisler, D.H., Xu, Z.Z., Loos, K., Youngquist, R.S. and Salfen, B.E. (1995) Characterization of ovarian follicular cysts and associated endocrine profiles in dairy cows. *Biology of Reproduction* 53, 890–898.

Hatler, T.B., Hayes, S.H., Laranja da Fonseca, L.F. and Silvia, W.J. (2003) Relationship between endogenous progesterone and follicular dynamics in lactating dairy cows with ovarian follicular cysts. *Biology of Reproduction* 69, 218–223.

Hollenstein, K., Janett, F., Bleul, U., Hässig, M., Kähn, W. and Thun, R. (2006) Influence of oestradiol on adrenal activity in ovariectomized cows during acute stress. *Animal Reproduction Science* 93, 292-302.

Hooijer, G.A., Lubbers, R.B., Ducro, B.J., van Arendonk, J.A., Kaal-Lansberger, L.M. and van der Lende, T. (2001) Genetic parameters for cystic ovarian disease in Dutch Black and White dairy cattle. *Journal of Dairy Science* 8, 286–291.

Humblot, P. (2001) Use of pregnancy-specific proteins and progesterone assays to monitor pregnancy and determine the timing, frequencies and sources of embryonic mortality in ruminants. *Theriogenology* 56, 1417–1433.

Jainudeen, M.R. and Hafez, E.S.E. (1993) Sheep and goats. In: Hafez, E.S.E. (ed.) *Reproduction in Farm Animals.* Lea and Febiger, Philadelphia, pp. 330–342.

Kinder, J.E., Bergfeld, E.G.M., Wehrmann, M.E., Peters, K.E. and Kojima, F.N. (1995) Endocrine basis for puberty in heifers and ewes. *Journal of Reproduction and Fertility, Suppl.* 49, 393–407.

Lamming, G.E. and Darwash, A.O. (1998) The use of milk progesterone profiles to characterise components of subfertility in milked dairy cows. *Animal Reproduction Science* 52, 175–190.

LeBlanc, S.J., Duffield, T.F., Leslie, K.E., Bateman, K.G., Keefe, G.P., Walton, J.S. and Johnson, W.H. (2002) Defining and diagnosing post-partum clinical endometritis and its impact on reproductive performance in dairy cows. *Journal of Dairy Science* 85, 2223–2236.

Leroy, J.L.M.R., Vanholder, T., Mateusen, B., Christophe, A., Opsomer, G., de Kruif, A., Genicot, G. and Van Soom, A. (2005) Non-esterified fatty acids in follicular fluid of dairy cows and their effect on developmental capacity of bovine oocytes. *Reproduction* 130, 485–495.

Lewis, G.S. (2003) Steroidal regulation of uterine resistance to bacterial infection in livestock. *Reproduction of Biology and Endocrinology* 1, 117.

López-Gatius, F., Labérnia, J., Santolaria, T., López-Béjar, M. and Rutllant, J. (1996) Effect of reproductive disorders previous to conception on pregnancy attrition in dairy cows. *Theriogenology* 46, 643–648.

Lucy, M.C. (2001) Reproductive loss in high-producing dairy cattle: where will it end? *Journal of Dairy Science* 84, 1277–1293.

Lussier, J.G., Matton, P. and Dufour, J.J. (1987) Growth rates of follicles in the ovary of the cow. *Journal of Reproduction and Fertility* 81, 301–307.

Mann, G.E. and Lamming, G.E. (2000) The role of sub-optimal preovulatory oestradiol secretion in the aetiology of premature luteolysis during the short oestrous cycle in the cow. *Animal Reproduction Science* 64, 171–180.

Mann, G.E., Lamming, G.E. and Payne, J.H. (1998) Role of early luteal phase progesterone in control of the timing of luteolytic signal in cows. *Journal of Reproduction and Fertility* 113, 47–51.

McCracken, J.A. (1998) Luteolysis. In: Knobil, E. and Neill, J.D. (eds) *Encyclopedia of Reproduction.* Academic Press, San Diego, California, pp. 1083–1094.

Opsomer, G., Coryn, M., Deluyker, H. and de Kruif, A. (1998) An analysis of ovarian dysfunction in high-yielding dairy cows after calving based on progesterone profiles. *Reproduction in Domestic Animals* 33, 193–204.

Peter, A.T. (2004) An update on cystic ovarian degeneration in cattle. *Reproduction in Domestic Animals* 39, 1–7.

Ribadu, A.J., Nakada, K., Morijoshi, M., Zhang, W.C., Tanaka, Y. and Nakao, T. (2000) The role of LH pulse frequency in ACTH-induced ovarian follicular cysts in heifers. *Animal Reproduction Science* 64, 21–31.

Schams, D. and Berisha, B. (2004) Regulation of *corpus luteum* function in cattle – an overview. *Reproduction in Domestic Animals* 39, 241–251.

Sheldon, I.M. and Dobson, H. (2004) Post-partum uterine health in cattle. *Animal Reproduction Science* 82–83, 295–306.

Shresta, H.K., Nakao, T., Higaki, T., Suzuki, T. and Akita, M. (2004) Resumption of post-partum ovarian cyclicity in high-producing Holstein cows. *Theriogenology* 61, 637–649.

Sirois, J. and Fortune, J.E. (1988) Ovarian follicular dynamics during the oestrous cycle in cattle monitored by real-time ultrasonography. *Biology of Reproduction* 39, 308–317.

Thatcher, W.W. (1998) Ruminants. In: Knobil, E. and Neill, J.D. (eds) *Encyclopedia of Reproduction.* Academic Press, San Diego, California, pp. 302–311.

Thorburn, G.D., Challis, J.R.G. and Currie, W.B. (1977) Control of parturition in domestic animals. *Biology of Reproduction* 16, 18–27.

Thun, R., Kündig, H., Zerobin, K., Kindahl, H., Gustafsson, B.K. and Ziegler, W. (1993) Die Uterusmotorik des Rindes während Spätgravidität, Geburt und Puerperium. III. Anwendung von Flunixin Meglumin und endokrine Veränderungen. *Schweizer Archiv für Tierheilkunde* 135, 333–344.

Thun, R., Kaufmann, C. and Janett, F. (1998) The influence of restraint stress on reproductive hormones in the cow. *Reproduction in Domestic Animals* 33, 255–260.

Vanroose, G., de Kruif, A. and Van Soom, A. (2000) Embryonic mortality and embryo–pathogen interactions. *Animal Reproduction S*cience 60–61, 131–143.

Williams, G.L., Gazal, O.S., Leshin, L.S., Stanko, R.L and Anderson, L.L. (2001) Physiological regulation of maternal behaviour in heifers: role of genital stimulation, intracerebral oxytocin release, and ovarian steroids. *Biology of Reproduction* 65, 295–300.

Sample Collection

2

2.1 Study Design

Chris Bartels and Franz J. Conraths

When the prevalence of an infection such as neosporosis in dairy cattle needs to be known, the best method is to test all the animals in the population: this is called census. If all data are recorded correctly, the only source of error relevant for this method will be related to the measurement (e.g. the use of a serological test with a certain sensitivity and specificity). For many reasons (practical, financial, time), a census is usually not possible and a sample of animals is taken and tested. The result of this testing is then used to draw conclusions concerning the population. In such a situation, there is additional error in the study due to sampling. Yet, a well-planned sample can provide virtually the same information as a census, at a fraction of the costs.

In this chapter we will discuss a number of issues that seem relevant for general practitioners who wish to test populations (i.e. herds) for diseases without sampling all the animals in the herd. For this reason, we will use a standard example of a veterinary practice with 100 dairy farms, each with, on average, 50 head of cattle and 35 head of young stock.

Sample Size Calculations

There are two different approaches to sampling a population:

© CAB International 2007. *Protozoal Abortion in Farm Ruminants*
(eds L.M. Ortega-Mora, B. Gottstein, F.J. Conraths and D. Buxton)

- A qualitative objective aiming at the detection of infection in animals or herds from a population.
- A quantitative objective aiming at estimating the prevalence of an infection within a herd or infected herds within a population.

The difference between these two approaches is important, as they lead to different conclusions and require different sample sizes. The first approach aims at detection of infection (or disease) above a certain threshold at a selected level of confidence. It does not attempt to estimate to what extent the infection is present above or below the stated threshold. A typical question is:

> *Does the prevalence infection with* Neospora caninum, *BVD,* Salmonella *or IBR in a particular dairy herd exceed 10% or not?*

In the second case one wants to estimate the prevalence of infection with a given level of precision at a selected level of confidence. A typical question is:

> *What is the prevalence of infection with* N. caninum, *BVD,* Salmonella *or IBR in the dairy herds?*

For the quantitative objective, the number of samples required is higher. In addition, the required precision influences the sample size, with a higher degree of precision requiring an increased sample size.

Software

There are a number of good software packages available for calculation of sample sizes. There is no need to use one's own calculator, rather, it is better to spend time on the study design so as to minimize any bias. It is not our intention to list all the available software. We only wish to illustrate its availability and use.

An overview of software available is given on the website 'EpiVetNet, website for veterinary epidemiologists': http://www.vetschools.co.uk/EpiVetNet/Sampling_software.htm. This website is maintained by the Royal Veterinary College, University of London, UK; it was originally created at the EpiCentre, Massey University, New Zealand. In Table 2.1.1, a list of the most common software packages available on the Internet is shown.

The information on WIN EPISCOPE and FREECALC/SURVEY TOOLBOX is given in more detail in the appendix at the end of this chapter, because these packages were used for the sample calculations presented in this chapter.

Statistical factors that relate to sample size calculations

In general, the required sample sizes depend on:

- Estimated prevalence: P

Table 2.1.1. Overview of software packages at http://www.vetschools.co.uk/EpiVetNet/Sampling_software.htm

Program	Current version	Operating system	Description
CSURVEY		MS-DOS	Design of cluster surveys
WIN EPISCOPE	2.0	MS-Windows	Software for sample size estimation, evaluation of diagnostic tests (incl. ROC analysis), cross-tabulation of data from various types of epidemiological studies, Reed-Frost simulation models
FREECALC	2.0	Windows	Epidemiological calculator for planning and analysis of surveys to detect disease or prove freedom from disease; calculates sample size requirements based on diagnostic test sensitivity and specificity, taking population size into account, and analyses the results of such surveys
G*POWER	2.0	MS-DOS, Macintosh	Statistical power analyses for the most common statistical tests: t-tests; *post hoc, a priori* and compromise power analyses
HERDACC	3.0	MS-Windows	Program for calculating herd (aggregate) sensitivity and specificity for a wide range of values of individual level sensitivity and specificity
PS	2.1.23	MS-Windows	Interactive program for performing power and sample size calculations. It can be used for studies with dichotomous, continuous or survival response measures. The alternative hypothesis of interest may be specified either in terms of differing response rates, means or survival times, or in terms of relative risks or odds ratios.
POWER PLANT	1.0 (beta)	MS-Windows	Power and sample size calculations for variety of hypothesis testing situations; probability calculations using electronic calculator; publication quality graphs (power curves); spreadsheet output of results
SAMPLE XS		MS-Windows	Sample size calculator for cross-sectional surveys using simple random, systematic or complex samples
SSIZE		Palm	Calculates sample size using either power or confidence interval widths for a variety of distributions with respect to linear models, and for a variety of specialized situations such as for logistic models and contingency tables
SURVEY TOOLBOX	1.03	Windows 95	Suite of software programs designed to make planning, conducting and analysing statistically valid, efficient and practical animal health surveys easier

It requires a rough estimate of the prevalence when one wants to calculate the sample size necessary to determine a certain prevalence. Such estimates can be obtained from other studies and from knowledge of the dynamics of the disease.

- Population size: N
  This is of particular interest in relation to the sample size. As a rule of thumb, when the sample size (n) is greater than 10% of the population, it needs to be corrected for the population size. The available computer programs perform this correction automatically. In most epidemiological studies, sampling is carried out without replacement. This means that a sampled element is not put back into the population and can thus not be sampled again. If the proportion of the sampled population is relatively high (> 10%), this can substantially increase the precision of the estimate over what would be expected from a population of infinite size. Consequently, the estimated variance can be adjusted downward by a *finite population correction* given as:

  $$FPC = N - n/N - 1 \quad (1)$$

- Required precision: L
  When prevalence needs to be estimated, the precision of the prevalence estimate needs to be selected. The more precise one wishes to be, the larger a sample size required. In general, precision is set at 5%; alternatives are 2 and 10%.
- Confidence level: CL
  Additionally, in descriptive studies, one has to decide how sure one wants to be that the confidence interval from the estimate will include the true population value. Values frequently used for confidence levels (expressed as 1 – a) are:

α	CL (%)	Z-value (two-sided testing)
0.10	90	1.65
0.05	95	1.96
0.01	99	2.56

Sample size for detection of infection in a population

A herd is regarded as being positive when at least a certain number of animals (D) are diseased. This number of diseased animals can be regarded as the detection level or threshold. The sample size (n) required to investigate whether a population (or herd) is infected depends on the desired level of confidence (CL), the population size (N) and the number of expected infected animals (D) in the population or the prevalence threshold to be detected:

$$n = (1 - (1 - CL)^{1/D}) * (N - (D - 1)/2) \quad (2)$$

Example:
In a herd (100 animals) with fertility problems, one wants to know with 95% confidence if the prevalence of N. caninum*-infected animals is below or above 10%. The required sample size is:*

$n = (1 - (1 - 0.95)^{1/10} * (100 - (10 - 1)/2) = 25$

(see Fig. 2.1.1).

This same formula can be used to determine the maximum number of positives (D) in a population, given that all samples tested test negative:

$$D = (1 - (1 - CL)^{1/n}) * (N - (n - 1)/2) \quad (3)$$

Example:
In the same herd of 100 animals, one has tested 20 animals negative. What is the maximum number of N. caninum*-infected animals in that herd with 99% level of confidence?*

$D = (1 - (1 - 0.99)^{1/20}) * (100 - (20 - 1)/2 = 19$

Expressed as prevalence; with 99% confidence one knows that the prevalence of N. caninum *infection is 19/100 = 19% or less* (see Fig. 2.1.2).

Estimation of the prevalence of an infection or disease

It is also possible to estimate the prevalence of disease in a population by sampling only a fraction of the population. The sample size depends on the desired precision of the estimate, the confidence level and the expected prevalence.

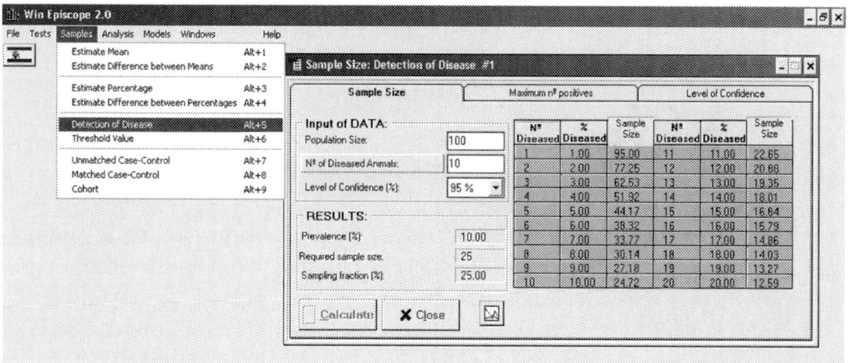

Fig. 2.1.1. Sample size calculation for detection of disease at 10% prevalence in a population of 100 animals with 95% confidence (from WIN EPISCOPE 2.0).

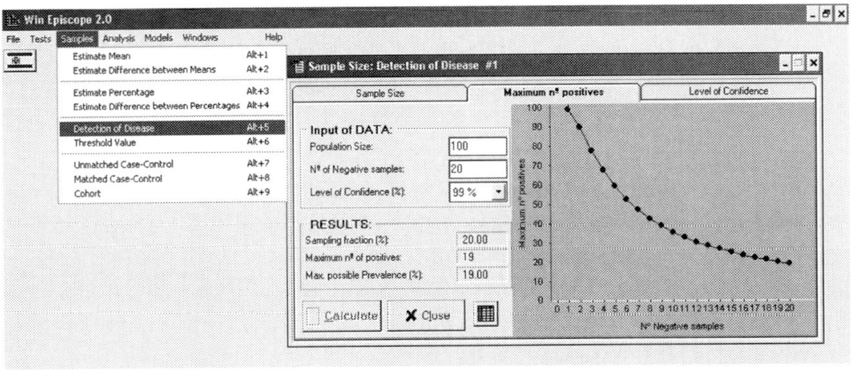

Fig. 2.1.2. Maximun number of positives at 99% confidence when 20/20 animals sampled tested negative from a population of 100 animals (from WIN EPISCOPE 2.0).

$$n = Z^2 * P(1 - P) / L^2 \quad (4)$$

where:
n = sample size
Z = student's t-value related to level of confidence
P = expected prevalence in population
L = required precision expressed as maximal acceptable error (proportion scale).

Example:
In a veterinary practice, the practitioner wants to evaluate the Neospora *herd prevalence to monitor the efficacy of a new control strategy over time. Nothing is known about the prevalence in the population. Therefore, a prevalence of 50% is assumed because this is the worst-case scenario concerning sample sizes. The sample size for 50% prevalence with 90% confidence level and with an acceptable error of 10% is calculated as:*

$$n = 1.65^2 * 0.5(1 - 0.5)/0.10^2$$

n = 68 herds.

The number of required herds to test is 68. However, the number of herds in a population is not infinite and this sample size needs to be corrected for the fact that there are only 100 herds available. As mentioned above, such corrections need to be applied when the sample size is around or greater than 10% of the population size. In the software programs this will be automatically taken into account.

$$n^* = n/(1 + ((n - 1)/N) \quad (5)$$

where:
n^* = adjusted sample size.
n = calculated sample size

N = population size.

For this example:

$n^* = 68/(1 + (67/100) = 41$ herds (see Fig. 2.1.3).

At this point one knows that 41 herds must be sampled. The next step is to decide when to call a herd infected. There are two ways to do this. First, one can test all animals within each herd and define a herd as being infected when at least one or more animals test positive. This might involve a lot of testing. Another possibility is to sample each herd in such a way that one can define a herd as being uninfected when the required number of animals sampled have all been tested negative. Strictly speaking, the herd might still contain infected animals, but the level of infection would be under a defined (and relevant) prevalence level.

For this example, we want to detect 10% prevalence. We have seen that we need to test 25 animals in a herd of 100 animals, but not all herds have 100 animals. Thus, we can define the number of animals that need to be tested for each herd size.

In Table 2.1.2, the required sample sizes for different herd sizes, confidence levels and detection levels are given. As one can see, the required number of samples increases only slightly when the herd size increases. The same applies when the confidence level is increased. This is different, however, when low prevalence levels are to be detected.

Sample size to estimate the mean of a parameter

It is not always sufficient to test simply for the absence or presence of an infection. It can also be determined whether an abnormally low level of a certain

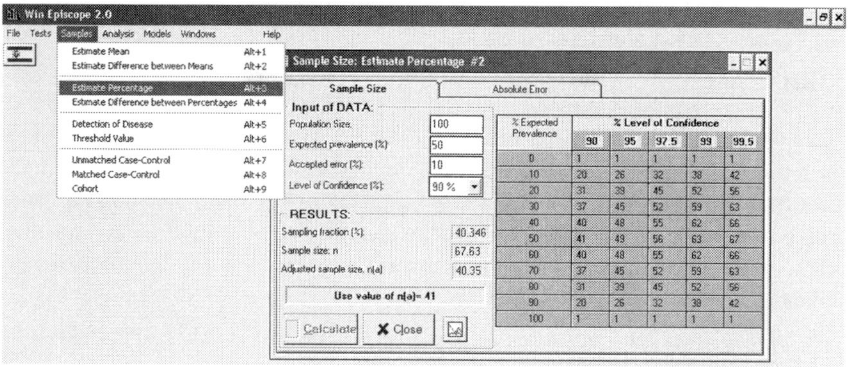

Fig. 2.1.3. Sample size calculation for estimation of 50% herd prevalence in a finite population of 100 herds with 95% confidence and 10% acceptable error (from WIN EPISCOPE 2.0).

Table 2.1.2. Sample size requirements for detection of disease at 10 and 25% with 90 and 95% levels of confidence.

Herd size	Confidence level	Animals to be tested for detection of 10% prevalence (*n*)	Animals to be tested for detection of 25% prevalence (*n*)
25	90	17	8
	95	20	9
50	90	18	8
	95	22	10
100	90	20	8
	95	25	10
200	90	21	8
	95	27	11

parameter is present in a population, e.g. a herd that requires intervention: for example, the selenium (Se) level in the blood of heifers with no Se supplementation and housed indoors.

The calculation of the sample size is, in this case, based on a normal distribution of the parameter. It is necessary to specify the naturally present variation and the maximal acceptable variation around the mean. This last information indicates the precision of the estimated mean from the sample. The sample size also depends on the population size:

$$n = (Z^2 * SD^2 * N)/((N - 1) * d^2 + (Z^2 * SD^2)) \quad (6)$$

where:

n = sample size
Z = Student's t-value for the expected confidence level
SD = expected standard deviation of parameter
N = population size
d = accepted error (precision) (proportion scale).

For example:
In a herd of 400 sheep, one expects a shortage of cobalt (Co). A sample is taken in order to decide whether sheep need to be treated. One wishes to take a decision based on a sample, at the 90% confidence level and with an accepted error of 25% around the mean. The normal range of Co in sheep is 400–4000 pmol/l. We assume that the normal range covers 95% of the normal distribution, that the mean is 2200 pmol/l and that the standard deviation is approximately the normal range divided by four standard deviations = (4000–400)/4 = 900. The number of animals that one minimally needs to sample is:

$$n = (1.65^2 * (900/2200)^2 * 400)/((400 - 1) * 0.25^2 + 1.65^2 * (900/2200)^2)$$

$n = 7$ animals.

The WIN EPISCOPE program calculates that eight samples are required. The difference is an effect of rounding-off to integers (see Fig. 2.1.4).

Sample size to calculate differences between proportions

To demonstrate the difference between two groups of animals with regard to a proportion, a minimum number of animals per group is required. The number per group is dependent on the expected difference in proportion between the groups.

In addition, it is necessary to indicate the level of confidence and to define the power of the calculation. The power of a study is the probability that a true difference between groups is demonstrated in the study (see Table 2.1.3).

A common value for the confidence level is 95%, while for power, values of 80 and 90% are widely used. This means that one usually tries to detect most existing associations, while accepting that some associations may be 'detected' that do not exist in reality.

The required sample size is determined by four values (Z_a, Z_b, P_1 and P_2) in the following formula:

$$n = [(Z_a * \sqrt{}\, 2 * P(1 - P) + Z_b * \sqrt{}\, (P_1 * (1 - P_1) + P_2 * (1 - P_2))]^2 / [P_2 - P_1]^2 \quad (7)$$

where:

Z_a = Student's t-value of confidence level
Z_b = student's t-value of power
$P = P_1 + P_2$
P_1 = proportion in group 1
P_2 = proportion in group 2.

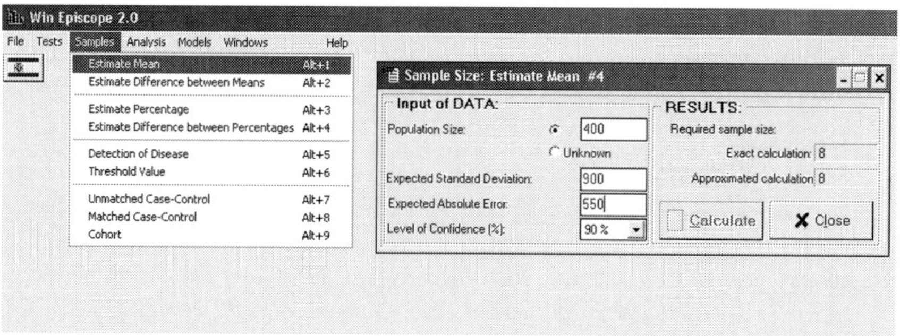

Fig. 2.1.4. Sample size calculation for estimation of the cobalt mean when the expected standard deviation is 900 pmol/l and the expected absolute error is 550 (25% of 2200 pmol/l), in a population of 400 animals with 90% confidence (from WIN EPISCOPE 2.0).

Table 2.1.3. Relationship between results of a statistical test and the true difference between possible outcomes.

	True status = null hypothesis, e.g. the prevalences in two herds do not differ	True status = not-null hypothesis, e.g. the prevalences between two herds do differ
Test outcome: accepting the null hypothesis; i.e. the prevalences do not differ	Probability (1 – P_a) of correctly accepting the null hypothesis; i.e. the prevalences do not differ and we conclude that they are not different	Type II error (P_b); probability of the not-null hypothesis; i.e. the prevalences do differ but we conclude that they are not different
Test outcome: rejecting the null hypothesis; i.e. the prevalences do differ	Type I error or alpha; probability (P_a) of a false rejection of the null hypothesis; i.e. the prevalences do not differ but we conclude that they are different	Power (1 – P_b); probability of correctly rejecting the not null hypothesis; i.e. the prevalences do differ and we conclude that they are different

For example:
In a dairy herd with a number of recent abortions, one wishes to determine whether there is an association between these abortions and neosporosis. It is expected that 80% of aborting cows are infected with N. caninum, *whereas only 10% of the non-aborting cows are infected. The calculations need a power of 80% and 90% level of confidence. The required sample size is:*

$$n = [(1.64 * \sqrt{2 * 0.9 * 0.1}) + (1.28 * \sqrt{0.9 * 0.1 + 0.8 * 0.2})]^2/[0.1 - 0.8]^2$$

n = 5 animals in each group (see Fig. 2.1.5).

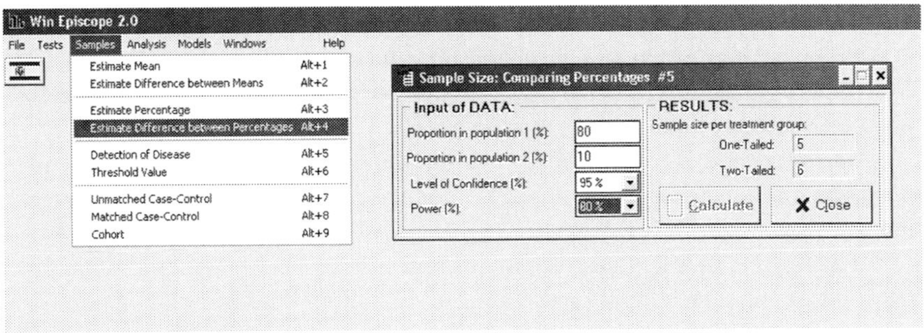

Fig. 2.1.5. Sample size calculation for estimation of differences between percentages in populations with 95% confidence and 80% power (from WIN EPISCOPE 2.0).

Effect of test quality on sample size

These sample sizes are based on the assumption that perfect tests – i.e. assays with 100% sensitivity and specificity – are used. Unfortunately, such tests do not exist in reality. It may therefore be necessary to adjust the sample size when imperfect tests are used. If the sensitivity is less than 100%, the sample size can be multiplied with the inverse of the test sensitivity.

> *Example: if sample size calculations indicate that 20 animals in a herd need to be tested for the detection of neosporosis and if the available test has a sensitivity of 80%, the adjusted sample size is 1/0.8 * 20 = 25 animals.*

Usually, the specificity of the test does not influence the sample size that much. Therefore, a correction of the sample size for tests with imperfect specificity is usually not required. When specificity is below 100% there are, in principle, two ways to adjust:

- Positive samples are confirmed using a different test, e.g. Western blot.
- A higher cut-off for defining a herd infected is used. For example, when all animals in a herd of 100 cattle are sampled with a test of 95% specificity, there is a high probability of testing one or more animals seropositive. It would be more relevant to define such a herd as seropositive when five or more animals test seropositive. A number of software programs (FREECALC, HERDACC) allow for the inclusion of test specificity and sensitivity when calculating sample size.

Accuracy and Precision

When a study is designed in which sampling is performed, accuracy and precision need to be considered.

Sampling is performed both in descriptive studies (surveys) and in analytical studies (observational studies). The first type aims at describing features of the population such as production level and prevalence of a disease, while the second type aims at testing hypotheses about an association between exposure and disease or infection outcome in a population.

The objective is to use the results obtained from the study population to make assumptions concerning the total population (i.e. all animals or all herds), the so-called target population.

This means that the study elements (animals or herds, etc.) are taken from the total population, e.g. all dairy cattle in a country. The study population is a subpopulation of the total population, consisting of the selected study elements (regardless of whether or not they were actually included in the study). Preferably, the study population equals the target population, but often the study population covers only part of the target population. The question to be answered is whether the study population is representative for the target population. This is particularly important for descriptive studies.

Accuracy

The accuracy of a result obtained from a sample is given by the extent to which it represents the true value in the population (also referred to as validity). In other words, accuracy describes the relative absence or presence of bias in results. Bias is any systematic error in the design, conduct or analysis of a study that renders results invalid.

Major types of bias are:

- Selection or sampling bias. As a result of selection bias, study elements (participants, animals) are not representative for the target population. This may be due to failure to use a random sampling method (convenience sampling) or the result of a high degree of a non-response of the sampling units that were selected by random sampling and where the non-response is linked to a particular factor.
  If, for example, a prevalence study on neosporosis is conducted and a large number of farmers do not wish to participate because their dog is not happy with visitors, the outcome of this study may be biased due to the non-response of dog-owning farmers.
- Information bias. In this situation, bias is due to incorrect classification or measurement of the study subjects (the animals or herds tested) or their exposure to certain factors (e.g. presence of possible risk factors). In the case of incorrect classification of categorical variables, the resulting bias is referred to as misclassification bias. If the variables of interest are continuous, the erroneous result is called measurement bias.
- Confounding. A confounder or confounding variable is a variable which is responsible for the difference (or for a proportion of the difference) observed between two groups, where this difference has been erroneously attributed to another factor. Age can act as a confounder. The proportion of animals giving a positive serological reaction will very often increase with age.

For example, when comparing two production systems (beef and dairy), one could conclude that the percentage of animals serologically positive for neosporosis was higher in dairy herds, whereas the difference could be partially or entirely due to age differences (the confounding factor) between the two production systems.

Several types of bias can be reduced by standardizing procedures before the survey and by exercising strict discipline and control during the study.

Precision

When a population is sampled, there is always variation in the results due to chance. If we assume that ten animals are selected at random from a farm consisting of 50 infected and 50 non-infected animals, the number of infected animals contained in a series of samples drawn from the herd would vary. Most often there would be five, but sometimes four, six or, less frequently, three or seven, etc. in the sample of ten animals. The results obtained in 100 random samples of ten animals each is shown in Fig. 2.1.6.

It can be seen that most often (26% of cases) the sample contains five infected animals; nevertheless, in 21% of cases it contains four or six and in 12% it contains three or seven infected animals. Thus, the values obtained by sampling 10 out of 100 animals are spread around (dispersed about) the average value, which in addition is equivalent to the true value in the population. The frequency with which an observed value occurs decreases with the distance from the true value of the parameter in the population.

The smaller the selected sample is, the greater the dispersion will be. For a sample of only two animals, the dispersion is shown in Fig. 2.1.7. In half of the cases, the sample will contain one infected animal, but in one out of four cases there will be none, and in one out of four cases there will be two infected animals. As a consequence, such a small sample size will lead to results that are widely dispersed around the true value, since in 25% of cases the conclusion would be that 'there are no infected animals' and in 25% 'only infected animals are present'.

The precision of an estimate obtained from a survey is determined by the size of the sample: increasing the sample size increases the level of precision. However, precision will not affect accuracy which is, in turn, determined by the study design and the efforts paid to reduce bias.

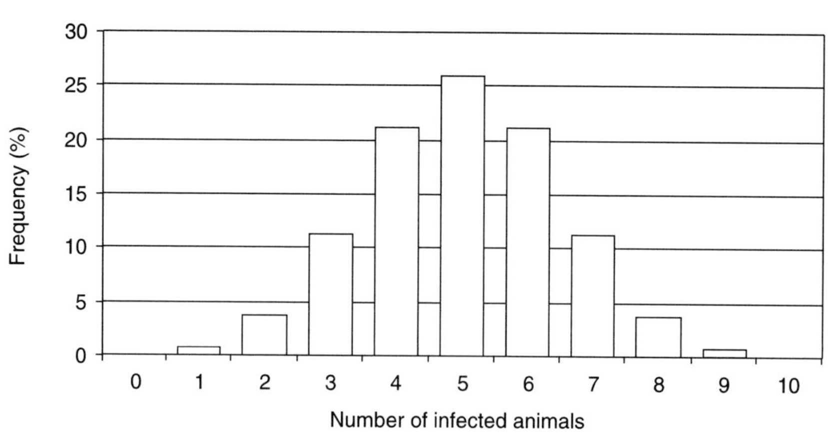

Fig. 2.1.6. Frequency distribution of number of infected animals observed in 100 random samples of ten animals each, drawn from a population of 100 animals, of which 50 are infected.

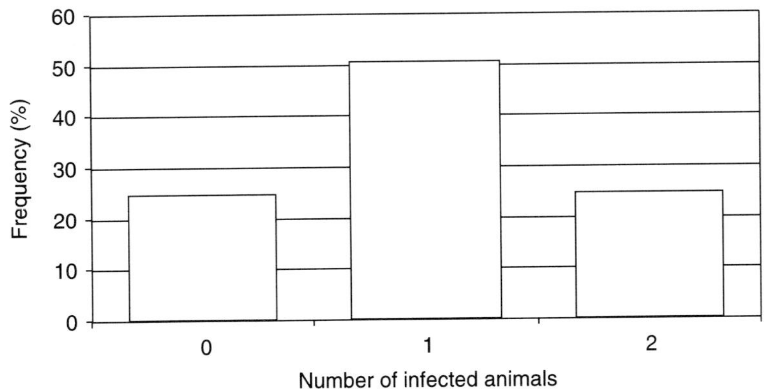

Fig. 2.1.7. Frequency distribution of number of infected animals observed in 100 random samples of two animals each, drawn from a population of 100 animals, of which 50 are infected.

Sampling Methods

Several sampling methods exist, which include the following approaches:

Non-random sampling or non-probability sampling

This method is not based on the random selection of subjects, but is based on convenience: for example, easily accessible herds or animals, farms whose owners are known or likely to agree to take part in a project. Such a sample method will not lead to a selection of units (animals, herds) that are representative for the study population, and hence it will not be justifiable to extrapolate results from the studied sample as valid for the target population. Moreover, it is not possible to define a probability with which a particular unit (animal, herd) will be included in the study and, consequently, it is not possible to calculate a required sample size with the help of the methods that are discussed later on.

Probability or random sampling

A probability sample is one in which every element in the population has a *known* probability of being included in the study. This probability must be greater than zero.

Simple random sampling

In this situation, a sample is selected at random from a population in which each member of the population has the same probability of being selected. A basic requirement for performing a simple random sampling is a list of all members of the population. Such a list is called the *sampling frame.* Random sampling can be based on drawing numbers from a hat, by using computer-generated random numbers, by using a random number table, by flipping a coin or by throwing dice.

Example:
We assume that someone wishes to estimate the prevalence of neosporosis in a beef population of 100,000 animals. A list (or identification) of all these animals is required. Then each head is numbered and 400 numbers are randomly selected for sampling.

In practice, simple random sampling is seldom used, because:

- The sampling frame is often not known.
- A list of all members of the population is not available.
- Other sampling methods might be less expensive, depending on the study objective and choice of sampling unit.
- The method is only applicable if the target population is homogeneous with regard to the studied problem. If the population is known to be heterogeneous with regard to the factors of interest, the results obtained from random sampling can lack precision.

If, for instance, the proportion of *N. caninum*-infected herds varies greatly between regions in a country (for example from 2 to 20%), a simple random sampling would produce an average result (in the order of 10%), which would not necessarily take into account the level of heterogeneity between the regions, whereas other sampling methods (see stratified random sampling) might be able to reflect these differences.

Systematic random sampling

This method does not require a sampling frame. Moreover, it is not necessary to know the exact population size prior to the investigation. Only the *sampling fraction* must be determined. The sampling fraction is the proportion obtained by dividing the required number of units in the sample by an estimate of the number of units in the population. Once this fraction is set at 1/X, it is sufficient to select at random a number Y, between 1 and X, and to select individuals starting with the one represented by the number Y and the individuals with the numbers Y + X, Y + 2X, Y + 3X … This can be illustrated by imagining that all animals of the population form a row and are numbered. The first animal (X) is randomly selected (e.g. by drawing a number from a hat) and, if Y is 6, every sixth animal in the row is also included in the sample.

Example:
Required number in sample = 50, total population = 2500. Sample fraction is 1/50. Random number selection between 1 and 50: 42.

Individuals included in the sample: 42, 92, 142, etc.

In the case of variables with a cyclical character, it is necessary to avoid a selection interval which is synchronized with the studied variable.

Stratified random sampling

This sample is obtained by random sampling within strata of the population that have previously been defined. In this context, a stratum is a subgroup within the population which is more homogeneous with respect to a given feature (region, farm size, gender).

Within each stratum, a random sample is drawn (by simple or systematic random sampling). Each sample is independent of the samples drawn from the other strata. The size of the sample drawn from each stratum can be a function of the number of units contained in each stratum (proportional sampling) or of the expected prevalence in each stratum. The advantages of stratified random sampling are:

- All strata are represented.
- The precision of overall estimates might be greater than those derived from a simple random sampling. The gain in precision results from the fact that the between-stratum variation is explicitly removed from the overall estimate of variance.
- This method produces estimates of stratum-specific outcomes, although the precision of these estimates will be lower than the precision of the overall estimate.

Example:

We assume that someone wants to estimate the prevalence of N. caninum *in cattle within a country. The prevalence may vary between different regions (provinces). Therefore, it is better to perform a stratified random sampling than a simpler systematic random sampling in this situation. In such a case, the sample size per province should be proportional to the number of animals in each province.*

Cluster sampling

A cluster is a natural or convenient collection of elements with one or more characteristics in common. In veterinary medicine, the best example of a cluster is a herd. In a cluster sample, the primary sampling unit (herd) is larger than the unit of concern (animal). In a cluster sample, every element within the cluster is included. Cluster sampling is done because it may be easier to get a list of clusters than to get a list of individuals. It is also frequently less expensive to sample a small number of clusters than to travel from herd to herd to collect information or samples from many different herds or individual animals.

However, caution is needed when interpreting the results from this sampling method when it is used to determine the disease status in individual animals. The information obtained will be less precise compared to sampling the same number of elements (animals) by simple or systematic sampling. The reason for this is that the prevalence of a disease, especially for infectious diseases, varies more between herds than within herds.

Multi-stage sampling

Multi-stage sampling is the selection of a sample in two or more stages. An example is the random selection of dairy herds in a region, after which a random sample is taken from the animals in each of the selected herds. Sample sizes for multi-stage sampling depend on four values: the within-herd variation of disease, the between-herd variation of the disease, the cost for sampling a herd and the cost of sampling animals. The ratio of the between-herd and within-herd variation indicates whether: (i) relatively more herds with a small number of animals per herd; or (ii) fewer herds with more animals per herd need to be sampled.

If there is a great variation between herds and a relatively small within-herd variation, more herds and fewer animals per herd need to be sampled. Most infectious diseases fall into this category. When there is little between-herd variation and large within-herd variation, fewer herds, but herds with many animals per herd, need to be sampled, e.g. to analyse production parameters (milk production, number of piglets per litter).

In addition, the relative cost of sampling herds compared to sampling animals within herds is taken into account to optimize the product term of variance times cost. For more details, please refer to the section, below, on further reading (listed books and software).

Clustering

The last two sampling methods involve the sampling of animals within groups. Animals within groups are usually more alike than animals chosen randomly from the population. From a statistical perspective, this means that observations made on individuals belonging to clusters are not independent. The lack of independence needs to be taken into account in the analysis. Discussion of appropriate methods is beyond the scope of this chapter. The reader is referred to the textbooks and software listed in the section, below, on further reading.

Choice of the sampling method

The sampling methods presented here are not mutually exclusive: the different methods can be combined in one survey design. The decision as to what type of sampling is to be used depends on the objectives of the survey and on available funds. Before a decision on the survey design is taken, it is essential to consult a statistician in order to establish which of the possible scenarios offers the best ratio of precision to cost.

Further Reading

A number of veterinary epidemiology textbooks explain the different aspects of study design and sampling in great detail. For compiling this chapter we referred to the following:

Applied Veterinary Epidemiology and the Control of Disease in Populations (1996) by B. Toma, B. Dufour, M. Sanaa, J.J. Bénet, F. Moutou, A. Louzã and P. Ellis, ISBN 92-9044-487-8, AEEMA.

Sampling Animal Populations (1996) by Y.H. Schukken, M.L. Beiboer and H.W. Barkema. Dutch manual, 1996.

Veterinary Epidemiological Research (2003) by Ian Dohoo, Wayne Martin and Hendrik Stryhn (available at http:// www.upei.ca/ver).

Veterinary Epidemiology (1995) 2nd edn by M. Thrusfield, Blackwell Science, Oxford, UK.

Appendix

In this appendix, information on WIN EPISCOPE and FREECALC/SURVEY TOOLBOX is given, since these packages were used when compiling this chapter.

WIN EPISCOPE

WIN EPISCOPE is software for quantitative veterinary epidemiology. It is suitable for both the design and analysis of epidemiological studies, and as an aid to the teaching of quantitative epidemiology. WIN EPISCOPE 2.0 has been programmed with BORLAND DELPHI© 1.0. The program can be run on IBM-compatible PCs with the Windows© 3.1 operating system, or higher. Modules on sample size calculations are available, providing the following calculations:

- To estimate means.
- To estimate differences between means.
- To estimate percentages (e.g. prevalence).
- To estimate difference between percentages.
- For detection of disease in a population.
- For threshold level.
- In unmatched case-control studies.

- In matched case-control studies.
- In cohort studies.

In addition, this software program provides modules on diagnostic tests, analytical observational studies and models.

FREECALC and SURVEY TOOLBOX

FREECALC is an epidemiological probability calculator designed to assist in the planning and analysis of surveys to demonstrate freedom from disease, or surveys to detect disease. FREECALC has two modules: Sample Size Calculation and Analysis of Results. Commonly used approaches to demonstrating freedom from disease have either: (i) failed to take into account the imperfect nature of laboratory tests (sensitivity and specificity not equal to 1); or (ii) assumed infinite population sizes (or sampling with replacement).

At the heart of FREECALC is a new probability formula that adjusts for imperfect tests and for population size, providing an exact result.

SURVEY TOOLBOX is a suite of software programs designed to make planning, conducting and analysing statistically valid, efficient and practical animal health surveys easier. These programs were written for veterinarians in developing countries, but many have uses for anyone involved in planning surveys.

The features included in this program are listed in Table 2.1.4.

2.2 Sample Collection

Gorka Aduriz and Raquel Atxaerandio

The clinical history is the first step for the diagnosis of abortion or early pregnancy losses problems in ruminants. Aspects like housing, nutrition, vaccination protocols, bio-security measures, reproductive performance, general management, replacement policy and feeding should be recorded in a questionnaire format to facilitate the investigation of the case (see Box 2.2.1). Moreover, it will be also very helpful to the laboratory personnel in order for them to select the most appropriate diagnostic tests.

On the farm, the clinical examination of the aborted female can give useful information because unspecific causes (i.e. pyrexia or endotoxaemia) can induce abortion. After this, a complete macroscopic examination of the aborted fetus should be carried out, even although gross lesions are not very common. An important aspect that should be included in the clinical history is the gestational age of the fetus. A pestivirus infection of the non-immune cattle, sheep or goats

Table 2.1.4. Contents of SURVEY TOOLBOX.

Random sampling from a sampling frame	Enables simple random sampling or probability proportional to size sampling, with or without replacement from a sampling frame. The frame may be in dBASE or Pardox formats
Random selection of animals in a village	Assists with the construction of a sampling frame and random selection of individual animals from a population made up of many small groups
Random geographic coordinate sampling	Tools to assist with a sampling technique that allows a true random sample to be drawn without the need for a sampling frame.
Two-stage prevalence survey design and analysis	This program calculates sample sizes for, and analyses results from, two-stage prevalence surveys using one of three sampling strategies: (i) probability proportional to size sampling at the first stage (when a good sampling frame with herd size data is available); (ii) simple random sampling (when no herd size data are available); or (iii) random geographic coordinate sampling (when no sampling frame is available)
True prevalence	Prevalence estimates from a survey can be misleading due to imperfect diagnostic tests used. This program converts the apparent prevalence from a survey to the true prevalence, taking false positive and false negative results into account
Comparison of two prevalences	The aim of prevalence surveys is usually to detect change. This program compares two prevalence estimates and their variance estimates, to determine if any difference is likely to be real or simply due to chance
Survival analysis sample size	This program calculates the sample size required when using survival analysis to compare the disease experience of two groups
Survival analysis	This program performs simple survival analysis of data with censored observations. It produces Kaplan-Meier survival curves, summary statistics and compares two survival curves using the log rank test and the hazard ratio. Although it may be applied to any type of survival data, it is specifically designed to analyse the results of retrospective disease outbreak surveys
Capture recapture	This program estimates the size of a population based on observations from two independent data sources. It may be used to calculate disease incidence based on a combination of disease reports or laboratory submissions and a structured disease survey
FREECALC – freedom from disease	The FREECALC program is used to help plan and analyse surveys to detect disease or demonstrate freedom from disease. It contains a sample size calculator, as well as an analysis module. The program may be used for surveys at the herd level or of large areas (state, province, national)

can induce an abnormal development of the fetus, with a reduced size for the corresponding gestational age. In cattle (Caldow and Gray, 2004), this age in days (x) can be calculated from the crown–rump length in centimetres (y), as follows:

$$x = 2.5\ (y + 21) \tag{1}$$

Fetuses with a different grade of autolysis are observed in the field. Grossly, they can be classified as fresh, autolytic, macerated or mummified. Fetal mummification is indicative of protozoal or pestiviral involvement. The presence of white plaques in the skin or placenta suggests a fungal involvement. White foci of necrosis in the cotyledons are seen in toxoplasmosis. Various other unspecific macroscopic changes like hydropericardia, consolidation of the lungs or hepatomegalia can also be detected during macroscopic examination. Laboratory tests are necessary to clarify the significance of these changes.

Box 2.2.1. Important background information when investigating an abortion outbreak in ruminants.

Herd/flock composition

Number of head (by productive or age groups; no. of males)
Production type (breed)

General management

Open *vs.* closed
Bio-security measures (quarantine and testing)
Sanitation
Stress-causing factors (space and sunlight)
Recent changes

Dietary management (quality and quantity)

Forage and silage
Concentrates
Mineral supplementation
Water supply and cleanliness

Case history

Actual and previous abortion rates (per female at risk)
No. of stillbirths and weak newborns
Age of the aborted female
Status of the fetus and placenta (macroscopic examination)
Clinical signs of the dam
Previous abortive causes in the herd/flock
Recent purchase of new stock
Method of service

Treatments and vaccinations

IBR (product and date of last administration)
BVD (product and date of last administration)
Parasite control (product and date of last administration)
Mastitis control (dry period and lactation)
Other (product and date of last administration)

At this juncture, the practitioner must decide to collect the appropriate samples for the specialized diagnostic laboratory. Unfortunately, diagnostic rate with abortion cases in cattle or sheep is low, varying between 23 and 46% (Kirkbride, 1990; Barr and Anderson, 1993). However, in about 90% of the cases in which a diagnosis is reached, an infectious aetiology is confirmed, due to the specialization of the veterinary diagnostic laboratories. A detailed list of the main infectious agents that cause reproductive problems in ruminants – and the corresponding diagnostic tests – is presented in Table 2.2.1.

Aborted Fetuses

In the face of an abortion problem in cattle (i.e. > 3–5% of aborted cows at risk in a short period of time) or in sheep, the practitioner should collect and deliver to the laboratory the following samples (see Fig. 2.2.1):

A fresh fetus or fetuses with placenta

Placenta is a very valuable sample because many of the agents reach the materno-fetal barrier and can be detected with a high probability in this tissue. When placenta is lost, a sample can be taken from the female. If this is not possible, a bovine placentome can be obtained from the uterine wall.

It is preferable to submit the entire fetus and the placenta chilled, in a hard, leak-proof container with ice packs to maintain a low temperature. Sometimes, however, a necropsy has to be conducted in the field. A sharp knife, pruning shears (to remove the rib cage), a saw (to remove the brain), some leak-proof plastic bags, a leak-proof container with 10% formalin and sterile syringes and needles comprise the basic equipment to carry out this task.

Although the best option can be to discuss the collection and transport of samples with the laboratory personnel, as a general rule the following samples are required:

- Formalin fixed: brain, heart, lung, liver, skeletal muscle, lip, eyelid, spleen, kidney and placenta.

Table 2.2.1. Suggested samples and methods for the diagnosis of reproductive problems in ruminants.

Case analysis		Sample(s)	Method(s)/agent(s)
Aborted fetus (bovine, ovine or caprine)	Pathology	Placenta, brain, lip, eyelid, liver, kidney, lung, heart, skeletal muscle, spleen (10% buffered formalin)	Histology (haematoxilin-eosin), immunocytochemistry; pestiviruses, neosporosis, toxoplasmosis, IBR
	Microbiology	Placenta, liver, stomach contents (chilled, NOT frozen)	Staining: modified Ziehl-Nielsen (*Chlamydia*, Q Fever) wet mount (fungi) Culture: general media, selective media (*Brucella* sp., *Campylobacter* sp., *Yersinia* sp.)
		Brain, kidney, lung, thyroid gland, spleen (chilled, NOT frozen)	FAT: pestiviruses, IBR, *Leptospira* ELISA (antigen): pestiviruses; Isolation of virus: pestiviruses, IBR PCR: neosporosis, toxoplasmosis, pestiviruses
	Serology	Thoracic fluid (chilled, NOT frozen)	IFAT: neosporosis, toxoplasmosis ELISA: Neosporosis, pestiviruses, IBR MAT: leptospirosis
Aborted female	Serology	Serum: aborted female and herd/flock profile (10–30%) Paired serum samples (chlamydiosis only) (plain tubes refrigerated)	Rose Bengal Test: brucellosis CFT: brucellosis, chlamydiosis, Q Fever ELISA: neosporosis, pestiviruses, IBR MAT: leptospirosis IFAT: toxoplasmosis
Bull/cow (infertility)	Microbiology	Preputial smegma/cervico-vaginal mucus (specific transport medium – refrigerated)	Selective media: campylobacteriosis, trichomonosis mycoplasmosis/ureaplasmosis
	Serology	As above	As above
	Biochemistry	Serum: 3–4 samples of each productive status (plain tubes refrigerated)	Enzymes: AST (GOT), γ-GT Metabolites: total protein, urea, NEFAs Minerals: Ca, Mg, Cu, Zn

(CNS, central nervous system; CFT, complement fixation test; IFAT, indirect fluorescent antibody test; FA, fluorescent antibody test; MAT, microagglutination test; NEFAs, non-sterified fatty acids).

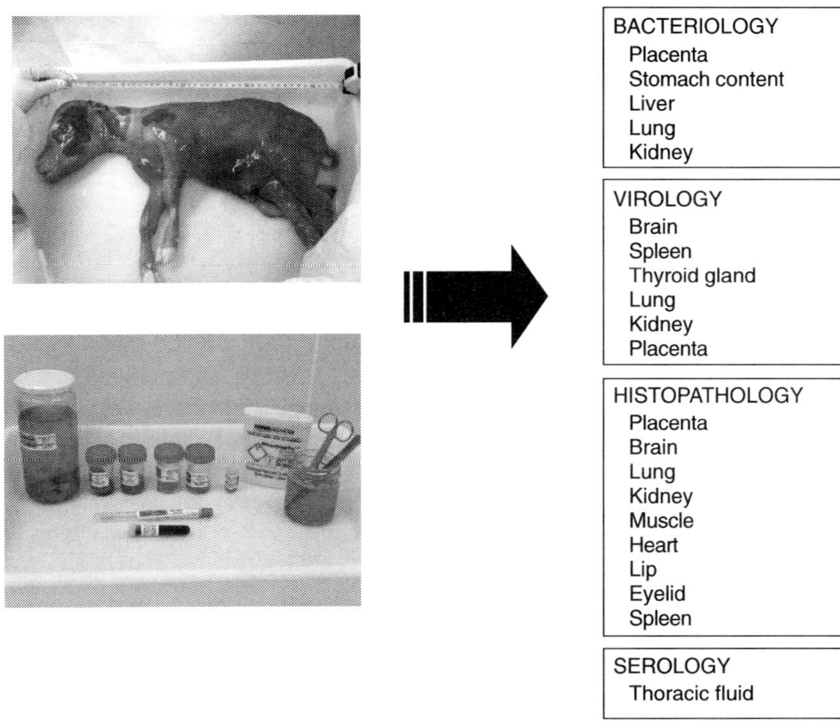

Fig. 2.2.1. Abortion investigation: collecting samples for the laboratory.

- Fresh: lung, kidney, spleen, thyroid gland, liver, 1–3 ml stomach contents (syringe), 1 ml thoracic fluid (syringe) and placenta.

In all cases, it is necessary to prevent the leakage of fluids because some of the agents causing abortion in ruminants have zoonotic potential (i.e. brucellosis, toxoplasmosis, chlamydiosis, Q fever).

Serum and blood samples

When serum of the aborted cow or sheep is the only sample submitted, a diagnostic rate of less than 15% has to be expected. In this case, an additional serologic profile of the herd or flock gives very useful information at a relatively low cost. It can be recommended, as a general rule, to take samples from 10% of each productive group (dairy herds), or 10–30% of the entire herd (beef) or flock. For specific agents (i.e. *Neospora*), a different approach should be adopted (see Chapter 3.1, this volume).

When abortion due to *Chlamydia* is suspected, paired serum samples, taken at the time of abortion and 3 weeks later, can be useful. Blood samples should be collected in vacuum plain tubes for serological tests, or in tubes with anticoagulant (EDTA) for virology (do not freeze whole blood).

Infertility

When infertility is the main reproductive problem and an infectious cause is suspected, the following samples should be taken:

Serum and blood samples

A serologic profile for the main infectious agents is the best option. Blood samples should be collected in vacuum plain tubes for serological tests or in tubes with an anticoagulant (EDTA) for virology (viral antigen detection). Serum samples can be used to assess the metabolic and mineral status of the herd/flock, although for this purpose good selection of the animals to be tested is necessary.

Vaginal and cervical swabs

When an abnormally high level of metritis or vaginitis is detected, samples from several problematic females should be collected. To avoid bacterial contaminants, the use of long swabs is recommended, like those designed for sampling mares. Different transport media should be used to investigate bacteria, viruses or mycoplasms/ureaplasms. In some cases (i.e. bovine trichomonosis), pyometra fluids or cervico-vaginal mucus can be layered directly onto a selective liquid transport medium (see Chapter 3.4, this volume).

Preputial smegma (beef herds)

This is the best sample to take when a venereal infection is suspected (trichomoniasis, campylobacteriosis or mycoplasmosis/ureaplasmosis). It is convenient to give the bull a sexual rest of 2–3 weeks before samples are taken. Once the bull is under light sedation, a long catheter is inserted to the fornix of the preputial cavity. A small quantity of sterile saline can be inoculated and aspirated to help in obtaining the smegma. Selective media for the different agents can be inoculated on the farm and carried to the laboratory under refrigeration.

References

Barr, B.C. and Anderson, M.L. (1993) Infectious diseases causing bovine abortion and fetal loss. *The Veterinary Clinics of North America: Food Animal Practice* 9, 343–368.

Caldow, G. and Gray, D. (2004) Fetal loss. In: Andrews, A.H. (ed.) *Bovine Medicine. Diseases and Husbandry of Cattle.* Blackwell Science Ltd., Oxford, UK, pp. 577–593.

Kirkbride, C.A. (1990) *Laboratory Diagnosis of Livestock Abortion,* 3rd edn. Iowa State University Press, Ames, Iowa.

Aetiological Diagnosis

3

3.1 Neosporosis

General Considerations

Franz J. Conraths and Bruno Gottstein

Neospora caninum is regarded as a major cause of bovine abortion worldwide (Dubey and Lindsay, 1996; Hemphill and Gottstein, 2000; Dubey, 2003a, b). The prevalence of the infection in cattle can vary relatively widely, as has been shown by standardized epidemiological studies in Europe (Bartels *et al.*, 2006).

In cattle, trans-placental transmission from infected dams to their offspring appears to be the major natural route of infection. Prenatally infected but healthy calves remain persistently infected and can pass the infection to their own offspring (Anderson *et al.*, 1997). This leads to endogenous trans-placental transmission of the infection through successive pregnancies and cattle generations (Björkman *et al.*, 1996; Schares *et al.*, 1998; Davison *et al.*, 1999).

Dogs and coyotes (*Canis latrans*) are definitive hosts for *N. caninum* (McAllister *et al.*, 1998; Gondim *et al.*, 2004a). Dogs can transiently shed oocysts upon ingestion of *N. caninum*-infected body tissues from intermediate hosts (McAllister *et al.*, 1998; Basso *et al.*, 2001; Dijkstra *et al.*, 2001b; Schares *et al.*, 2001b; Gondim *et al.*, 2002). Post-natal bovine infection (horizontal transmission) through peroral

© CAB International 2007. *Protozoal Abortion in Farm Ruminants* (eds L.M. Ortega-Mora, B. Gottstein, F.J. Conraths and D. Buxton)

ingestion of oocysts shed by dogs (Gondim *et al.*, 2002) or perhaps other, so far unidentified, definitive hosts is likely (McAllister *et al.*, 2000; Schares *et al.*, 2002a, b).

The role of other modes of horizontal transmission, e.g. by means of infected placenta or lochial fluid ingestion, is still under discussion. In cattle herds with epidemic abortion, point-source exposure to *N. caninum*, e.g. through oocyst-contaminated fodder or drinking water, is regarded as the most probable cause of infection (McAllister *et al.*, 2000; Dijkstra *et al.*, 2002c; Schares *et al.*, 2002a).

With some exceptions (Barling *et al.*, 2001; Rodriguez *et al.*, 2002; Fischer *et al.*, 2003), previous studies identified dogs – kept together with a cattle herd – as a major risk factor for bovine *N. caninum* infection (Paré *et al.*, 1998; Mainar-Jaime *et al.*, 1999; Ould-Amrouche *et al.*, 1999; Schares *et al.*, 2003) and *N. caninum*-associated abortion (Bartels *et al.*, 1999).

A study conducted in Texas suggested that thc abundance of wild canids, i.e. grey fox (*Urocyon cinereoargenteus*) and coyote (*Canis latrans*), was associated with the *N. caninum* seroprevalence in cattle (Barling *et al.*, 2000). As European foxes (*Vulpes vulpes*) failed to shed *N. caninum* oocysts upon experimental infection (Schares *et al.*, 2002b), it seems unlikely that a similar situation exists in Europe regarding the potential risk that wild carnivores may pose to cattle herds with respect to *N. caninum* infections.

However, the findings that coyotes can excrete *N. caninum* oocysts in their faeces (Gondim *et al.*, 2004a) and that white-tailed deer (*Odocoileus virginianus*) are natural intermediate hosts of the parasite (Dubey *et al.*, 1999; Lindsay *et al.*, 2002) suggest the existence of a sylvatic cycle of neosporosis in North America (Rosypal and Lindsay, 2005).

A number of techniques are available for investigation of *N. caninum* infection in bovine abortion cases. Reliable diagnostic conclusions regarding the role of *N. caninum* in causing abortion in a bovine animal or a whole herd can be obtained by primarily examining aborted fetuses for infection with *N. caninum* and by – complementarily – testing dams and their progeny for antibodies to the parasite. To assess the situation in a herd with regard to abortions due to infection with *N. caninum*, it will therefore be necessary to apply a combination of diagnostic techniques in most cases.

Fetal tissue can be histologically examined or the parasite can be specifically detected by PCR or immunohistochemistry (see Table 3.1.1). The histologically detectable lesions caused by *N. caninum* infection – in brain, heart and liver – are typical of protozoal infection (Wouda *et al.*, 1997b). However, a definite diagnosis is possible only by applying *N. caninum*-specific detection techniques such as immunohistochemistry or PCR.

A comparison of several direct parasite detection techniques in a ring trial revealed that immunohistochemistry has a limited sensitivity, meaning that certain PCR techniques can be considered as the method of choice for direct parasite detection (van Maanen *et al.*, 2004). It must be stressed, however, that the

Table 3.1.1. Diagnosis of *Neospora caninum* infection in aborted fetuses.

Target	Test	Sensitivity	Specificity
Disease	Histology	+/++ (?)	?
Parasite	Immunohistochemistry	+/++ (?)	?
Parasite	PCR	+++	+++
Parasite	Cell culture	(+) (?)	+++
Parasite	Infection of experimental animals	(+) (?)	+++
Parasite-specific antibodies	Immunofluorescent antibody test	+++	+++
Parasite-specific antibodies	ELISA	+++ (?)	+++
Parasite-specific antibodies	Immunoblot	+++	+++

question whether an abortion was caused by *N. caninum* cannot be answered just by demonstrating the presence of the parasite in an aborted fetus. For establishing *N. caninum*-infection as the cause of death, the judgement of a pathologist is required.

Tissue culture techniques and infection of experimental animals can be used to isolate *N. caninum* from infected animals for further propagation and characterization of the isolates, but these methods are mainly used for research purposes. Serological techniques, such as the indirect immunofluorescent antibody test (IFAT) and immunoblotting, can also be used to obtain evidence for an *N. caninum* infection in aborted fetuses by detection of specific antibodies in fetal fluids. However, the diagnostic sensitivity is strongly dependent upon the age of the fetus, with elder fetuses being more likely to be seropositive than younger ones.

In live cattle under routine conditions, this infection can so far only be diagnosed by detection of *N. caninum*-specific antibodies in serum or milk. Although *N. caninum* has also been detected by PCR in the blood and semen of cattle (Caetano-da-Silva *et al.*, 2004; Okeoma *et al.*, 2004, 2005; Ferre *et al.*, 2005), these techniques are not suitable for routine application, due to the intermittent presence of *N. caninum* in blood (Ferre *et al.*, 2005).

Generally speaking, the risk of abortion is increased two- to fourfold for seropositive dams as compared with seronegative cows (Paré *et al.*, 1997; Moen *et al.*, 1998). IFAT – a microagglutination test (MAT), immunoblotting and a considerable number of ELISAs have been described so far. These techniques can vary widely in their characteristics, qualities and in the objectives of testing. Some are optimized for the detection of infected animals; others preferentially recognize cattle that have aborted due to infection with *N. caninum*. A number of ELISAs have also been modified to study the avidity of antibodies found in infected cattle (Björkman *et al.*, 1999; Maley *et al.*, 2001; Schares *et al.*, 2002a; Sager *et al.*, 2003).

A comparison of avidity ELISAs was performed by Björkman and co-workers (2006). Antibodies of low avidity may indicate recent infection and are frequently found in cases of epidemic abortions, i.e. largely horizontal transmission, while antibodies of high avidity can usually be detected in animals that had contracted the infection a long time previously. Consequently, high avidity antibodies can be expected in trans-placentally infected cattle, i.e. in cases of vertical transmission.

Individual serological tests for the detection of bovine antibodies to *N. caninum* were compared with smaller panels of sera (Dubey *et al.*, 1997; Wouda *et al.*, 1998a; Schares *et al.*, 1999a; Jenkins *et al.*, 2000; Reichel and Pfeiffer, 2002; Álvarez-García *et al.*, 2003). A comparison of several tests with a large panel of well-documented sera was recently published (von Blumröder *et al.*, 2004).

To assess the herd status and the main transmission pattern within herds, the serological status of cow–calf pairs can also be established, and the role of endogenous and exogenous trans-placental transmission analysed (Parè *et al.*, 1996; Schares *et al.*, 1998). In these studies it is of utmost importance that the calves are pre-colostrally bled to avoid the maternal antibodies directed against *N. caninum* becoming confused with the specific immune response raised by the calves themselves. Alternatively, umbilical cord blood could be used for the same purpose (Michael Hässig, Zürich, 2003, personal communication).

Relatively little is known about the role of natural *N. caninum* infections in sheep (Buxton, 1998). The first *N. caninum* isolation from the brain of a naturally infected pregnant ewe was obtained by inoculation of immunodeficient mice (Koyama *et al.*, 2001). The same group also found *N. caninum* tissue cysts in the brains of twin fetuses and of the ewe – including histopathological findings such as focal gliosis with mononuclear cell cuffing (Kobayashi *et al.*, 2001).

Insight into the pathogenesis of ovine neosporosis has been provided by experimental infections of pregnant ewes (Buxton *et al.*, 1998), leading to fetal death and respective resorption or abortion. Immunohistochemical examination showed that the parasite had predominantly invaded the placentae and fetal brains of all cases examined (Buxton *et al.*, 1998). In the field, however, ovine neosporosis is regarded as an infrequent cause of abortion (Owen and Trees, 1999; Hurtado *et al.*, 2001), reports being restricted to a few cases (Helmick *et al.*, 2002; Hässig *et al.*, 2003).

Even less is known about the putative role of *N. caninum* as an abortifacient organism in goats. Some authors found the parasite associated with abortion in goats (Barr *et al.*, 1992; Eleni *et al.*, 2004). A granulomatous encephalitis has been histologically detected in a neurologically impaired goat kid: lesions were associated with the presence of degenerating *N. caninum* tissue cysts (Corbellini *et al.*, 2001).

In terms of diagnosis, the same tools (PCR and histology/immunohistochemistry) could be applied to detect *N. caninum* directly in abortion cases. Serological techniques – mainly IFAT and ELISA – were adapted from the bovine system to the ovine and caprine systems.

Biology, Transmission and Clinical Signs

Willem Wouda

Neosporosis is a disease of animals caused by the apicomplexan parasites *Neospora caninum* and *N. hughesi*, which are obligate intracellular protozoa closely related to *Toxoplasma gondii*. *N. caninum* was first described in dogs in 1984 (Bjerkås *et al.*, 1984) and later in calves with encephalomyelitis (O'Toole and Jeffrey, 1987; Parish *et al.*, 1987), although it was not isolated and named until 1988 (Dubey *et al.*, 1988a, b). *N. hughesi* has been described only in horses (Marsh *et al.*, 1998). This review will focus on neosporosis caused by *N. caninum*.

Life cycle and infectious stages

Neospora caninum has a heteroxenous life cycle (see Fig. 3.1.1). To date, only dogs (McAllister *et al.*, 1998) and coyotes (Gondim *et al.*, 2004a) have been identified as definitive hosts of *N. caninum*. A wide range of domestic and wild animals, including cattle, water buffaloes (Rodrigues *et al.*, 2004) and white-tailed deer (Gondim *et al.*, 2004b; Vianna *et al.*, 2005) can act as intermediate hosts (Dubey and Lindsay, 1996).

There are three infectious stages of the parasite: tachyzoites, bradyzoites and sporozoites.

Tachyzoites and bradyzoites are asexually proliferating stages that occur in tissues of infected hosts (intermediate and definitive), whereas sporozoites are present in sexually produced oocysts that are excreted in the faeces of definitive hosts. Tachyzoites are ovoid, lunate or globular, depending on the stage of division and are approximately 5–7 μm in length and 1–2 μm wide (Dubey and Lindsay, 1996; Hemphill, 1999). They may actively invade a wide variety of nucleated cells, becoming enclosed in parasitophorous vacuoles, where they divide rapidly by endodyogeny. Proliferating tachyzoites cause host cell lysis and newly formed tachyzoites infect neighbouring cells. Tachyzoites may be spread throughout the body by the bloodstream, with trans-placental transmission to the fetus frequently occuring in pregnant animals.

Bradyzoites are slowly replicating stages of the parasite that are capable of forming intracellular tissue cysts. Bradyzoites are slender, approximately 6–8 μm long and 1–2 μm wide (Hemphill, 1999; Dubey *et al.*, 2002). Tissue cysts have a distinct cyst wall up to 4 μm in thickness, which protects the parasites from immunological and physical reactions on the part of the host. The thickness of the cyst wall depends on how long the infection has persisted (Jardine, 1996).

Tissue cysts may vary considerably in size (up to 100 μm), depending on the number of bradyzoites within them (Dubey *et al.*, 2002). Tissue cysts occur mostly in the central nervous sysem (CNS), but have also been reported in

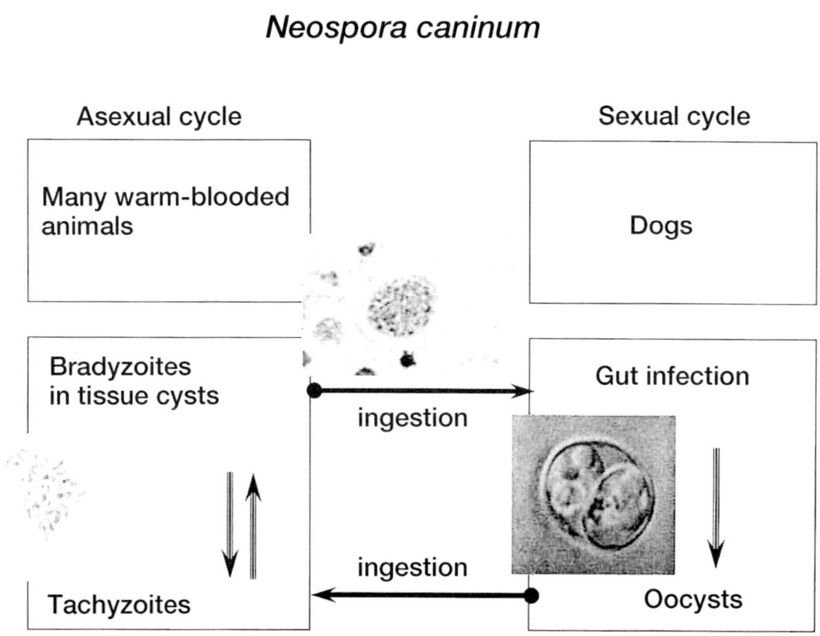

Fig. 3.1.1. Life cycle of *Neospora caninum.*

skeletal muscle (Peters *et al.*, 2001a). Tissue cysts can persist in infected hosts for several years without causing significant clinical manifestations. Definitive carnivore hosts can acquire the infection by ingestion of tissues containing tissue cysts. The cyst wall is resistant to acid pepsin digestion in the stomach, which allows bradyzoites to be released in the gut.

Unsporulated *N. caninum* oocysts, approximately 10 × 12 μm, are excreted in canine faeces. Sporulation then occurs so that each oocyst contains two sporocysts, each of which contains four sporozoites (McAllister *et al.*, 1998). The schizogonic and gametogenic stages that are presumed to precede the formation of oocysts in the intestines of dogs have not yet been documented.

Experimentally, dogs have shed oocysts after ingesting naturally infected tissues from cattle (Dijkstra *et al.*, 2001b; Gondim *et al.*, 2002), water buffaloes (Rodrigues *et al.*, 2004) and white-tailed deer (Gondim *et al.*, 2004b). *N. caninum* oocysts have infrequently been identified in the faeces of naturally infected dogs (Basso *et al.*, 2001; McGarry *et al.*, 2003; Schares *et al.*, 2005), but on one occasion protracted or repeated shedding of oocysts has been documented (McGarry *et al.*, 2003). *N. caninum* oocysts constitute a reservoir of the infection in the environment, although little is known of their viability and persistence.

Transmission routes

There are two transmission routes for *N. caninum* in cattle (see Fig. 3.1.2). The first, and probably most important, route is usually called vertical transmission or congenital transmission, or more precisely endogenous trans-placental transmission. This refers to the trans-placental passage of organisms from a persistently infected cow to her calf *in utero*. The infection originates from reactivated bradyzoites within tissue cysts in the dam's tissues during a so-called recrudescence of infection, which is enhanced by a decrease of cell-mediated immunity (Entrican, 2002).

The second route is referred to as horizontal or post-natal transmission, which occurs when cattle ingest sporulated *N. caninum* oocysts. If a pregnant cow is infected by ingestion of oocysts, the infection may be passed on to her calf trans-placentally.

Recently, the terms endogenous trans-placental transmission and exogenous trans-placental transmission have been proposed to describe more precisely the origin and route of infection of the fetus (Trees and Williams, 2005). Endogenous trans-placental transmission occurs in a persistently infected dam after recrudescensce of the infection during pregnancy, while exogenous trans-placental transmisson occurs after a primary, oocyst-derived, infection of a pregnant dam.

Vertical, i.e. endogenous trans-placental transmission, may lead to abortion but, in the majority of cases, a healthy, congenitally infected calf is born.

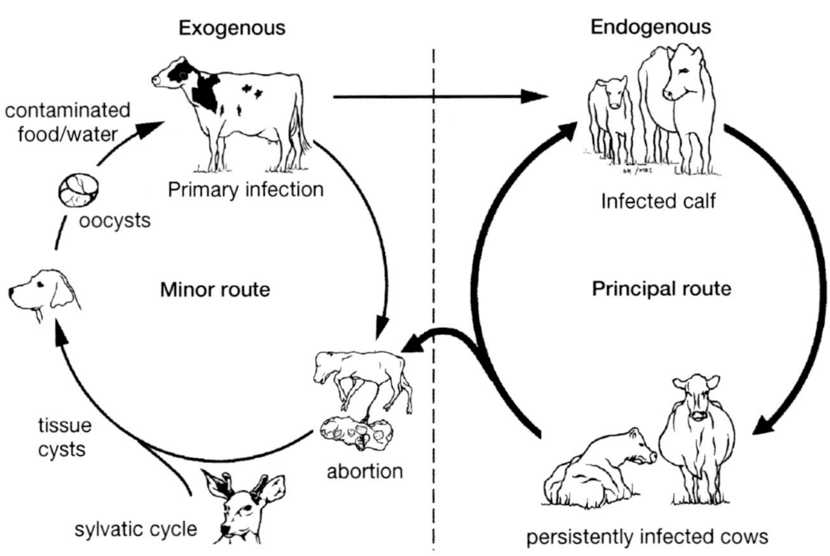

Fig. 3.1.2. Transmission routes for *Neospora caninum* in cattle.

Therefore, endogenous trans-placental transmission contributes significantly to the maintenance of the infection in a herd.

Cows may remain infected with *N. caninum* for life (Trees *et al.*, 1999) and may transmit the infection to their offspring in several consecutive pregnancies (Fioretti *et al.*, 2003) or intermittently (Boulton *et al.*, 1995; Wouda *et al.*, 1998b; Guy *et al.*, 2001). Congenital infection rates are high, varying from 66 to 95% (Paré *et al.*, 1996; Davison *et al.*, 1999; Dijkstra *et al.*, 2003).

In two separate studies, the congenital infection rate decreased with the increasing parity of the dam, suggesting that cows may, over time, develop a degree of immunity that prevents endogenous trans-placental transmission (Romero *et al.*, 2002; Dijkstra *et al.*, 2003). In one of these, 20 Costa Rican dairy herds were investigated and the daughters born to dams with six or more parturitions were shown to have a significantly decreased probability of being seropositive when compared with daughters born to dams with one or two parturitions (Romero *et al.*, 2002).

In the other study, researchers in the Netherlands analysed 500 dam–daughter pairs in 21 dairy herds with a history of neosporosis and discovered a congenital infection rate of 80% in heifers, 71% in second parity cows, 67% in third parity cows and 66% in fourth parity and older cows (Dijkstra *et al.*, 2003).

Despite the efficiency of endogenous trans-placental transmission, it is evident from theoretical modelling that *N. caninum* infection cannot be sustained in herds without horizontal transmission (French *et al.*, 1999). Moreover, epidemiological studies and field observations are providing increasing evidence for the occurrence of horizontal infection in cattle, leading to exogenous trans-placental transmission to the fetus in infected dams: particularly, epidemic abortion outbreaks involving 20% or more of the animals at risk, have been associated with horizontal transmission of *N. caninum* (McAllister *et al.*, 1996a, 2000; Thurmond *et al.*, 1997; Patitucci *et al.*, 1999; Waldner *et al.*, 1999; Dijkstra *et al.*, 2001a).

Important observations in this respect are: (i) a characteristic epidemic curve of a given outbreak of *N. caninum*-associated abortions (sudden progression, high amplitude, skewed to the right), suggesting a point-source exposure (Yaeger *et al.*, 1994; McAllister *et al.*, 1996a; McAllister *et al.*, 2000); (ii) an increasing seropositivity with age (Dyer *et al.*, 2000; Romero *et al.*, 2002); and (iii) the lack of an association between the seropositivity of dams and daughters in infected herds (Thurmond *et al.*, 1997; Mainar-Jaime *et al.*, 1999; Waldner *et al.*, 1999; Dyer *et al.*, 2000; Dijkstra *et al.*, 2001a).

One study showed remarkable differences in seroprevalence among age groups in herds with *N. caninum*-associated abortion epidemics (Dijkstra *et al.*, 2001a). Animals in seropositive age groups had either seronegative dams or seronegative offspring. The first situation indicates horizontal infection of the daughters, whereas the second situation indicates horizontal infection of the dams after the birth of their seronegative daughters.

Further, an avidity-ELISA (Björkman *et al.*, 1999) has been used to provide evidence of horizontal infection in a herd, as the presence of many seropositive cows with low avidity suggests a recent infection of the herd (McAllister *et al.*, 2000; Dijkstra *et al.*, 2002a).

The presence of dogs on a farm has been shown to be a risk factor for infection of cattle (Paré *et al.*, 1998; Bartels *et al.*, 1999; Wouda *et al.*, 1999b). The introduction of a new dog to a farm with endemic bovine neosporosis appears to be a risk factor for horizontal transmission in the herd (Dijkstra *et al.*, 2002c).

Dogs may become infected by eating placenta from freshly calved infected cows (Dijkstra *et al.*, 2001b, 2002b). Other sources of infection for dogs may be bovine carcasses (Trees and Williams, 2002) or raw meat (Basso *et al.*, 2001). Aborted fetuses appear to be a less likely source of infection (Bergeron *et al.*, 2001a).

Coyotes – and possibly other wild canids – may play a role in a sylvatic cycle, with cervids (deer) as intermediate hosts. Transmission of *N. caninum* between wild and domesticated animals has been postulated. Hunting by humans favours the transmission of *N. caninum* from deer to canids, because deer carcasses are usually eviscerated in the field. Infection of canids in turn increases the risk of transmitting the parasite to domestic livestock (Gondim *et al.*, 2004b).

At present, there is no evidence of cow-to-cow transmission of *N. caninum.* For example, in one study 25 heifers, seronegative for *N. caninum*, were housed from birth with 25 heifers seropositive for the parasite and their progeny were monitored serologically for *N. caninum* infection. The seronegative heifers remained seronegative and gave birth to calves not infected with *N. caninum*, while the seropositive heifers remained clinically normal but gave birth to congenitally infected calves. Seven of the latter, congenitally infected calves (four of which were recumbent) were necropsied and all had histological evidence of *N. caninum* infection (Anderson *et al.*, 1997).

Theoretically, *N. caninum* may be excreted in the milk or uterine discharges of infected cattle. It has been speculated that lactogenic transmission of tachyzoites – or ingestion of fetal membranes or uterine fluids containing tachyzoites – may contribute to such infection (Davison *et al.*, 2001). However, it seems currently that these routes are not important.

Additionally, it is unlikely that *N. caninum* is transmitted venereally in cattle. Although *N. caninum* DNA has been found in the semen of naturally exposed bulls (Ortega-Mora *et al.*, 2003; Caetano da Silva *et al.*, 2004; Ferre *et al.*, 2005), the results suggest that viable organisms, if present, are few and infrequent.

Basic prevalence data

Serological surveys indicate that *N. caninum* occurs in dairy and beef cattle throughout the world, but that seroprevalence varies greatly between countries (Wouda *et al.*, 2000; Dubey, 2003a). It is difficult, however, to compare the results of various studies, because they were obtained by using different serological techniques and study designs.

Recently, herd and animal prevalences for *N. caninum* in beef and dairy cattle within four European countries were compared using a standardized cross-sectional study design and harmonized serological techniques (Bartels *et al.*, 2006). The results of this study, conducted in Germany, the Netherlands, Spain and Sweden, show that the importance of *N. caninum* infection varies greatly within Europe. Herd prevalences for dairy herds were estimated to be 16% in Sweden, 50% in Germany, 63% in Spain and 80% in the Netherlands.

The lowest animal true prevalence was estimated in dairy cattle in Sweden: 0.5%, while the highest animal true prevalence was estimated in dairy cattle in Spain: 16.2%. Within-herd prevalences varied greatly, with very few farms in Sweden having more than 10% seropositive animals, while in Spain more than 10% of the herds had within-herd prevalences between 50 and 100%. Recent studies in Italy (Otranto *et al.*, 2003; Rinaldi *et al.*, 2005) and Portugal (Canada *et al.*, 2004) provide further evidence that *N. caninum* prevalence in cattle is relatively high in southern Europe, which may be related to climatic factors.

In a study in Costa Rica, herd prevalence was 100% and within-herd prevalences were between 25.0 and 70.5% in dairy cattle (Romero *et al.*, 2002). The probability of horizontal infection in these herds was calculated to be 0.22% (Romero and Frankena, 2003). This is in line with modelling studies, which show that a high within-herd prevalence is probably due to a high external exposure (French *et al.*, 1999).

Climatic factors such as the mean temperature in July, the mean temperature in January and the total yearly precipitation in districts and cities contributed to the variability in the regional prevalences in a German study (Schares *et al.*, 2003). In addition, prevalences were higher in regions with an advanced degree of urbanization, which is related to dog density.

Also, researchers in Italy, using geographic information systems and remote sensing, found climatic and environmental factors – such as the minimum temperature during spring and the vegetation index during summer, to be related to *N. caninum* prevalence (Rinaldi *et al.*, 2005). These results provide evidence that, in addition to risk factors related to management on individual farms, and also to risk factors related to the farm location such as climatic factors and dog density in the surrounding area, are important in the epidemiology of bovine neosporosis.

Clinical signs

Neosporosis is a major cause of reproductive failure in dairy and beef cattle in many countries, causing considerable economic losses (Trees *et al.*, 1999). The infection may cause fetal resorption, mummification, abortion, premature birth and stillbirth. However, most fetuses infected during pregnancy survive and are born congenitally infected but clinically healthy. Rarely, congenitally infected calves may have nervous signs varying from mild ataxia to tetraparalysis (Dubey, 2003a). The hind limbs and/or the forelimbs may be flexed or hyperextended. Exophthalmia or an asymmetrical appearance in the eyes has been reported and, occasionally, birth defects have included scoliosis, hydrocephalus and a narrowing of the spinal cord (O'Toole and Jeffrey, 1987; Parish *et al.*, 1987; Barr *et al.*, 1993; Dubey and Lindsay, 1996).

Abortion is the main clinical manifestation of bovine neosporosis in cattle. Most abortions occur during mid-gestation. Fetuses dying *in utero* between 3 and 8 months of gestation are usually expelled showing moderate autolysis, but fetuses dying before 5 months' gestation may be mummified and retained in the uterus for several months; those dying at an early stage of gestation may be reabsorbed, causing repeat breeding (Anderson *et al.*, 1991; Barr *et al.*, 1991; Dubey, 2003a).

Abortions may be epidemic or endemic. In some herd outbreaks as many as 33% of dairy cows have been reported to abort over just a few months (Thilsted and Dubey, 1989; Yaeger *et al.*, 1994; McAllister *et al.*, 1996a, 2000; Moen *et al.*, 1998; Wouda *et al.*, 1999a; Dijkstra *et al.*, 2001a; Schares *et al.*, 2002a). Abortions have been defined as epidemic if more than 10 or 12.5% of cows at risk abort within 6–8 weeks (Wouda *et al.*, 1999a; Schares *et al.*, 2002a).

A small proportion (< 5%) of cows may have repeated abortion due to neosporosis (Anderson *et al.*, 1995). Most abortions due to neosporosis are probably the result of reactivation of a latent infection. Pregnancy-induced immunosuppression may cause reactivation of encysted parasites and subsequent parasitaemia.

Other factors causing immunosuppression, like mycotoxicosis (Bartels *et al.*, 1999) or BVDV infection (Björkman *et al.*, 2000) may be involved in triggering recrudescence of a latent infection. *N. caninum*-infected cows have a 2- to 3.5-fold higher risk of abortion than non-infected herd mates (Paré *et al.*, 1997; Moen *et al.*, 1998; Davison *et al.*, 1999; Stenlund *et al.*, 1999). In congenitally infected heifers, the risk of abortion is even higher (Thurmond and Hietala, 1997). Aborting cows show no general signs of illness, returning to oestrus and possibly becoming pregnant again without delay.

Neosporosis in other animal species

Neosporosis has been sporadically associated with abortion in goats (Barr *et al.*, 1992; Eleni *et al.*, 2004) and sheep (Dubey and Lindsay, 1996; Hässig *et al.*, 2003). Sheep have been successfully used for experimental reproduction of the disease as a model for neosporosis in cattle (McAllister *et al.*, 1996b; Buxton *et al.*, 1997). Abortions due to *N. caninum* have also been documented in a horse (Dubey and Porterfield, 1990), water buffaloes (Guarino *et al.*, 2000) and camelids (Serrano-Martinez *et al.*, 2004).

Naturally occurring neosporosis has also been found in a 2-month-old fawn of a wild black-tailed deer (*Odocoilos hemionus columbianus*, Woods *et al.*, 1994), in a stillborn fawn of a captive elk (*Cervus eldi siamensis*, Dubey *et al.*, 1996b), in stillborn calves of captive antelopes (*Tragelaphus imberbis*, Peters *et al.*, 2001b) and in a calf of free-ranging white rhinoceros (*Ceratotherium simum*, Williams *et al.*, 2002).

Diagnostic Techniques

Histology

Willem Wouda and David Buxton

Histological examination of fetal tissues can be of great value as a diagnostic procedure, because the lesions caused by *N. caninum* are often distinctive. Most aborted fetuses infected with *N. caninum* have randomly scattered foci of mononuclear inflammatory cells and foci of necrosis in many tissues, including brain, heart, muscle, liver, kidney, lung, adrenal gland and placenta (Barr *et al.*, 1990).

The following tissues should be sampled routinely for histological examination: brain (ensuring adequate coverage of the cerebral cortex, brainstem and cerebellum), myocardium, liver, lung and placenta (if available). The latter two tissues are less relevant for the positive diagnosis of neosporosis, but are useful for ruling out other infectious causes of abortion.

Procedure

1. Fix tissues in 10% formol saline (ensure that samples are immersed in 10 × their volume of fixative) and trim selected blocks.
2. Dehydrate through graded alcohols.
3. Clear in xylene (or an equivalent).
4. Embed in paraffin wax.
5. Cut 4 μm sections with a microtome.
6. Mount sections on glass histology slides.
7. Stain with haematoxylin and eosin (H & E).
8. Mount under a coverslip.
9. Histopathological examination by a pathologist.

Below, a brief description of the histopathological lesions in the tissues recommended for sampling is given.

Histopathological findings

BRAIN The most characteristic finding is a multifocal, non-suppurative encephalitis in which small inflammatory foci are scattered throughout the brain,

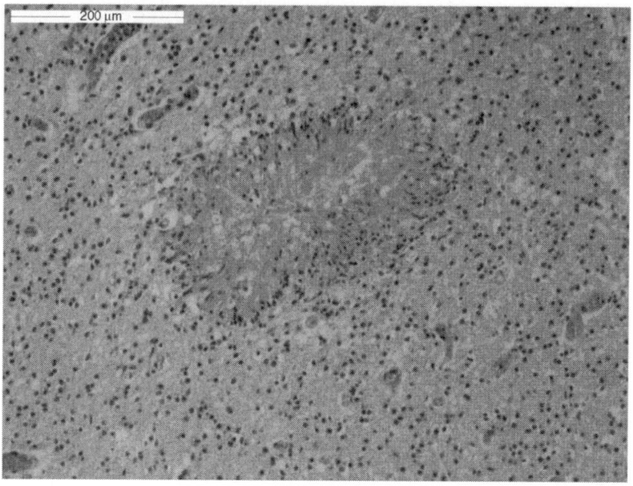

Fig. 3.1.3. *Neospora caninum* lesions in fetal brain: necrotic centre surrounded by glial cells.

typically consisting of a necrotic centre surrounded by a rim of glial cells (see Fig. 3.1.3). However, focal necrosis without a cellular reaction and focal gliosis without necrosis (see Fig. 3.1.4) may also occur.

The lesions in the brain are often accompanied by some vascular endothelial hyperplasia and associated mild infiltration by mononuclear cells.

In some cases, small foci of leukomalacia may be seen in the cerebral white matter. These are not normally associated with an inflammatory response and

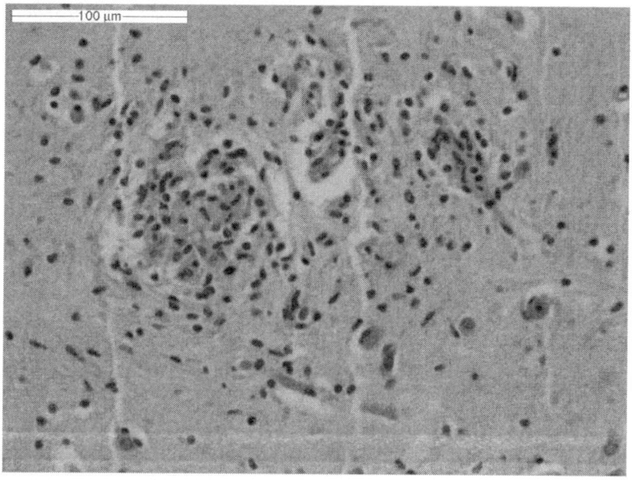

Fig. 3.1.4. *Neospora caninum* lesions in fetal brain: focal gliosis.

are presumably the result of severe fetal anoxia immediately prior to parturition. Their presence is not necessarily specific to neosporosis, but rather an indication of placental dysfunction.

Parasites are not easily found in H & E stained sections, but occasionally tachyzoites may be observed in the lumen of blood vessels or lying in clusters within the brain parenchyma (Barr *et al.*, 1990; Wouda *et al.*, 1997b). Small tissue cysts may also be observed.

MYOCARDIUM Focal or diffuse myocarditis is a frequent finding in infected fetuses (Barr *et al.*, 1990; Wouda *et al.*, 1997b), with lesions characterized by varying numbers of mixed mononuclear cell infiltrates in the epicardium, myocardium and endocardium (see Fig. 3.1.5). In acute lesions, necrosis of cardiomyocytes may be observed, but lesions are often obscured by autolysis. Rarely, aggregations of tachyzoites can be found within cardiac muscle cells (Wouda *et al.*, 1997b). Similar lesions may be found in skeletal muscle.

LIVER Two types of lesion can be encountered in the liver: periportal hepatitis (see Fig. 3.1.6) and multifocal hepatocellular necrosis (see Fig. 3.1.7). Both types of lesions may occur simultaneously. Periportal hepatitis, characterized by infiltration of mononuclear cells in portal areas, is the most common lesion. Multifocal hepatocellular necrosis, with associated intra-sinusoidal deposits of fibrin, has been associated with epidemic abortions (Wouda *et al.*, 1997b). This type of lesion may be related to an acute infection or a high parasite load.

PLACENTA Descriptions of lesions in the placenta are limited, as relatively few samples of placenta become available for examination and those that do are very

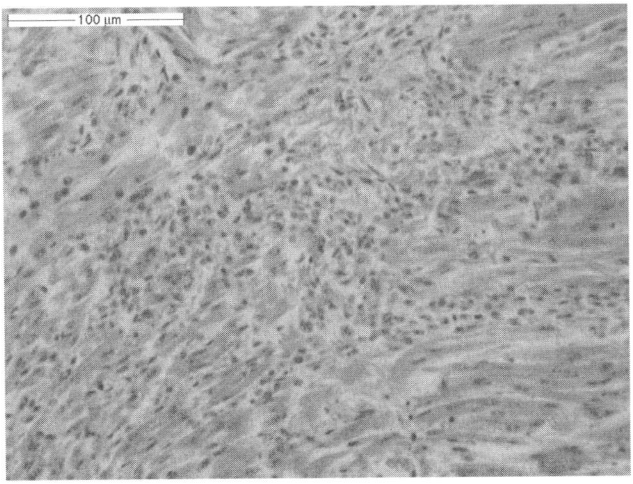

Fig. 3.1.5. *Neospora caninum* lesions in fetal heart: myocarditis.

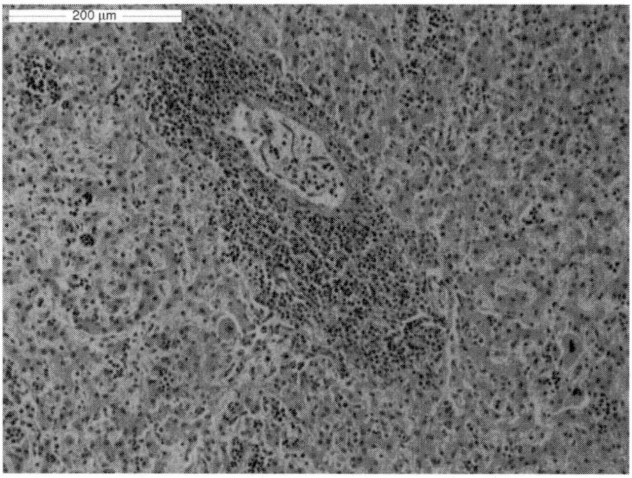

Fig. 3.1.6. *Neospora caninum* lesions in fetal liver: periportal hepatitis.

often autolytic. A non-suppurative placentitis has been reported (Otter *et al.*, 1995; Bergeron *et al.*, 2001b), and experimentally induced lesions confirm that in the second half of pregnancy a non-suppurative placentitis may occur, sometimes with mineralization (Maley *et al.*, 2003). In experimentally induced disease earlier in gestation, there may be acute necrosis of fetal placental villi (Macaldowie *et al.*, 2004).

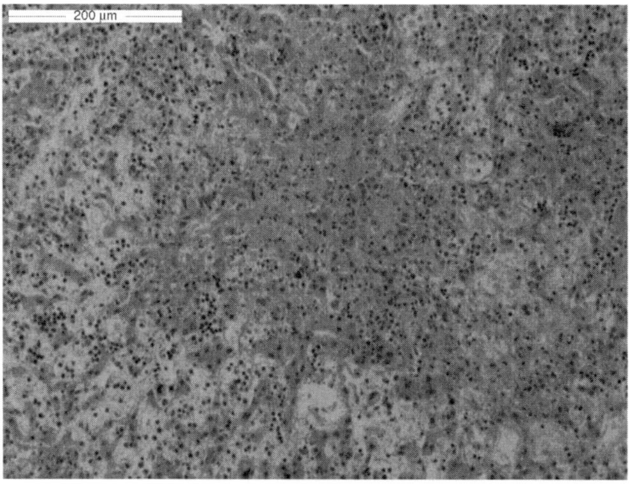

Fig. 3.1.7. *Neospora caninum* lesions in fetal liver: hepatocellular necrosis.

Immunohistochemistry

Willem Wouda and David Buxton

A definitive diagnosis depends on the identification of *N. caninum* in fetal tissues associated with the lesions described in the previous section. This can be achieved by immunohistochemistry, in which specific antibody binds to *N. caninum* antigen in the tissue section and is then visualized by applying reagent(s) that locate on the bound antibody and produce a colour reaction. The method labels intact *N. caninum* or antigenic debris in tissue sections from normally fixed paraffin wax-embedded tissues (including archive blocks).

Several immunohistochemical procedures are suitable, including the peroxidase-antiperoxidase (PAP) technique, the Vectastain avidin biotin complex method (Vector Laboratories, California, USA) and the DAKO EnVision+ System (DAKO Corporation, California, USA). Polyclonal antibodies against *N. caninum* tachyzoites are more commonly used as a primary antibody in diagnosis, as they will recognize a number of different parasite antigens (Lindsay and Dubey, 1989), but certain monoclonal antibodies may also be appropriate (Cole *et al.*, 1993). For each technique, the instructions of the manufacturer should be strictly followed. The procedure for the PAP technique is briefly outlined below.

Reagents, materials and equipment

Tris buffer solution (pH 7.6)
Methanol 100%
Ethanol 100%
Ethanol 96%
Ethanol 70%
Xylene (or an equivalent)
H_2O_2 30%
Haematoxylin
Anti-*Neospora* antiserum (polyclonal or monoclonal), prepared in (non-ungulate) animal A
Anti-animal A-IgG antibodies prepared in (non-ungulate) animal B
Peroxidase–antiperoxidase complex
Diamino benzidine (DAB) solution (DAKO)
Demineralized water
Histological slides
Coverslips
Usual equipment of an (immuno)histological laboratory

Procedure

Note: In each laboratory, optimization of antiserum dilutions should be performed using a positive control sample before application of the test for routine use. In each test session, duplicate negative control sections (not treated with specific antiserum) should be prepared together with the test sections, as well as positive control sections from a confirmed positive case.

1. Prepare unstained paraffin wax sections as detailed in the previous section.
2. Dewax sections via xylene, reverse graded alcohols and water.
3. Incubate for 30 min in 1% H_2O_2 in methanol to inhibit endogenous peroxidase.
4. Wash with water and tris buffer solution, respectively.
5. Incubate with anti-*Neospora* serum of animal A for 30–45 min at room temperature (for negative control use tris buffer solution).
6. Wash with tris buffer solution.
7. Incubate with anti-animal A-IgG antibodies.
8. Wash with tris buffer solution.
9. Incubate with peroxidase–antiperoxidase complex.
10. Wash with tris buffer solution.
11. Incubate with DAB solution.
12. Wash with tris buffer solution.
13. Stain lightly with haematoxylin.
14. Dehydrate via graded alcohols.
15. Clear with xylene.
16. Mount under a coverslip.
17. Histopathological examination by a pathologist.

An advantage of immunohistochemistry, compared with the PCR, is that the presence of parasites, or parasite antigen, can be related to the lesions. However, immunohistochemistry has a lower sensitivity than the PCR. Also, due to methodological differences, the experience of the operator and the time devoted to screening, the sensitivity of the procedure may vary considerably between laboratories (Van Maanen *et al.*, 2004). It is important to be aware that there may be some cross-reactivity with *Toxoplasma gondii* (Van Maanen *et al.*, 2004), necessitating that the specificity of antisera be checked against tissues containing *Toxoplasma.*

Brain is the tissue of choice for making an immunohistochemical diagnosis of *N. caninum* in bovine fetuses, although other tissues such as the liver and heart may also be useful (Wouda *et al.*, 1997b). *Neospora caninum* tachyzoites or parasite antigen are usually found within focal lesions of necrosis and/or inflammation. Diffuse soluble antigen may be found in neurons and neuronal processes as well as in other cells. Parasites often occur in small groups, but sometimes they are sparse and not easily found. Occasionally, tachyzoites may be found intravascularly or in close proximity to blood vessels. *Neospora caninum* tissue cysts may be found in fetal brain without associated cellular response.

Polymerase chain reaction

Jens G. Mattsson and Norbert Müller

Introduction

In the last few years, the diagnosis of neosporosis has been much improved by the development of PCR tests that allow a fast and methodically highly sensitive identification of the parasite through the amplification, and subsequent demonstration, of parasite-specific DNA sequences. As can be concluded from the current literature, the NC5- (Müller *et al.*, 1996) and the ITS1- (Holmdahl and Mattsson, 1996) PCRs are generally accepted tests for routine diagnostic and epidemiological applications in the field of neosporosis.

The high diagnostic value of these two PCRs has been demonstrated in a European inter-laboratory evaluation (van Maanen *et al.*, 2004), where the two DNA amplification reactions proved to be much more reliable than immunohistochemistry. However, this study also clearly demonstrated that a standardization of these two PCR assays, as well as regular quality control rounds, are needed. It is evident that lack of standard protocols, as well as a variability in the quality of reagents and equipment, influence the efficient inter-laboratory dissemination of the PCR methodology. In order to achieve international uniformity in the molecular diagnosis of neosporosis, the following standardized procedures for the application of the NC5- and ITS1-PCR are suggested.

Materials

SAMPLE PREPARATION DNeasy™ Tissue Kit (Qiagen)

DNA AMPLIFICATION Oligonucleotide primers for the NC-5 PCR:

Np21plus (5'CCCAGTGCGTCCAATCCTGTAAC3')

Np6plus (5'CTCGCCAGTCAACCTACGTCTTCT3')

Oligonucleotide primers for the ITS1 PCR:

PN3 (5'TAC TAC TCC CTG TGA GTT G3')

PN4 (5'TCT CTT CCC TCA AAC GCT A3')

AmpliTaq DNA polymerase (Applied Biosystems, Foster City, California, USA)

10 × Gene Amp™ PCR buffer (Applied Biosystems, Foster City, California, USA)

Deoxynucleoside triphosphates (dNTPs) 10 mM of each, (dATP, dGTP and dCTP and dTTP or 0.4 mM dUTP instead of dTTP (Amersham Biosciences (now GE Healthcare), Little Chalfont, UK)

Uracyl DNA glycosylase (UDG), heat-labile (Roche Diagnostics, Basel, Switzerland)

Nuclease-free water
Recombinant internal positive control (inhibition control, IC)
DNA thermal cycler.

GEL ELECTROPHORESIS DNA grade agarose (Amersham Biosciences (now GE Healthcare), Little Chalfont, UK)

50 × TAE buffer: 242 g Tris base, 57.1 ml glacial acetic acid and 100 ml 0.5 M EDTA (pH 8.0)l, final concentration 2 M Tris-acetate 50 + mM EDTA

Ethidium bromide, 10 mg/ml (Bio-Rad, Hercules, California, USA) (*wear gloves when handling*)

6 × Gel loading solution (GLS): 0.25% bromophenol blue, 0.25% xylene cyanol FF and 40% (w/v) sucrose in water

Molecular weight marker PhiX174 RF DNA HaeIII fragments (Invitrogen, Carlsbad, California, USA); alternatively, the 100-bp marker from Amersham Biosciences can be used

Agarose gel electrophoresis equipment
UV-transilluminator

Methods

SAMPLE PREPARATION Genomic DNA is extracted from different diagnostic samples (e.g. tissue, blood) by using the DNeasy™ Tissue Kit according to the instructions (handbook) provided by the manufacturer. Note that volumes and quantities of starting materials can be scaled up or down according to sample type and availability.[1]

Store the DNA preparations at –20°C for later PCR analysis.

POLYMERASE CHAIN REACTION: THE NC5 REGION Detection of the parasite is by amplification of a 337 bp sequence of the NC5 region of *N. caninum* (Müller *et al.*, 1996).

1. For PCR, set up a 50 µl volume reaction for each sample containing:
5 µl 10 × Gene Amp™ PCR buffer
0.2 mM each dATP, dGTP and dCTP
0.4 mM dUTP (Pharmacia (Pfizer), New York, USA)
20 pmol each of *N. caninum*-specific primers Np21plus and Np6plus
1.25 units of AmpliTaq™ DNA polymerase
0.5 units of UDG
0.2–2.0 µl of sample DNA.
In parallel, the same PCR is run in the presence of approximately ten molecules of recombinant IC (see following section on serology).
2. Transfer the reaction tubes to a thermal cycler and amplify the 337-bp diagnostic PCR product and the 534 bp IC PCR product over 40 cycles using the following conditions: a 5 min denaturation of the DNA at 95°C is followed

by 40 cycles with denaturation (94°C, 1 min), annealing (63°C, 1 min) and primer extension (74°C, 3.5 min). After the last cycle, a primer extension is continued for 10 min at 74°C.
3. After the PCR, the samples can be analysed at this stage or stored frozen.

POLYMERASE CHAIN REACTION: ITS1 Detection of the parasite is by amplification of a 249 bp sequence of ITS1 of *N. caninum* (Holmdahl and Mattsson, 1996; Uggla *et al.*, 1998).

1. Set up a 50 μl volume reaction for each sample containing the following:
5 μl 10 × Gene Amp™ PCR buffer
0.2 mM each dATP, dGTP,dCTP and dTTP (Amersham Biosciences (now GE Healthcare), Little Chalfont, UK)
20 pmol each of *N. caninum*-specific primers PN3 and PN4
1.0 units of AmpliTaq™ DNA polymerase
0.2–2.0 μl of sample DNA.
2. Transfer the reaction tubes to a thermal cycler and amplify the 249-bp diagnostic PCR product over 30 cycles using the following conditions: a 2.5 min initial denaturation of the DNA at 94°C, followed by denaturation (94°C, 40 s), annealing (54°C, 1 min) and primer extension (72°C, 30 s). After the last cycle, a primer extension is continued for 3 min at 72°C.
3. After the PCR, the samples can be analysed at this stage or stored frozen for analysis at a later date.

PCR CONTROLS Any PCR detection method is susceptible to false negative and false positive results. Because of the sensitivity of the PCR, false positives usually result from cross-contamination of samples with minute quantities of target DNA, and careful steps,[2] as well as the UDG-decontamination system,[3] need to be taken to ensure this does not occur. It is essential to run appropriate controls in all experiments. The negative control should consist of distilled water (to ensure that none of the PCR reagents have become contaminated with parasite DNA). In addition to the negative control, a positive control – consisting of parasite DNA (DNA equivalents from approximately ten parasites) – should be added to ensure the integrity of the PCR.[4]

Finally, in order to monitor false negative results possibly caused by sample-related inhibitory compounds in the PCR tests, approximately ten molecules from a recombinant internal positive control (inhibition control, IC) are added.[5] The ITS1 PCR has not been evaluated using the UDG-decontamination system. However, that system could probably be directly incorporated into the ITS1 PCR system.

PRODUCT DETECTION Using control *N. caninum* DNA, sensitivity can be achieved to an approximative DNA equivalent of one parasite by ethidium bromide detection. The ITS1 amplicon can also be analysed by DNA sequencing.[6]

AGAROSE GEL ELECTROPHORESIS
1. Withdraw 15 μl of the completed PCR reaction and mix with 3 μl of GLS.

2. Load all samples, including controls, on a 2.0% agarose gel containing 0.5 μg/ml ethidium bromide and 1 × TAE buffer.
3. Separate the samples at 5 V/cm for about 40 min.
4. Analyse and record the gel image under UV illumination.

Endnotes

1 In low-level infections, the ability to detect *N. caninum* in any sample depends largely upon sampling strategy. If the parasite is very sparsely disseminated, the more tissue that can be collected and processed the greater chance there will be of detection by the PCR. However, it becomes impractical to process very large quantities of tissue routinely and so detection thresholds will always be dependent on how much tissue is collected. When processing tissue, small quantities of tissue can be pooled from different parts of an organ (such as the brain) to maximize the chance of parasite detection. The ITS1 PCR can also be used as part of a nested system to increase the sensitivity when analysing clinical samples. In the first step, the external primers F6 and 5.8B (5'CAGGTCTGTGATGCCC3' and 5'TCGCGTTTTGCTGCGTTCTTC3') are used. The conditions are as described above, except that 5 pmol of primers are used at an annealing temperature of 50°C, the number of cycles is reduced to 25 and the 72°C extension is 1 min per cycle. For the second step, 1 μl of a 1:5 dilution of the first PCR is used as a template for the internal PCR (as above).

2 General steps to avoid DNA cross-contamination should include the following: (i) the use of pipette tips with integral filters; although more expensive, filter tips are highly effective at preventing contamination of the pipette barrel with DNA; (ii) preparation and freezing of batches of the PCR reagents; each PCR experiment is then run with the same batch of materials and excess material discarded. By this method, any contamination of stock solutions (such as dNTPs) is contained within a single experiment and can be eliminated by repeating with a fresh batch of reagents; (iii) amplified DNA products (i.e. any material downstream of the PCR reactions) should be processed away from PCR preparation areas. In practice, this usually means keeping the product analysis equipment separate, preferably in a different area of the laboratory, or in a different room. Laminar flow-hoods are useful for setting up PCR reactions in a clean environment. Pipettes used for loading gels with product DNA should never be used for sample preparation or for setting up PCR reactions. Ideally, dedicated sets of pipettes should be used for the different processes. If a nested PCR is used, special precautions need to be instituted to avoid contamination problems. It is strongly recommended that a special area and a set of dedicated pipettes be used for the transfer of material from the first PCR to the second PCR reaction.

3 Carry-over PCR contaminations can be controlled by the following two steps as previously described (Longo *et al.*, 1990): (i) incorporating dUTP in all PCR products (by substituting dUTP for dTTP); and (ii) treating all subsequent fully pre-assembled starting reactions with uracil DNA glycosylase (UDG), followed by thermal inactivation of UDG. UDG cleaves the uracil base from the

phosphodiester backbone of uracil-containing DNA, but has no effect on natural DNA. If the contaminant contains uracils, carry-over contaminations of PCR can be controlled effectively through UDG-incubation of the mixes prior to the amplification reaction.

4 Positive control *N. caninum* DNA can be obtained by extraction of DNA from cell culture-derived *N. caninum* tachyzoites by using the DNeasy™Tissue Kit. Use DNA equivalents from approximately ten parasites as a positive control.

5 In order to avoid false negative results, possibly caused by sample-related inhibitory compounds in the PCR tests, a recombinant internal positive control (IC) was developed. This construct was the result of a PCR which used plasmid Bluescript KS plus (pBS+) (Stratagene, La Jolla, California, USA) as a template and was performed as follows: as a first step, chimeric primers were designed that contained the Np21plus-forward primer sequence plus a pBS^{+}-specific sequence (chimeric forward primer:
5'-CCCAGTGCGTCCAATCCTGTAACGCCAGTGAATTGTAATACG-3') and the Np6plus-reverse primer sequence plus a pBS^{+}-specific reverse sequence
(5'-CTCGCCAGTCAACCTACGTCTTCTGAGTGAGCTGATACCGC-3').
The PCR (profile: five cycles at 94°C, 30 s; 33°C, 30 s; 72°C, 2 min 30 s plus 30 cycles at 94°C, 30 s; 55°C, 30 s; 72°C, 2 min 30 s; followed by a primer extension cycle at 72°C, 15 min) with these chimeric primers produced a 534 bp pBS+ fragment with the *Neospora*-specific primer sequences incorporated at the ends. This fragment was then cloned into the pGEM™-Teasy vector (Promega, Madison, Wisconsin, USA) according to the instructions of the manufacturer. Internal positive control recombinant DNA (inhibition control, IC) can be provided by one of the authors (NM). ITS1 sequences have been extensively used to analyse coccidian phylogenies. There is some evidence that there might be polymorphisms in ITS1 when comparing *N. caninum* from various sources. A sequence analysis of ITS1 amplicons, combined with GenBank searches, could therefore aid in the study of the population structure of *N. caninum*.

Serology

Camilla Björkman, Heinz Sager and Gereon Schares

Introduction

Neospora caninum antibody assays are vital tools in support of clinical examinations and they are indispensable in epidemiological studies. A variety of antibody assays, including indirect fluorescent antibody tests (IFAT), enzyme-linked immunoassays (ELISA), immunoblotting (IB) and agglutination tests have been developed over the years (reviewed by Björkman and Uggla, 1999; Atkinson *et al.*, 2000; Jenkins *et al.*, 2002).

Several ELISA kits based on different antigen preparations, an agglutination test kit and reagents for *N. caninum* IFAT are now commercially available. Each test system has its own advantages and disadvantages, which should be carefully considered when tests are chosen for different applications. It is also essential that each laboratory carefully validates the performance of its *Neospora* tests.

In this section, we focus on the analysis of bovine and canine samples, give a general presentation of each test principle and describe some of the test protocols that are successfully used in our laboratories.

Tests to detect infection

IFAT The IFAT is based on the principle of attaching intact *N. caninum* tachyzoites to microscope slides. The parasites on the slides are then incubated with diluted test sera and, in a second step, with fluorescein-labelled antibodies directed against immunoglobulins of the animal species under investigation (Trees *et al.*, 1994; Dubey *et al.*, 1996a). Reactions are read under a fluorescence microscope (Paré *et al.*, 1995b).

Performing the test requires training and experience, and it must be stressed that IFAT results are, to a certain degree, subjective. It is also important to bear in mind that the IFAT reagents and protocol (method of antigen preparation, origin of conjugate, conjugate dilution, etc.) affect the test result. Consequently, each laboratory has to establish its own cut-off value. IFAT titres reported by different laboratories are not necessarily equivalent and can therefore not be directly compared.

Reagents

- Cell culture medium: Dulbecco's Modified Eagle Medium (DMEM), supplemented with 5% horse serum, 2 mM glutamine, 50 U/ml penicillin and 50 mg/ml streptomycin.[1]
- Phosphate-buffered saline (PBS): 58 mM Na_2HPO_4, 17 mM NaH_2PO_4, 68 mM NaCl, pH 7.3–7.4; alternative PBS recipes may also be used.
- Formaldehyde, 37%.
- Ethanol.
- Methanol.
- Evans blue solution: 0.01% in PBS.
- Conjugate: FITC-conjugated rabbit anti-bovine IgG (Sigma) diluted approximately 1:200 in Evans blue solution. The optimal dilution is decided for each new batch.[2]
- Mounting fluid: Fluoprep (BioMérieux, Marcy l'Étoile, France).

Procedure for the antigen preparation

1. Vero cells used for the propagation of *N. caninum* tachyzoites are maintained in cell culture medium at 37°C/5% CO_2.[1]

2. Add two to four parasites per Vero cell, to a cell culture that has been seeded the previous day and culture at 37°C/5% CO_2.

3. Harvest the parasites when about 60–80% of the Vero host cells have been lysed overnight (typically after 3–4 d). Parasites are harvested by vigorously striking the side of the flask several times. Alternatively, the cultures can be scraped off using a plastic cell scraper.
4. Wash the preparation twice in PBS (centrifugation at 2000 *g* for 10 min).
5. Filter through a 3 µm polycarbonate or 5 µm PVDF (e.g. Millipore, Billerica, Massachusetts, USA) filter.[3]
6. Dilute tachyzoites to 4×10^7 tachyzoites/ml in PBS and mix carefully.
7. Add formaldehyde to a final concentration of 0.2%.
8. Keep in a rotator for 24 h at 4°C.
9. The tachyzoite suspension can either be used directly for preparation of slides or stored at –20°C for later use.

Preparation of slides
1. Wash Teflon-printed microscope slides (with ten wells each of 7 mm diameter) with 96% ethanol.
2. Thaw the tachyzoite suspension, pipette 30 µl into each well of the microscope slide and allow to dry.
3. The slides can be used immediately or stored frozen at –20°C until used.

Test procedure
1. Warm the required number of slides to room temperature.
2. Fix the parasites by soaking the slides in methanol for 10 min.
3. Wash the slides twice for 10 min in PBS.
4. Add 30 µl diluted serum (1:640 in PBS) per well. Positive and negative control sera should be included in each test series.[4,5]
5. Incubate the slides in a humid chamber at room temperature for 30 min.
6. Gently rinse the slide with PBS using a washing bottle and then soak them twice for 10 min in PBS.
7. Drain the slides. They may be allowed to dry slightly but should be visibly damp.
8. Add 30 µl conjugate to each well and incubate for 30 min in the dark in a humid chamber at room temperature.
9. Gently rinse the slides and then soak them twice for 10 min in PBS.
10. Drain the slides. They may be allowed to dry slightly but should be visibly damp.
11. Place two to three drops of mounting fluid on each slide and top with a coverslip.
12. View the slide in a fluorescence microscope at 100–150 × magnification. Confirmation can be done at 400 × magnification. A bright, unbroken peripheral fluorescence of the tachyzoites is required for a valid positive reaction, whereas sole fluorescence of the apical part of the tachyzoites ('cap' or 'polar' staining) is regarded as non-specific (Paré *et al.*, 1995b) (see Fig. 3.1.8).
13. A serum is deemed negative if no positive reaction is detected at the 1:640 dilution.[5]

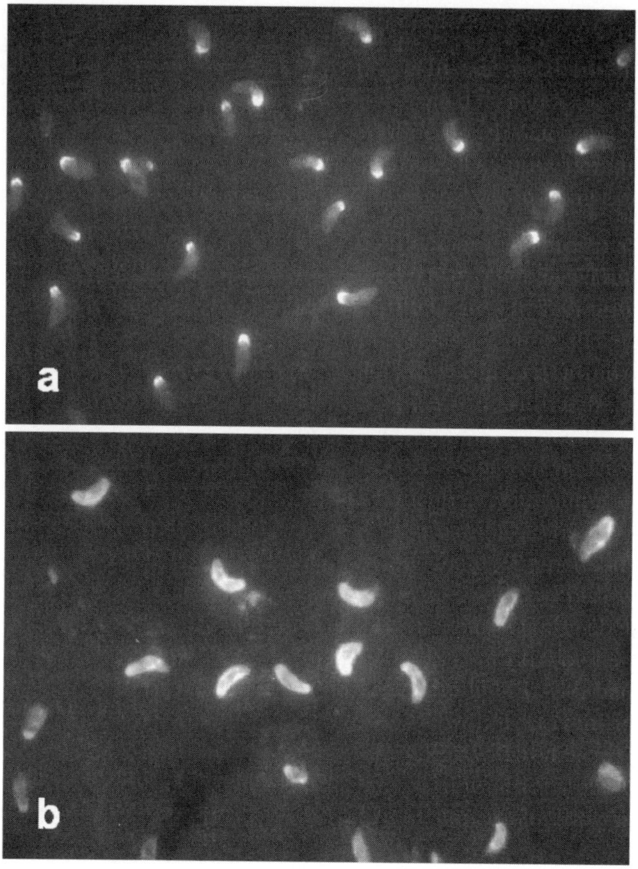

Fig. 3.1.8. Non-specific polar (a) and specific peripheral (b) immunofluorescence of *Neospora caninum* tachyzoites.

ELISA In indirect *N. caninum* ELISA systems, the particular antigen preparation is coated onto the plastic surface of multiwell microtitre plates. After incubation with the diluted test sera, an enzyme-labelled, species-specific anti-immunoglobulin antibody (conjugate) is applied. In a final step, a substrate is added which is converted into a coloured reaction product by the enzyme conjugated with the antibody. The absorbance or optical density (OD) is measured by a spectrophotometer. When compared with the IFAT, the ELISA has the advantage that the measurement of the given reaction is objective. Moreover, the assay can readily be automated. The ELISA is therefore a technique well suited to the screening of large numbers of samples, e.g. in serological surveys.

In addition to the indirect ELISA described here, a competitive inhibition ELISA has also been described for the analysis of sera from different animal species. Some bovine *N. caninum* ELISA tests have been modified to enable the analysis of milk samples for specific antibodies, and are used to test individual or bulk tank milk samples. Some tests have also been utilized to analyse serum samples from other ruminant species such as sheep, goats and water buffaloes.

In a multi-centred study, various *N. caninum* ELISAs used for demonstrating antibodies in bovine serum were compared (von Blumröder *et al.*, 2004). Most tests showed a high level of agreement in the interpretation of the test results (positive or negative).

The test system described below is an indirect ELISA based on a soluble *N. caninum* tachyzoite antigen (Gottstein *et al.*, 1998). In addition, indirect ELISA based on purified antigens, *N. caninum* iscoms or Nc-p38 are described. The tests employing purified antigens were developed to overcome problems with non-specific binding and cross-reactivities that are observed with some sera (Björkman *et al.*, 1997; Schares *et al.*, 2000).

Reagents

- *Neospora caninum* tachyzoites from Vero cell cultures (see the first four steps in IFAT, antigen preparation, above).
- PBS.
- Coating buffer (50 mM sodium carbonate/hydrogen carbonate, pH 9.6): 4.29 g $Na_2CO_3 \cdot 10H_2O$, 2.93 g $NaHCO_3$, add distilled water up to 1000 ml
- PBS-Tween: 0.3% Tween 20 in PBS.
- Horse serum (Amimed 2-05 F00-G).
- Conjugate: Monoclonal anti bovine IgG-AP conjugate (Sigma).
- Substrate: Alkaline phosphatase substrate (Sigma 104® Phosphatase Substrate).
- Stop solution: 3M NaOH.

Coating of plates

1. Purify the *N. caninum* tachyzoites on a PD-10 TM column filled with Sephadex G25M.[3]
2. Resuspend tachyzoites in PBS to a final concentration of 1×10^8 tachyzoites/ml.
3. Lyse the tachyzoites by three freeze–thaw cycles (–50°C/+37°C) and ultrasonication (3 × 20 s at 65 W).
4. Centrifuge the solution at 10,000 *g* at 4°C for 30 min.
5. Collect the supernatant and use it directly or store at –80°C.
6. Coat Polysorp 96-well microtitre plates (NUNC, Roskilde, Denmark) with the soluble *N. caninum* antigen diluted 1:2000 in coating buffer (100 μl/well).
7. Incubate for 16–24 h at 4°C.

Test procedure

1. Discard the coating buffer.
2. Wash the plate twice with PBS-Tween.

3. Add 200 µl PBS-Tween containing 1% horse serum to each well.
4. Incubate for 30 min at 37°C.
5. Remove blocking solution. *Do not wash.*
6. Add serum samples diluted 1:200 in PBS-Tween (100 µl/well). Include a high positive and a negative control serum on each plate.
7. Incubate for 90 min at 37°C.
8. Wash twice with PBS-Tween.
9. Add conjugate appropriately diluted in PBS-Tween with 1% horse serum.
10. Incubate for 45 min at 37°C.
11. Wash twice in PBS-Tween.
12. Add 100 µl alkaline phosphatase substrate per well.
13. Incubate for 15 min at room temperature.
14. Stop the reaction by adding 100 µl stop solution per well.
15. Read OD at 405 nm (reference wavelength 630 nm) in a microplate spectrophotometer.[6]

IMMUNOBLOT In the immunoblot (IB), tachyzoite proteins are separated by electrophoresis in a SDS-polyacrylamide gel and then transferred (electroblotted) on to a nitrocellulose or PVDF membrane (Immobilon-P, Millipore, Billerica, Massachusetts, USA), where they bind in the same pattern as formed in the gel. Antigen-coated membrane strips are first incubated with the test sera and then with enzyme-labelled anti-species (e.g. bovine) immunoglobulin. A substrate is then added and the antigen pattern recorded that is recognized by the antigen-specific antibodies present in the tested serum.

As IB combines the resolution of gel electrophoresis with the specificity of immunochemical detection, the method has a high specificity. However, it is also a time-consuming technique and different laboratories reported varying sensitivities. It should also be noted that serological reactions against specific antigens are not an unambiguous indication that an animal is truly infected. Previous infections, and also exposure to related parasite species exhibiting cross-reactive epitopes on their antigens, must be considered as possible explanations, in particular when weak reactions against single antigens are observed. For routine diagnosis, recognition of a certain number of defined immunodominant antigens (IDA) is used to classify a sample as positive, negative or inconclusive.

The IB system presented here is based on the methods described by Schares *et al.* (1998, 2001b).

Reagents

1. Antigen preparation

Sample buffer (5×): 3.75 g tris(hydroxymethyl)aminomethane, 10 g sodium dodecyl sulphate (SDS), 35 ml glycerol, 25 mg bromophenol blue.

Adjust to pH 6.8 with HCl and add distilled water up to 100 ml.

2. SDS-PAGE
- Solution A (stacking gel buffer): 6.06 g tris(hydroxymethyl)aminomethane, 4 ml 10% (w/v) SDS. Adjust to pH 6.8 with HCl and add distilled water up to 100 ml.
- Solution B (separation gel buffer): 18.17 g tris(hydroxymethyl)-aminomethane, 4 ml 10% (w/v) SDS. Adjust to pH 8.8 with HCl, add distilled water up to 100 ml.
- Solution C (acrylamide stock solution): 30 g acrylamide, 0.8 g N,N'-methylen-bis(acrylamide), add distilled water up to 100 ml.
- Electrophoresis buffer (5×): 15.2 g tris(hydroxymethyl)aminomethane, 72.1 g glycin, 5 g SDS, add distilled water up to 1000 ml.
- Ammonium persulphate 40% (w/v) in distilled water.
- N,N,N',N'-tetramethylethylenediamine (TEMED).
- LMW Electrophoresis Calibration Kit (17-0446-01; GE Healthcare Life Sciences, Little Chalfont, UK).

3. Transfer of antigen to nylon membranes.
- Anode buffer I (to soak six sheets of filter paper): 36.3 g tris(hydroxylmethyl) aminomethane, 200 ml methanol and distilled water up to 1000 ml.
- Anode buffer II (to soak three sheets of filter paper): 3.03 g tris(hydroxyl-methyl)aminomethane, 200 ml methanol and distilled water up to 1000 ml.
- Cathode buffer (to soak nine sheets of filter paper): 40 mM 6-aminohexanoic acid 5.2 g, 0.01 % SDS (w/v) 0.1 g, 20% (v/v) 200 ml methanol and distilled water up to 1000 ml.
- India Ink (Hancock and Tsang, 1983).
- PBS-T-G: PBS supplemented with 0.05 % (v/v) Tween 20 and 2% (v/v) fish gelatine liquid (Serva No. 22156).

4. Immunoblot
- PBS-T-G: PBS with 0.05 % (v/v) Tween 20 and 2% (v/v) fish gelatine liquid (Serva No. 22156).
- PBS-Tween: 0.05% Tween 20 in PBS.
- Conjugate: anti bovine IgG[H+L] peroxidase (Dianova, Hamburg, Germany) or anti-dog IgG[H+L] peroxidase (Dianova, Hamburg, Germany) are used for testing cattle or dog sera, respectively.
- Substrate: 30 mg 4-chloro-1-naphthol (Sigma, C-6788), 10 ml methanol, 30 ml PBS and 40 µl 30% H_2O_2.

Equipment

The described procedure is appropriate to the following equipment:

- Minigel SDS-PAGE system (e.g. Mini Protean II equipment, Bio-Rad Laboratories, Hercules, California, USA).

- Semidry blotting system for antigen transfer (Multiphor II, NovaBlot from GE Healthcare Life Sciences, Little Chalfont, UK).
- Block heater (e.g. Thermomixer, Eppendorf, Hamburg, Germany).

Procedures

1. Antigen preparation

- Purify *N. caninum* tachyzoites by filtration using 3–5 μm polycarbonate or PVDF filter units. Wash the parasites five times (centrifuge for 10 min at 1500 *g* and 4°C, and resuspend in PBS).[3]
- Add 80 μl distilled water and 20 μl sample buffer (5 ×) to a pellet containing 4 × 10^7 tachyzoites. This amount is sufficient for one Minigel (60 × 70 × 1 mm).
- Heat in block heater for 10 min at 94°C.

2. SDS-PAGE

- Prepare the separation gel solution (Table 3.1.2).
- Cast the separation gel, overlay it with distilled water and let it stand for approximately 30 min (i.e. until the gel has polymerised).
- Prepare the stacking gel solution (see Table 3.1.2).
- Pour on the stacking gel solution and insert the comb into the solution to form the slots.
- Wait at least 30 min (i.e. until the gel has polymerized) before the comb is removed.
- Submerge the gel in 1 × electrophoresis buffer.
- Load the tachyzoite antigen into the preparative slot (of about 60 mm length) and the marker protein into the narrow slot (about 2 mm).
- Run the gel at 120 volts until the bromophenol blue front leaves the gel.

Table 3.1.2. Composition of separation and stacking gels. The volumes are sufficient for two Minigels (60 × 70 × 1 mm).[a]

	5% (w/v) separation gel	12.5% (w/v) separation gel	Stacking gel
Solution A (ml)	–	–	1.5
Solution B (ml)	2.5	2.5	–
Solution C (ml)	1.7	4.2	0.7
H_2O (ml)	5.8	3.3	3.6
APS (40% w/v) (μl)	25	25	8
TEMED (μl)	10	10	16

[a] The 12.5% (w/v) SDS-polyacrylamide separation gel is used for demonstration of *N. caninum*-specific antibodies in bovine and canine sera. To investigate canine sera for antibody reactions towards the 152 kDa antigen, a 5% (w/v) SDS-polyacrylamide separation gel is used.

3. Transfer of antigen to nylon membranes

- Transfer the separated antigens to PVDF membranes (Immobilon-P, Millipore, Billerica, Massachusetts, USA). We prefer an electrophoretic transfer by semidry Western blotting (Multiphor II, NovaBlot, GE Healthcare Life Sciences, Little Chalfont, UK) using a discontinuous buffer system. This protocol is described here. Other transfer systems or a continuous buffer system may also work.
- To assemble the transfer unit follow the sandwich assembly illustrated in Fig. 3.1.9. Use, for instance, gel blotting paper from Schleicher & Schuell GmbH (Dassel, Germany) GB 001 to assembly the transfer unit.
- Run the electrophoresis at a current of 1.5 mA/cm^2 gel for 90 min and a discontinuous buffer system.
- Stain the section of the membrane containing the molecular weight marker and a small part of the antigen section with India Ink.
- Incubate the remaining part of the membrane with PBS-T-G for at least 30 min to block all remaining protein-binding sites.
- The membrane is dried overnight and can then be stored frozen (–20°C) until used.

4. Antibody detection

- Cut the required number of 3–4 mm-wide strips from the membrane and label each strip.
- Incubate the strips for 10–30 min with PBS-T-G.
- Remove PBS-T-G and incubate each strip with a test serum diluted 1:100 in PBS-T-G (or undiluted fetal fluid), appropriately diluted positive and negative control serum or PBS-T-G for 1 h (500–1000 μl serum dilution required, depending on the slot size of the incubation

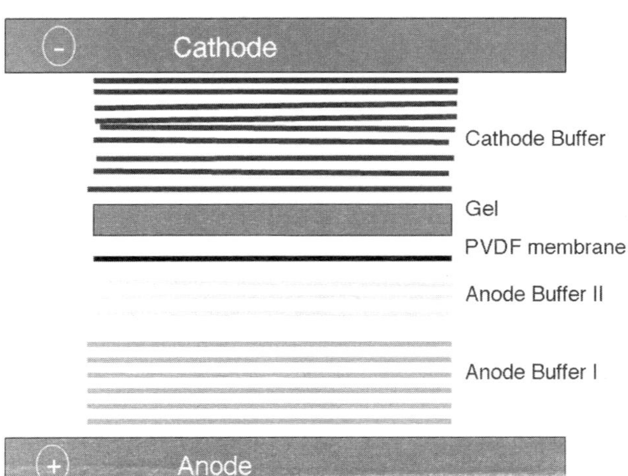

Fig. 3.1.9. Sandwich assembly for semi-dry Western blotting.

tray). A positive and a negative control – as well as PBS-T-G – must be included in each test run.

- Wash the strips four times with PBS-Tween.
- Incubate the strips with conjugate for 1 h (300–1000 μl required per strip).
- Wash the strips three times for 3 min with PBS-Tween and twice for 3 min with PBS.
- Incubate the strips with substrate for 20 min (500–1000 μl required per strip).
- Wash the strips once for 3 min with PBS and dry them on filter paper.
- Read results.[7]
- When bovine and canine sera are tested for *N. caninum*-specific antibodies, reactions against five IDAs with relative molecular masses of 17−19, 29, 30, 33 and 37 kDa are recorded (see Fig. 3.1.10). The results are interpreted as follows:
  - Two or more than two IDAs recognized: positive;
  - one IDA recognized: inconclusive;
  - no IDA recognized: negative.
- When canine sera are tested for antibody reactions against the 152 kDa antigen, a serum is judged positive if a band at this relative molecular mass is visible (an example of a positive reaction is displayed in Schares *et al.*, 2001b).

AGGLUTINATION TEST The agglutination test is based on the fact that intact, formalin-treated tachyzoites agglutinate in the presence of specific antibodies (Packham *et al.*, 1998; Romand *et al.*, 1998).Tachyzoites are placed on the bottom of a microtitre well, test serum is added and the agglutination reaction is read after an incubation period. Mercaptoethanol is included in the assay to destroy specific and non-specific IgM, so that IgG antibodies only will be detected. A critical step in this assay is the reading of the agglutination reaction. This is done either by direct visual inspection or by using a microplate spectrophotometer coupled to a computer with agglutination-reading software.

As no species-specific conjugate is required for this test system, it can be used to analyse sera from a variety of animal species. Nevertheless, the method must be appropriately validated for each separate species. When we applied a commercially available agglutination test in our laboratories, we had considerable problems in judging whether certain bovine sera were positive or negative. We have also had false positive results when sera of different origin were analysed, especially when the samples were haemolysed.

Tests to estimate duration of infection

It is often desirable to distinguish between acute and chronic infections with *N. caninum*. To this end, indirect *N. caninum* ELISAs have been modified to measure

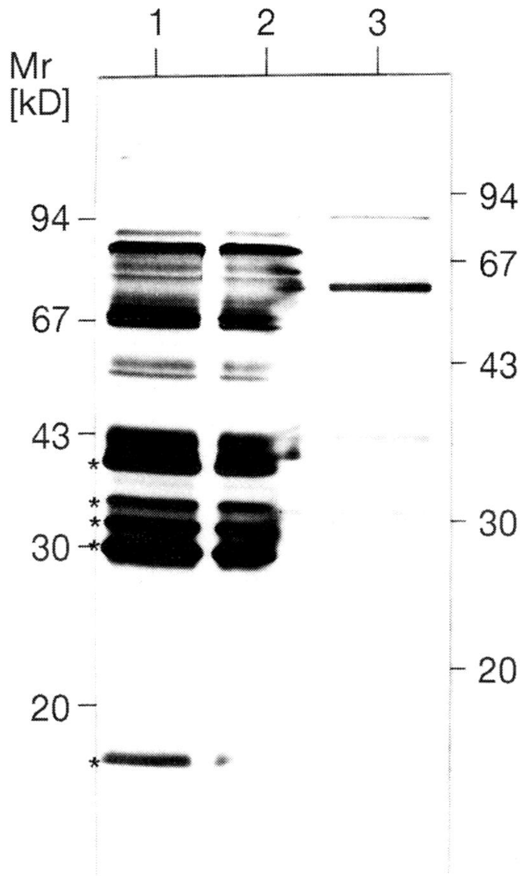

Fig. 3.1.10. Reactivity of a *Neospora caninum*-positive serum against *N. caninum* tachyzoite antigens separated under non-reducing (Lane 1) and reducing (Lane 3) conditions. Note that non-reduced antigens were also loaded in Lane 2. However, during the gel run, the reducing agent (2-mercaptoethanol) diffused from Lane 3 into Lane 2 and led to a partial reduction of the antigens. Since exactly the same antigen concentrations had been loaded into each lane, the result clearly demonstrated that the reactivity against the IDAs (marked with *) is strongly reduced when the antigens are denatured by reducing agents.[7]

IgG avidity (functional affinity). Avidity assays are based on the fact that the first antibodies synthesized after antigenic challenge or primary infection have a lower affinity for the antigen than those produced later on.

In the avidity ELISA, antibodies are allowed to bind as usual, but low-affinity antibodies are then eluted by incubation in 6–8 M urea, while high-avidity antibodies remain unaffected (i.e. bound) by this treatment. The antibody titres obtained with and without incubation with urea are then used to calculate the IgG avidity values. For more information about the *N. caninum* avidity tests and their applications, refer to Björkman *et al.* (1999, 2003), Schares *et al.* (2002a), Sager *et al.* (2003) and Aguado-Martínez *et al.* (2005). A comparison of the avidity tests used in several laboratories has been performed by Björkman *et al.* (2006).

Validation of N. caninum *antibody tests*

Before a laboratory starts to use any *N. caninum* antibody test, feasibility should be investigated by analysing positive and negative sera, as recommended in the OIE Manual of Standards for Diagnostic Tests and Vaccines (http://www.oie.int/eng/normes/mmanual/A_00013.htm, accessed 28 March 2006). This manual also provides general guidelines for further validation to establish that a test consistently provides results that correctly identify animals as seropositive or seronegative.

There is no gold standard for *N. caninum* infection: different combinations of tests and animals have therefore been used in attempts to overcome this difficulty (Williams *et al.*, 1999; Schares *et al.*, 2000; Baszler *et al.*, 2001). Also, alternative statistical methods, such as latent class models or techniques based on Bayes' theorem, have been applied (Frössling *et al.*, 2003). We recommend readers with a special interest in test evaluation and validation to consult a special issue of *Preventive Veterinary Medicine*, dedicated to different aspects of test validation (Gardner and Greiner, 2000).

Endnotes

[1] Different continuous and primary cell lines can be used as feeder cells for *N. caninum*. Cell culture medium may differ depending on the feeder cells. Feeder cells can be maintained with fetal calf serum (FCS), but it is important that the serum is free from *N. caninum* antibodies. Freedom from *N. caninum*-specific antibodies should be tested on each individual batch of FCS. When it is planned to use the antigen for bovine serology, horse serum should be used instead of FCS to grow the feeder cells.

[2] For testing canine sera, FITC-conjugated rabbit anti-dog IgG (Sigma), diluted approximately 1:100 in Evans blue solution, is used. The optimal dilution is determined for each new conjugate batch.

[3] Different methods can be used to purify tachyzoites from cell debris. The mixture may be filtered through a PD-10 TM column filled with Sephadex G25M (Amersham Biosciences, Uppsala, Sweden) or can be layered on top of 30% isotonic Percoll (Amersham Biosciences, Uppsala, Sweden) and centrifuged for 10 min at 2000 *g*. Some laboratories omit this purification step and use a mixture of tachyzoites, Vero cells and cell debris as antigen.

4 If it is desirable to determine IFAT titres rather than distinguishing just between positive and negative, two-fold serial dilutions can be prepared in PBS.

5 Serum dilution depends on a range of factors such as conjugate characteristics and microscope properties, and can be arbitrarily selected by varying the dilution of the conjugate. Thus, serum dilutions and cut-off titres differ between laboratories but are often set to 1:160–1:640 for adult cattle serum and 1:20–1:10 for bovine fetal sera and fetal fluid. Canine sera are often judged negative if no fluorescence is detected at the dilution 1:20 or 1:40.

6 A cut-off value should be established by each laboratory. The cut-off we apply is the mean absorbance + three standard deviations of 30–40 negative sera ($\mathrm{Mean}_{[neg]} + 3 \times \mathrm{StDev}_{[neg]}$). A correction factor (CF) is used to compensate for any test-to-test variation. When starting with a new antigen batch the CF is calculated:

$$\mathrm{CF} = \frac{\mathrm{Mean}_{[neg]} + 3 \times \mathrm{StDev}_{[neg]}}{\mathrm{NC[plate1]}} \quad (1)$$

$NC_{[plate1]}$ is the absorbance of the negative control on the first test plate. For the following tests, the cut-off is calculated by multiplying the OD of the negative control serum (NC) on each plate with the correction factor (CF). It is a prerequisite for the validity of each test series that the OD of the high positive control serum included on each test plate falls within certain established limits.

7 *Neospora caninum*-infected intermediate hosts develop an antibody response against non-reduced IDA (Barta and Dubey, 1992; Bjerkås *et al.*, 1994; Paré *et al.*, 1995a). At the FLI in Wusterhausen, Germany, reactions against the 17–19, 29, 30, 33 and 37 kDa IDA are recorded. The major epitopes on these IDA are destroyed if the antigens are denatured under reducing conditions (i.e. by using 2-mercaptoethanol or dithiothreitol in the SDS-PAGE sample buffer). In Fig. 3.1.10, immunoblot reactions against immunodominant antigens of *N. caninum* are shown. This figure also illustrates how the use of reduced antigens can diminish the reactivity of sera in the IB. Nevertheless, reduced antigens have been successfully used for diagnostic purposes (see, e.g. Álvarez-Garcia *et al.*, 2002). In dogs, antibody reactions against a 152 kDa tachyzoite antigen have been observed that might help to identify dogs that have shed *N. caninum* oocysts in the past (Schares *et al.*, 2001b).

T-cell proliferation and IFNγ detection

Elisabeth A. Innes

Introduction

Following infection with the parasite, the host mounts a variety of humoral and cell-mediated immune responses (Innes *et al.*, 2002) that may be useful markers and/or tools in aiding the diagnosis of infection. The tachyzoite stage of the parasite invades and multiplies within the cells of the host and the bradyzoite stage is maintained long-term within tissue cysts (Dubey, 2003a).

T-cells and immune cytokines are thought to be important components of protective immunity against the parasite, where they act by either directly lysing the parasite-infected cells (Staska *et al.*, 2003) or indirectly affecting intracellular growth of the tachyzoites through the action of cytokines (Innes *et al.*, 1995). The pro-inflammatory cytokine, interferon gamma (IFNγ), is known to be important in host protection and therefore may be a useful diagnostic marker.

Lymphocytes from peripheral blood or secondary lymphoid organs, including spleen and lymph nodes, from *N. caninum*-infected cattle, will proliferate *in vitro* when stimulated with specific antigen, and supernates from these cultures have been shown to contain IFNγ (Lunden *et al.*, 1998; Marks *et al.*, 1998; De Marez *et al.*, 1999; Williams *et al.*, 2000; Andrianarivo *et al.*, 2001; Innes *et al.*, 2001; Trees *et al.*, 2002; Bartley *et al.*, 2004).

Aims

Several techniques have been published on how to measure lymphoproliferative and IFNγ responses in cattle. This section will describe how to quantify lymphoproliferative and IFNγ responses in *N. caninum*-infected cattle using techniques developed and routinely used in our laboratory.

Materials and methods

PREPARATION OF BOVINE PERIPHERAL BLOOD MONONUCLEAR CELLS

Materials

- Preservative-free heparinized evacuated blood collection tubes (Vacutainer, Becton Dickinson, Oxford, UK)
- Hanks Balanced Salt Solution (HBSS)
- Fetal calf serum (FCS) (Labtech International, Sussex, UK)
- Penicillin, Streptomycin (Northumbria Biologicals, Cramlington, UK)
- Lymphoprep (Nycomed, Solihull, UK)
- Jouan CR422 bench centrifuge
- Iscove's modified Dulbecco's medium (Sigma, Poole, UK)

Method

1. Venous blood is collected into preservative-free heparinized evacuated blood collection tubes.
2. Blood samples are diluted 1:2 in sterile PBS and centrifuged for 20 min at 450 *g*.

3. The buffy coat is removed and diluted 1:3 in HBSS supplemented with 2% fetal calf serum (FCS), 100 U/ml penicillin and 50 μg/ml streptomycin (wash medium).
4. The cell suspension is layered carefully over 5 ml lymphoprep in a 10 ml plastic centrifuge tube and centrifuged at 550 *g* for 30 min.
5. Peripheral blood mononuclear cells are carefully harvested from the interface and added to 20 ml of wash medium in a plastic universal tube.
6. The cells are then washed three times by repeated centrifugation (300 *g*) and resuspension using wash medium.
7. An aliquot of cells (50 μl) is removed and mixed 1:1 with a solution of Nigrosin dye to enable viable cells to be counted using an improved Neubauer haemocytometer chamber.
8. Cells are adjusted to a concentration of 2 × 106/ml in Iscove's modified Dulbecco's medium supplemented with 10% FCS, 100 U/ml penicillin and 50 μg/ml streptomycin (culture medium).

PREPARATION OF *N. CANINUM* ANTIGEN

Materials

N. caninum tachyzoites

- Vero cells
- 25 and 75 cm^2 tissue culture flasks (Costar, Corning, New York, USA)
- Fetal calf serum (FCS) (Labtech International, Sussex, UK)
- Horse serum (Northumbria Biologicals, Cramlington, UK)
- Penicillin, Streptomycin (Northumbria Biologicals, Cramlington, UK)
- Iscove's modified Dulbecco's medium (Sigma, Poole, UK)
- BCA reagent (Pierce, Rockford, Illinois, USA)
- Monarch 2000 spectrophotometer (Instrumentation Laboratories, Palo Alto, California, USA)
- Humidified 5% CO_2 incubator
- Sterile cell scraper (Costar, Corning, New York, USA)
- Phosphate buffered saline, pH 7.4

Methods

1. Vero cell monolayers are maintained by twice weekly passage (1:10 split) in IMDM supplemented with 5% horse serum or FCS, 100 IU/ml of penicillin and 50 μg/ml of streptomycin in 25 cm^2 or 75 cm^2 tissue culture flasks. Cells were cultured in a 5% CO_2 humidified incubator at 37°C.
2. *N. caninum* tachyzoites are maintained by twice weekly passage within Vero cells. Parasites are added to Vero cells, seeded 24 h previously, at a tachyzoite:cell ratio of 2:1. Parasites are cultured within the cell monolayers for 3–4 d until free tachyzoites are observed in the culture flask. The infected cells

are then disrupted using a sterile cell scraper and the contents of the flask are collected into a sterile universal tube.

3. The parasites are washed three times in sterile, 0.3 M phosphate-buffered saline by repeated centrifugation at 500 *g*.

4. The resultant pellet, typically containing approximately $1–2 \times 10^9$ tachyzoites, is resuspended in 2 ml of sterile distilled water and further disrupted by three cycles of freezing and thawing, followed by 7 × 15 s cycles of sonication on ice.

5. The sonicated tachyzoites are centrifuged at 10,000 *g* for 30 min at 4°C to remove insoluble debris, and the supernatant containing the water-soluble antigen fraction is collected.

6. The protein concentration of the antigen is assayed using BCA reagent, the antigen is aliquoted into sterile eppendorf tubes and stored at –20°C until required.

CELL PROLIFERATION ASSAYS

Materials

- 96-well, round-bottomed tissue culture plates (Nunc, Roskilde, Denmark)
- Concanavalin A (Sigma, Poole, UK)
- Humidified 5% CO_2 incubator
- ^{3}H thymidine (Amersham, Little Chalfont, UK)
- Fibreglass filters (Canberra, Packard, Meriden, Connecticut, USA)
- Gas proportional counter (Canberra, Packard, USA)
- Peripheral blood mononuclear cells
- *N. caninum*-specific antigen
- Culture medium

Method

1. Bovine peripheral blood mononuclear cells are prepared as described above and adjusted to a final concentration of 2×10^6/ml. *N. caninum*-specific antigen is added to the cells at a predetermined optimal concentration (in our experience, this ranges from 1–10 μg/ml final concentration). In addition, cells are also stimulated with the T-cell mitogen concanavalin A at 5 μg/ml as a positive control, and are cultured in media alone as a background control.

2. Equal volumes of cells and test antigens are set up in triplicate in 96-well, round-bottomed tissue culture plates and are cultured in a humidified 5% CO_2 incubator for 5 d.

3. Cells are pulsed for the final 18 h of the assay with 18.5 kBq ^{3}H thymidine per well.

4. The cells are then harvested on to fibreglass filters and the cell-associated radioactivity is quantified using a gas proportional counter.

5. Cell proliferation can be quantified directly to the counts per minute in each sample or can be expressed relative to the background proliferation and expressed as a stimulation index. In general, a stimulation index of greater than 3 (i.e. cell counts three times above background) is considered to indicate a positive proliferation response.

DETECTION OF GAMMA INTERFERON

Materials

- Bovine peripheral blood mononuclear cells
- *N. caninum*-specific antigen
- Bovigam™ Bovine gamma interferon test (CSL Veterinary, Parkville, Australia)
- Microplate reader (450 nm filter)

Methods

1. Similar cell and *N. caninum*-specific antigen cultures are set up as described above, and cell-free supernatants are harvested after 96 h incubation.
2. Other samples may include whole blood cultures stimulated with *Neospora* antigen (18–96 h) or plasma samples separated from whole blood.
3. Samples can be tested straight away or stored at –20°C prior to testing.
4. Samples are tested for the presence of IFNγ using the Bovigam ELISA kit following the manufacturer's instructions. The kit comes with a negative IFNγ control and a positive bovine IFNγ control. Typically, the mean adsorbance range of the negative control is < 0.130 and the positive control is > 0.700. Samples with adsorbance levels of > twofold the negative control are generally considered to be positive.
5. To quantify the levels of IFNγ in test samples, the ELISA can be calibrated using standards of recombinant bovine IFNγ (in-house research tool) prepared from doubling dilutions of a known quantity of reagent. The mean optical density values are then plotted against the unit/ml of recombinant IFNγ. A regression line is calculated and the quantity of IFNγ present in the samples is determined from the standard curve. We define a unit of IFNγ as the lowest concentration of IFNγ that will inhibit the cytopathic effect of Semliki Forest Virus on target cells by 100% (Entrican *et al.*, 1992).

Concluding remarks

Diagnostic techniques for detection of *N. caninum*-infected animals based on host immune responses currently rely on detection of specific antibody. However, antibody responses in pregnant animals infected with *N. caninum* are known to fluctuate and may even fall to levels that would be considered seronegative by some assays (Conrad *et al.*, 1993b; Dannatt, 1997; Stenlund *et al.*, 1999; Quintanilla-Gozalo *et al.*, 2000).

Quantification of cell-mediated immune responses has largely been used for research purposes and these techniques have not been extensively evaluated for diagnostic purposes. However, these techniques may prove useful, as the intracellular location of the parasite within the host is more likely to trigger a cell-mediated immune response and hence this may be a useful marker to detect persistently infected animals.

While the use of specific cell-mediated immune responses is not currently used as a diagnostic test for *N. caninum*-infected cattle, the demonstration of antigen-specific IFNγ responses *in vitro* has been shown to have application in the diagnosis of several other intracellular infectious diseases of cattle, including: (i) tuberculosis (*Mycobacterium bovis*) (Wood *et al.*, 1991); (ii) Johne's disease (*M. avium* ssp. *paratuberculosis*) (Billman-Jacobe *et al.*, 1992); (iii) brucellosis (Weynants *et al.*, 1995); (iv) infectious rhinotracheitis (BHV-1) (Godfroid *et al.*, 1995); and (v) bovine herpesvirus-4 infection (Godfroid *et al.*, 1996).

The cell-mediated assays described in this chapter would have to be evaluated against current diagnostic tests as a gold standard to assess their specificity and sensitivity. The assay could be further refined by using a panel of different antigens, for example specific tachyzoite and/or bradyzoite antigens which may increase the sensitivity of the tests and may allow discrimination between recently infected animals and persistently infected animals.

Oocyst detection and differentiation

Gereon Schares

Introduction

Among the family of Sarcocystidae, four genera are known to have the dog as a definitive host, *Sarcocystis*, *Cystoisospora*, *Hammondia* and *Neospora*. Recently, *Toxoplasma gondii* and *Hammondia hammondi* oocysts have been observed in canine faeces (Schares *et al.*, 2005). These findings suggest that coprophagia by dogs, with a subsequent intestinal passage, plays an important role in the dissemination of coccidian parasites for which cats are definitive hosts (Schares *et al.*, 2005).

The oocysts of *Sarcocystis* sp. have a unique morphology. They are thin-walled and sporulate within the intestine of the definitive host. Because of their thin wall the oocysts of *Sacrocystis* sp. are often ruptured, and thick-walled sporocysts are released. While the oocysts of *Cystoisospora* spp. are much larger than those of *N. caninum* and *H. heydorni*, the oocysts of *N. caninum*, *H. heydorni*, *T. gondii* and *H. hammondi* are morphologically very similar.

The unambiguous discrimination between *N. caninum* and *H. heydorni* became possible when molecular methods were applied to isolates of both parasite species, i.e. *N. caninum* NC-1 and *H. heydorni* Alabama 1 (Ellis *et al.*, 1999). Prior to the demonstration that the dog is a definitive host of *N. caninum* (McAllister *et al.*, 1998), isolates that were designated *H. heydorni* may, in fact, have been *N. caninum*. This may also be true for the isolate used to describe the biology of *H. heydorni* (Heydorn, 1973; Schares *et al.*, 2001a).

To date, dogs (*Canis familiaris*) and coyotes (*Canis latrans*) are the only known definitive hosts of *N. caninum* (McAllister *et al.*, 1998; Gondim *et al.*, 2004a). After experimental infections of dogs and red foxes (*Vulpes vulpes*), only dogs shed *N. caninum* oocysts, suggesting that the red fox is not a definitive host for *N. caninum* (Schares *et al.*, 2002b). *H. heydorni* also uses dogs as a definitive host and *Hammondia*-like oocysts have been observed in other canids, including foxes and coyotes (Dubey *et al.*, 2002).

Morphologically, the oocysts of *N. caninum, H. heydorni, T. gondii* and *H. hammondi* are very similar (see Fig. 3.1.11). Sporulated oocysts consist of two sporocysts, each containing four sporozoites. In water, the sporulated oocysts are 11.7 ± 0.43 μm long by 11.3 ± 0.38 μm wide (10.6–12.4 μm by 10.6–12.0 μm; *n*, 25). The length/width ratio is 1.04 (1.00–1.09; *n*, 25) (Lindsay *et al.*, 1999a). The first *H. heydorni* isolate that was characterized by molecular genetics (Ellis *et al.*, 1999) was named Alabama 1 (Dubey *et al.*, 2002). It measured 11.3–13.5 μm by 10.8–13.5 μm (mean 12.6 × 11.9 μm) (Blagburn *et al.*, 1988).

For diagnostic purposes, it has to be taken into account that oocysts are detected by flotation using concentrated salt or sucrose solutions. In concentrated solutions, oocysts are smaller than in water. Schares *et al.* (2005) compared the sizes of *H. heydorni, T. gondii, H. hammondi* and *N. caninum* oocysts in concentrated sucrose solution. *N. caninum* oocyst isolates were significantly smaller in length, with the 75th percentiles ≤ 10.7 μm when compared with oocyst isolates of the other species. *N. caninum* oocysts also showed smaller length:width ratios, with the 75th percentiles ≤ 1.06. It may thus be possible to develop criteria for a preliminary identification of *N. caninum* in dog faeces based on oocyst morphology (Schares *et al.*, 2005).

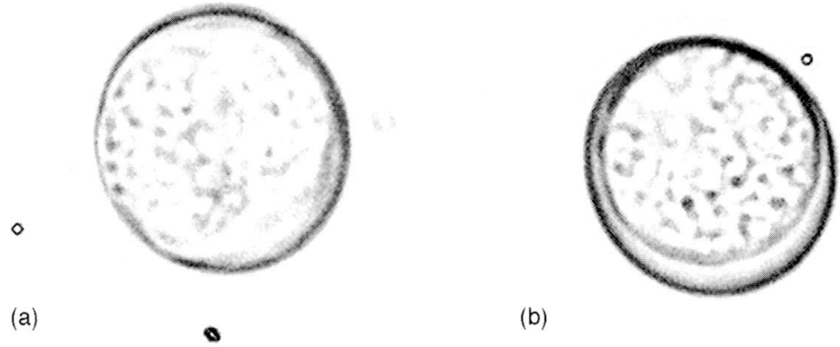

Fig. 3.1.11. Unsporulated *Neospora caninum* (a) and unsporulated *Hammondia heydorni* (b) oocysts in concentrated sucrose.

The oocysts of *Cystoisospora* spp. are much larger than those of *Neospora*, *Hammondia* or *Toxoplasma* and can thus readily be differentiated. The oocysts of, e.g., *C. burrowsi* are reported to measure about 21 × 18 μm.

Diagnostic examination of faeces for Hammondia- *or* Neospora-*like oocysts*

Many methods for the examination of canine faeces for *N. caninum*- or *H. heydorni*-like oocysts have been described. Techniques developed to diagnose other coccidian oocysts can also be applied to the examination of faeces for *N. caninum* and *H. heydorni*. In the following, details of the method used at the Friedrich-Loeffler-Institut, Federal Research Institute for Animal Health (FLI), in Wusterhausen, Germany, are described. The method consists of an initial sedimentation followed by flotation with concentrated sucrose.

EQUIPMENT
- balance
- spatula
- tea strainer (500–800 μm mesh)
- 200 ml beaker
- 250 ml conical measure or sedimentation cone
- centrifugation tubes (15 ml, 50 ml)
- centrifuge
- microscope slides
- flotation fluid
- right-angled wire loop (diameter of the loop about 2–3 mm)
- Bunsen burner
- coverslips
- light microscope.

FLOTATION FLUID Concentrated sucrose solution (specific gravity 1.3): dissolve 550 g sucrose in 450 ml tap water. Prepare fresh, store refrigerated for only a few days.

PROCEDURE

Sedimentation

1. Weigh out a 10 g faecal sample in a 200 ml beaker.
2. Place about 50 ml tap water in the beaker with the faecal material.
3. Mash the sample thoroughly with a spatula until the material is homogeneously suspended in the water.
4. Pour the mixture through a strainer into a 250 ml conical measure, stirring all remaining material in the strainer while pouring.

5. Add a small volume of water to the newly emptied 200 ml beaker and rinse the material clinging to the sides and bottom, then pour this mixture through the strainer, stirring the material in the strainer while pouring.
6. Carefully rinse the material in the strainer with tap water.
7. Press the remaining material in the strainer to extract as much of the fluid as possible; discard the remnant.
8. Fill the 250 ml conical measure completely and allow sedimentation for at least 30 min. To recover a higher proportion of oocysts, the sedimentation time might be extended up to 12 h.
9. After sedimentation, discard the supernatant.

FLOTATION

1. Stir the sediment and place 1.5 ml of the sediment in a 15 ml tube.
2. Fill the tube with flotation fluid up to a volume of 14 ml and mix.
3. Centrifuge the tube at 1600 *g* for 10 min.
4. With a loop, take four drops from the surface of the flotation fluid and place on a glass microscope slide.
5. Place a coverslip on the specimen.
6. Examine the specimen with a light microscope at a magnification of × 200–400.[1]

Quantitative isolation of Hammondia- *and* Neospora-*like oocysts from faeces*

Depending on the size of the faecal sample, the diagnostic sedimentation and flotation protocol (see above) may have to be scaled up. 2 l beakers, 2.5 l conical measures and large sieves (diameter 12–13 cm) are used at the FLI to perform the sedimentation. For flotation, 50 ml tubes are filled with 5 ml sediment and 45 ml concentrated sucrose. After centrifugation, 2 ml tap water are carefully pipetted on top of the flotation fluid (see Fig. 3.1.12 (2)). Using a shortened 1 ml pipette tip (see Fig. 3.1.12 (1)) fixed to a pipette, the water phase is stirred (see Fig. 3.1.12 (3)). The water phase and 3–4 ml of the flotation fluid are removed to recover the floating oocysts (see Fig. 3.1.12 (4)).

The supernatants of two flotation tubes (10 ml) are transferred into one fresh 50 ml tube. This tube is filled with 40 ml tap water and centrifuged without the use of the brake for 8 min, 1600 *g* (first wash). After centrifugation, 45 ml of the supernatant are carefully removed. The sediments of two first-wash tubes are combined in a single 50 ml tube which is filled up to 50 ml with tap water and centrifuged again without use of the brake (second wash). Third and fourth washes are performed in the same fashion and the sediments are combined to a final volume of 1–50 ml (depending on the size of the sample, i.e. 10–500 g).

The number of isolated oocysts is estimated by the microscopic examination of 10 μl samples using a Neubauer haemocytometer and counting of the oocysts in the haemocytometer chamber.

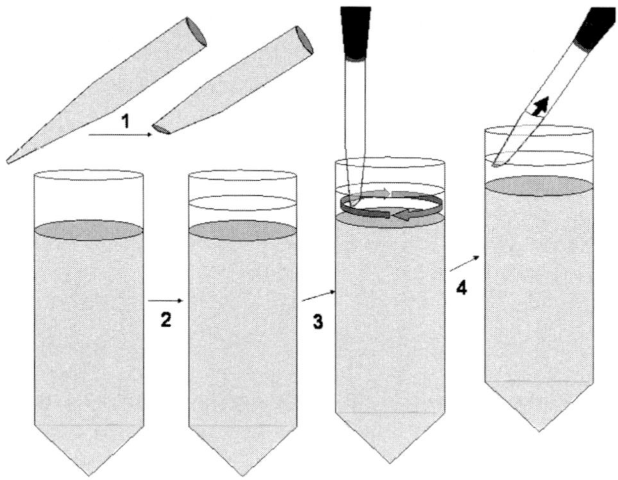

Fig. 3.1.12. Quantitative recovery of floating oocysts from a sucrose solution.

Oocyst detection in fodder, drinking water and environmental samples

For epidemiological studies on *N. caninum* infections, it may be desirable to examine not only the faeces passed by definitive hosts but also fodder, drinking water and environmental samples, for the presence of *N. caninum* oocysts. As yet, no protocols have been described to detect *N. caninum* oocysts in fodder, drinking water or environmental samples. However, procedures have been elaborated for *T. gondii* oocysts (reviewed by Dumetre and Darde, 2003, 2005) and it is likely that, after modification, these protocols will also be applicable to *N. caninum* oocysts.

Identification of N. caninum *oocysts by gerbil bioassay and PCR*

Once *Neospora-* or *Hammondia*-like oocysts have been obtained (see above), it is necessary to determine the species which is or are present in the isolated oocysts, with sensitive tools. In the following, a bioassay is described and a protocol for extracting DNA from oocysts presented. The latter is a prerequisite for the molecular characterization of isolated oocysts.

GERBIL BIOASSAY Gerbils (*Meriones unguiculatus*) are highly susceptible to infections with oocysts of various *N. caninum* isolates (Dubey and Lindsay, 2000; Basso *et al.*, 2001; Dijkstra *et al.*, 2001a; Schares *et al.*, 2001a, b, 2004a, 2005; Trees *et al.*, 2002). Dubey and Lindsay (2000) observed at 16 d post-infection (p.i.) parasite stages and lesions in the tissues of a gerbil inoculated with approximately one oocyst. Infection with five sporulated *N. caninum* oocysts was sufficient to induce a *N. caninum*-specific antibody response as determined by immunoblotting on day 31 p.i. (Schares *et al.*, 2002b).

Trees *et al.* (2002) observed that the inoculation of approximately one *N. caninum* oocyst was sufficient to induce a *N. caninum*-specific antibody response and a positive PCR signal using DNA extracted from the brain of the inoculated gerbil. The parasites could also be recovered from gerbil brain into cell culture. In the following, a brief description of the gerbil bioassay protocol used at the FLI is presented (Schares *et al.*, 2005).

Sporulation of oocysts

A prerequisite for the gerbil bioassay is the presence of sporulated, i.e. infectious, oocysts in the sample (see Fig. 3.1.13). To facilitate sporulation, the sample containing oocysts is incubated at room temperature (18–22°C, 3–5 d). With small samples of up to 5 ml, we use 50 ml tubes for this incubation. For larger volumes, Petri dishes are recommended to ensure that oocysts are supplied with enough oxygen. To prevent the growth of bacteria and fungi, potassium dichromate is added to a final concentration of 1–2%.

Sporulation can be monitored by examination under a light microscope (see Fig. 3.1.13). After sporulation, the oocysts must be stored in a refrigerator (4–8°C) until required. Prior to inoculation, the potassium dichromate is removed by washing with tap water (four times centrifugation (1100 *g*, 7 min, without use of the brake) and resuspending the pellet in 15 ml tap water; after the final wash the pellet is resuspended in the amount of water required to reach the inoculation volume of 0.5 ml).

Inoculation, blood sampling, assessment of the bioassay and cell culture isolation

There is no information indicating whether the age or sex of the gerbils can influence the result of the bioassay. To inoculate the animals orally, a bulb-ended curved needle (gavage) is used. The inoculation volume should not exceed 0.5 ml.[2] Prior to the inoculation, blood is taken from the lateral tail vein or the

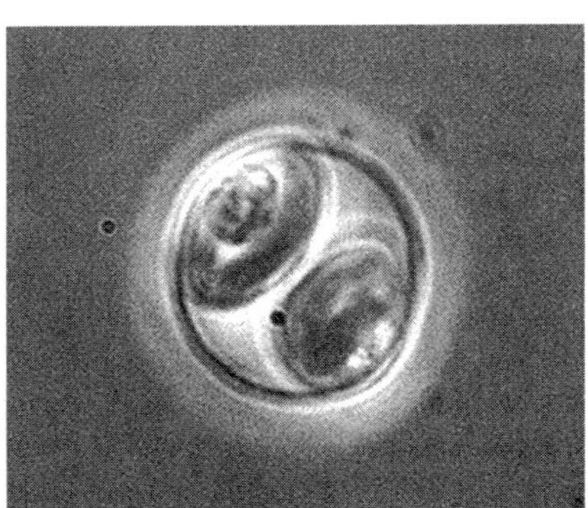

Fig. 3.1.13. Sporulated oocyst of *Neospora caninum*.

retro-orbital sinus of the animal in order to ensure that it was serologically negative prior to inoculation.

The gerbil bioassay is assessed by: (i) serological examination of the inoculated animal 3 weeks p.i.; (ii) by PCR of tissues recovered in the course of the post-mortem examination; and (iii) ideally by the post-mortem isolation of parasites from the inoculated gerbil.

For serology, the inoculated gerbil is bled on day 21 p.i. and tested for specific serological reactions against *N. caninum* antigens by immunoblotting (Schares *et al.*, 1998); a 1:100 serum dilution is used. A mouse-IgG/IgM peroxidase conjugate (Dianova, Hamburg, Germany) at a dilution of 1:1000 is applied to detect the reactions against immunodominant antigens of *N. caninum* tachyzoites (for details see the previous section on serology).

For the detection of *N. caninum* DNA, the gerbil is euthanized and necropsied. DNA is extracted from samples of brain, heart, lung, liver, spleen, kidney and skeletal muscle, and examined by specific primers (for details, see previous section on PCR).

Homogenized samples of organs from infected gerbils can be used to directly isolate the parasite into tissue culture.[3] Since the tachyzoite load of gerbil tissues is usually too low to obtain a reasonable cell culture infection (Schares, unpublished observation), it may be more suitable to inoculate gamma interferon knock-out (GKO) mice (C.129S7(B6)-*Ifng*^{tm1Ts}/J, The Jackson Laboratory, USA) intraperitoneally with homogenized gerbil tissues. Also, direct inoculation with sporulated oocysts may induce neosporosis in GKO mice (Schares *et al.*, 2005).

GKO mice are not able to control tachyzoite multiplication and consequently develop higher parasite loads in their tissues.[4] On the day that the first clinical signs in the inoculated animals are seen, they are necropsied and peritoneal exudate or homogenized organs used to infect cell cultures.[5] Many host cells are suitable for the isolation of *N. caninum* into cell culture (e.g. VERO, CPAE, COS-1, M617, HCT-8, HS68, bovine monocytes, equine kidney cells, mouse sarcoma cells).

EXTRACTION OF OOCYST DNA The specific detection of parasite DNA by PCR is regarded as very sensitive and specific. However, prior to amplification it is necessary to purify the parasite DNA, including the isolation of oocysts from the faeces, several incubation and washing steps to remove debris and contaminants, and a step to disintegrate the oocysts and the sporocyst wall. These procedures cause some loss of parasites (Hill *et al.*, 2001).

Together with PCR inhibitory effects of remaining contaminants (Hill *et al.*, 2001) and the fact that usually only an aliquot of the extracted DNA can be examined by PCR at one time, this causes PCR detection limits to be higher than those published for the gerbil bioassay (Slapeta *et al.*, 2002a, b; Schares *et al.*, 2005). At FLI, the sensitivity of the PCR is about four to eight oocysts per PCR analysis (Schares, unpublished data). Slapeta *et al.* (2002a) reported a sensitivity of two oocysts per analysis.

Depending on the volume of the DNA preparation (usually 50–100 µl), the volume of the template used for amplification (usually 2.5–10 µl) and the analytical sensitivity of the PCR, 10 to 400 oocysts have to be present in a sample that is subjected to DNA extraction in order to achieve a positive PCR reaction.

In the following section, a protocol is described that is used at FLI to extract DNA from isolated oocysts. This protocol consists of lysis steps, utilizing freezing and thawing cycles, treatment with saline saturated buffers, proteinase K and cety-trimethyl ammonium bromide. The method is partially based on the recommendations of Zhao *et al.* (2001).[6]

Reagents

- PBS: 300 mM NaCl, 2.7 mM KCl, 10 mM Na_2HPO_4, 1.7 mM NaH_2PO_4
- Sodium hypochlorite, aqueous solution, ≥ 4% as active chlorine (Aldrich 23,930-5)
- OOC-lysis buffer (pH 9.5): 600 mM EDTA, 1.3% (v/v) N-lauroylsarcosine, 2 mg/ml proteinase K
- OOC-CTAB buffer: 2% (w/v) cetyl-trimethyl ammonium bromide, 1.4 M NaCl, 0.2 % (v/v) mercapthoethanol, 20 mM EDTA, 100 mM tris(hydroxymethyl)aminomethane
- phenol/chloroform/isoamyl alcohol (25/24/1) (Sigma P3803)
- 4 M NaCl
- 96% (v/v) ethanol
- 70% (v/v) ethanol

Procedure

1. Isolate oocysts by sucrose flotation.
2. Use oocysts directly after isolation or store them in 1–2% $K_2Cr_2O_7$ in a refrigerator at 4–8°C.
3. Wash oocysts four times by centrifugation (1100 *g*, 7 min, without the use of the brake) in 15 ml PBS in a 15 ml centrifugation tube.
4. Incubate the pellet with oocysts and the remaining contaminants in 2 ml 5.75% sodium hypochlorite (30 min, 37°C).
5. Add double-distilled H_2O up to 15 ml.
6. Centrifuge supernatant in a 15 ml tube (1100 *g*, 7 min, without the use of the brake) and resuspend the pellet with PBS. Wash the pellet three times with PBS (1100 *g*, 7 min, without brake).
7. After a last centrifugation, resuspend the pellet in 1 ml PBS, transfer into a 1.5 ml reaction tube and spin down (1100 *g*, 7 min, without brake).
8. Carefully remove as much of the supernatant as possible and apply three freeze (10 min, –20°C)–thaw (2 min, room temperature) cycles to the pellet.
9. Resuspend the pellet in 100 µl OOC lysis buffer (45 min, 65°C).
10. Add 400 µl OOC-CTAB buffer (60 min, 60°C).

11. Mix with 500 µl phenol/chloroform/isoamylalcohol (25/24/1) by inverting 50 times. Centrifuge for 7 min at 13,000 *g*.
12. Transfer the supernatant to a fresh tube and add 0.04 volumes of 4 M NaCl and 2–3 volumes of –20°C 96 % (v/v) ethanol to precipitate DNA (keep at least 20–30 min at –20°C).
13. Centrifuge for 15 min at 13,000 *g*. Decant the supernatant.
14. Wash the pellet using 70% (v/v) ethanol and centrifuge for 15 min at 13,000 *g*.
15. Discard the ethanol solution and air-dry the pellet.
16. Resolve DNA in double distilled water for at least 12 h at 4°C.
17. Use 2.5–10 µl aliquots for PCR.

Endnotes

[1] Autofluorescence is reported for coccidia oocysts (Berlin *et al.*, 1998). The examination, using ultraviolet light, might help to locate oocysts in contaminated samples.

[2] Dubey and Lindsay (2000) reported that a dose of 1000 oocysts of the strain NC-Liv was sufficient to cause disease in gerbils. At the FLI, we have never observed disease in gerbils inoculated with various *N. caninum* isolates. However, the observed pathogenicity might be related to either the parasite strain or the strain of *Meriones unguiculatus*. Consequently, it might be wise to inoculate a single animal with no more than 500 oocysts to avoid fatal disease in the gerbil.

[3] Brain is the organ in *N. caninum*-infected gerbils that is found to be positive most frequently by PCR, and may thus be most suitable for isolation attempts (Dubey and Lindsay, 2000; Schares *et al.*, 2001a).

[4] Other immunodeficient mouse strains or mice treated with cortisone may be suitable.

[5] Direct cell culture isolation of *N. caninum* from oocysts might also be possible, but not all reported attempts were successful (Gondim *et al.*, 2002).

[6] Other protocols for the isolation of DNA from oocysts employ grinding with glass beads to disintegrate the oocysts (e.g. Hill *et al.*, 2001; Ruttkowski *et al.*, 2001; Slapeta *et al.*, 2002b).

Diagnostic Applications and Recommended Diagnostic Schemes

Cattle

Gema Alvarez-Garcia, Luis-Miguel Ortega-Mora, Thomas Dijkstra and Willem Wouda

Abortion is the main clinical manifestation of neosporosis. Fetuses dying *in utero* between 3 and 8 months of gestation are usually expelled showing moderate autolysis, but fetuses dying before 5 months' gestation may be mummified and retained in the uterus for several months. Fetuses dying at an early stage of gestation may be reabsorbed, causing repeat breeding. Abortions may be sporadic, endemic or epidemic (Dubey and Lindsay, 1996; Dubey, 1999; Dubey, 2003a, b).

The clinical history and epidemiological data can suggest a *N. caninum* infection, but for a more definite diagnosis the assistance of a veterinary diagnostic laboratory is needed. The examination of aborted fetuses in combination with maternal serology is recommended (see Tables 3.1.3 and 3.1.4).

Other infectious agents responsible for reproductive failure in cattle, such as bacteria, fungi, viruses or other protozoa, need to be ruled out, although mixed infections of *N. caninum* and BVDV have occasionally been described (Caldow, 1998; Bjorkman *et al.*, 2000; Mainar-Jaime *et al.*, 2001). Non-infectious causes of abortion should also be excluded.

Fetal diagnosis

Ideally, a number of diagnostic tests needs to be employed to increase the chance of detecting infection (Mainar-Jaime *et al.*, 2001; Sager *et al.*, 2001; Pereira-Bueno *et al.*, 2003). A combination of fetal histology and direct (PCR) or indirect (fetal serology) agent detection could be the best choice.

Detection of lesions and parasites in fetal tissues

Samples

Preferably, the whole fetus (with a sample of the placenta) should be submitted to a diagnostic laboratory so that a pathologist can collect relevant samples, not only for neosporosis but for other abortifacients as well.

Table 3.1.3. Individual diagnosis of *Neospora caninum*-associated abortions.

	Aborting dam		Aborted fetus			
	Serum		Serum, body fluids		Tissue samples: brain, heart, liver, placenta	
	IFAT, ELISA and WB		IFAT, ELISA and WB		Histological survey	
	Negative	Positive	Negative	Positive	Negative	Positive
Conclusions	Infection unlikely	Exposure to the parasite	Infection can not be ruled out	Infection	Infection can not be ruled out but implication in abortion is less likely	Compatible lesions with *Neospora* infection
Remarks	Antibody fluctuations possible; immunotolerance may occur	Is *N. caninum* responsible for abortion?	Absence of infection should be confirmed by histology and herd diagnosis	Implication in abortion should be confirmed by histology		Are other parasitic agents implicated in the abortion?
	Sequential serum samples should be analysed to confirm the absence of the infection	Low avidity results indicate a recent infection (*N. caninum*-induced abortion in an outbreak) High avidity results provide no additional information and complementary diagnosis on a herd level is required				The presence of *Neospora* should be confirmed by direct detection of parasite by means of IHC or PCR

Table 3.1.4. Herd diagnosis of *Neospora caninum*-associated abortions.

Calves		Breeding cattle			
Pre-colostral serum samples		Serum		Bulk milk	
IFAT, ELISA and WB		IFAT and ELISA		ELISA	
Negative	Positive	Negative	Positive	Negative	Positive
Conclusions					
Congenital infection unlikely	Congenital infection	Infection can not be ruled out	Infection	Infection can not be ruled out, as less than 10–15% prevalence in herd not detected	Prevalence of infection in herd is expected to be higher than 10–15%
Remarks					
Is there incidence of post-natal infection in the herd?	Is vertical transmission the route mainly responsible for *N. caninum* propagation in a herd?	At least two sequential samples should be analysed to avoid false negative results related to antibody fluctuations	Statistical comparisons between aborting and non-aborting cattle can be done	Bulk milk tests have a limited sensitivity; the infection can thus not be ruled out completely by bulk milk testing	Statistical comparisons between aborting and non-aborting cattle can be done
Samples collected at 6 months of age should be tested by serological methods	Paired samples from infected dams and their progeny should be tested		Avidity ELISA in aborting cattle can help to determine the route of infection (vertical or horizontal transmission)		Avidity ELISA in aborting cattle can help to determine the route of infection (vertical or horizontal transmission)
In the case of a positive result, post-natal infection should be confirmed by low avidity results	High avidity results are expected in aborting and non-aborting dams				

The brain is the target organ to be analysed by the different tests for neosporosis, although the probability of diagnosing the infection may increase when additional tissues, such as placenta, heart, lung and liver, are analysed (see previous section on histology). Tissues should be stored at –20°C for PCR analysis or fixed in 10% buffered formol saline for histological and immunohistochemical examination and stored at room temperature. PCR can also be applied in fresh, formalin-fixed and paraffin-embedded fetal tissues (Dubey and Lindsay 1996; Dubey, 1999, 2003b).

Histological examination

Routine sampling for histological examination of the brain heart, liver and placenta (Otter *et al.*, 1995) is an essential step in arriving at a diagnosis in a case of abortion, as it may rule in one cause while ruling out others. The histological lesions caused by *N. caninum* in these tissues are often distinctive, particularly those in the brain, and even in severely autolysed or mummified fetuses they can often be recognized under the microscope. The finding of multifocal encephalitis in a bovine fetus, together with myocarditis and/or hepatitis, is highly suggestive of neosporosis. The placenta is not of much help in diagnosing neosporosis, but is crucial for diagnosing a fungal or chlamydial abortion.

Immunohistochemistry and PCR methods

The diagnosis of *N. caninum* infection often relies on immunohistochemistry (IHC) or PCR techniques. While the sensitivity of the PCR is greater when compared with IHC, it should be performed in combination with a histological examination to confirm that any detected DNA is associated with relevant lesions, as the parasite may be present in fetuses that died from other causes.

IHC and PCR methods are highly specific. However, with IHC it is sometimes possible to get cross-reactivity with *Toxoplasma gondii* using certain antibody preparations, and care should therefore be taken to avoid this (Gottstein *et al.*, 1998; Van Maanen *et al.*, 2004). For this reason, the PCR may be more suitable for some laboratories than others.

Although the efficiency of detection of *N. caninum* by PCR may be dependent on the particular laboratory, the agreement among the PCR tests developed by different laboratories has been shown to be moderate to good, indicating that there is not a clear relationship between the PCR format (single or nested) and diagnostic sensitivity. Other factors influencing the performance of the PCR tests include the degree of autolysis of fetal tissues and sampling procedures. Special care should be taken to avoid cross-contamination and carry-over, which will lead to false positive results.

However, the most critical step appears to be the DNA extraction procedure, according to the observations reported by Van Maanen *et al.* (2004). Homogenization of samples is a crucial step, since the parasite is not homogeneously distributed in the CNS and, consequently, an appropriate sample preparation before DNA extraction can be important to ensure proper PCR results.

DETECTION OF SPECIFIC ANTIBODIES AGAINST *N. CANINUM* IN FETAL FLUIDS The bovine fetus develops the ability to produce antibodies during the fifth month of gestation (Osburn, 1988). Therefore, fetal serology can be applied to aid in the diagnosis of *N. caninum* infection in fetuses of 5 months or older. Fetal fluids from the abdominal or thoracic cavity can be examined for antibodies using IFAT or ELISA methods (Barr *et al.*, 1995; Wouda *et al.*, 1997a). However, the sensitivity of fetal serology may be low, due to low antibody titres in fetal fluids and post-mortem degradation of immunoglobulins.

An appropriate selection of the cut-off points different from the ones employed for adult cattle should be taken into consideration. An IFAT cut-off value between 1:16 (Alvarez-Garcia *et al.*, 2003) and 1:25 (Wouda *et al.*, 1997a) should be appropriate for fetuses. The western blot (WB) may be an alternative tool. Although it is not recommended in routine diagnosis, it improves fetal serology, as higher sensitivity values are obtained (Söndgen *et al.*, 2001). It should be kept in mind that the presence of antibodies to *N. caninum* in fetal fluids does not prove that the infection was the cause of the abortion; it shows only that the fetus has been exposed to the parasite.

ISOLATION OF *N. CANINUM* FROM PLACENTA AND FETAL TISSUES Isolation of *Neospora* in cell culture or by bioassay in mice is not suitable for routine diagnosis as the success of isolation depends on the presence of viable organisms, because most parasites in bovine fetuses rapidly die (Conrad *et al.*, 1993a; Davison *et al.*, 1997; Yamane *et al.*, 1998). However, these techniques may be suitable for certain research projects on molecular epidemiology.

KEY POINTS

- Samples from fetal tissues, including brain, heart, liver and placenta for histological examination and PCR, and from pleural or peritoneal fluids or abomasal contents for antibody detection in fetuses of over 5 months' gestation.
- At least two blocks of tissue from the brain (for example, two coronal slices through the brain, one from the cerebrum at the level of the thalamus and a second from the cerebellum, through the cerebellar peduncles and pons) as well as samples of liver, heart, lung and placenta should be investigated in order to increase the chance of detecting lesions and parasites.
- Specific anti-*Neospora* antibodies can be tested in fetuses older than 5 months by IFAT, ELISA or WB. When employing IFAT, a cut-off point in the range of 1:16–1:25 should be considered. WB increases sensitivity when it is carried out in abomasal content.
- The analysis of brain tissue, together with other organs such as heart, liver and placenta, improves the sensitivity of the histological procedures.
- Detection of characteristic or compatible lesions of *N. caninum* infection by histological examination (hematoxylin-eosin stained sections) is of great value as an initial diagnostic procedure.

- *N. caninum* may be detected in sections of fetus and placenta by IHC and PCR assay.
- An advantage of IHC is that the presence of parasites, or parasite antigen, can be related to the lesions. PCR offers greater sensitivity compared with IHC.
- To ensure an appropriate outcome of PCR methods prior to DNA extraction, fetal tissues should be carefully homogenized.

Post-natal detection of infection

Serological methods are usually employed to examine animals *in vivo* and thereby provide information on the stage of the infection.

NEWBORN CALF Trans-placental transmission of infection is the main route responsible for the maintenance of the parasite in a herd, and rates of endogenous (vertical) transmission in the range of 66–95% have been reported (Dijkstra *et al.*, 2003). Congenitally infected calves normally remain clinically healthy: only a small proportion of them develop clinical signs. Therefore, *in vivo* diagnosis depends mainly on serological assays (Jenkins *et al.*, 2002).

Clinical signs

Rarely, congenitally infected calves may show nervous signs varying from mild ataxia to tetraparalysis. Clinical signs have been reported only in calves younger than 2 months of age (Dubey and Lindsay, 1993).

Detection of specific antibodies against *N. caninum* in sera from newborn calves

To determine whether a newborn calf was infected *in utero*, a pre-colostral serum sample needs to be examined to avoid false positive test results due to colostral antibodies (mainly IgG1). When no specific antibodies are detected in newborn calves, infection with *N. caninum* is unlikely. Mostly, pre-colostral antibody concentrations in congenitally infected calves are high (Wouda *et al.*, 1998a).

IFAT and ELISA can be applied by considering a low cut-off point to maximize sensitivity, as the aim of the serological investigations carried out in calves is to provide evidence of the infection. Because maternal colostral antibodies in the calf may persist for several months, a post-natal infection in calves cannot be ruled out or confirmed before 6 months of age.

Detection of *N. caninum* in tissues from the newborn calf

IHC and PCR methods cannot be employed in the routine diagnosis of neosporosis in congenitally infected live calves. However, these techniques have proved to be valuable tools in the investigation of routes of transmission of *N. caninum* and in assessing the efficacy of pharmacological treatments in the brain and spinal cord from experimentally infected calves (Uggla *et al.*, 1998; De Marez *et al.*, 1999; Davison *et al.*, 2001; Innes *et al.*, 2001; Kritzner *et al.*, 2002).

KEY POINTS

- Precolostral sera from newborn calves
- Detection of specific antibodies to *N. caninum* by IFAT or ELISA
  - Pre-colostral serology to investigate endogenous trans-placental transmission of the infection
  - Serology to demonstrate post-natal transmission not before 6 months of age
- Post-mortem diagnosis by PCR and/or IHC in brain tissues
- Histopathology of defined tissues.

ADULT CATTLE In adult cattle, reproductive loss is the only clinical sign, so *in vivo* diagnosis largely depends on serological tests.

Clinical signs

Infection may cause fetal resorption, mummification, abortion, premature birth or stillbirth of calves.

Samples

The detection of specific antibodies is usually carried out with serum, but samples of milk may also be suitable (Schares *et al.*, 2003), although to date no comparison of the titres expected in milk has been published. Antibody concentrations in milk samples tend to be lower than in sera. Whenever possible, a sample should be taken shortly after abortion, since specific antibodies may decrease below the cut-off value within 2 months of abortion (Jenkins *et al.*, 2002).

Detection of specific antibodies against *N. caninum* in sera from individual cows

Cows aborting due to an infection with *N. caninum* often have high levels of antibodies shortly after the abortion. Therefore, post-abortion serology is a useful tool to aid in the diagnosis of neosporosis as a cause of abortion in individual cows (Wouda *et al.*, 1998a). However, the mere presence of antibodies to *N. caninum* does not prove that the infection caused the abortion, as many chronically infected cows are serologically positive.

If the infection status of a cow needs to be determined, e.g. for trade reasons but not related to an abortion, it is worth noting that antibodies in infected cattle may fluctuate substantially and may even drop below the cut-off value of the serological test. This may be particularly relevant if a test is used with a cut-off optimized to detect dams that have aborted due to infection with *N. caninum*. Consequently, the most sensitive tests available should be used for trade purposes or the cut-off of the test used adjusted appropriately (Schares *et al.*, 1999a; Alvarez-Garcia *et al.*, 2003).

A pattern of antibody response in naturally infected cattle has been reported by several investigators (Stenlund *et al.*, 1999; Quintanilla-Gozalo *et al.*, 2000). These investigations have revealed an association between recrudescence of infection and trans-placental (vertical) transmission during the third trimester of pregnancy. In aborting cattle, recrudescence around mid-pregnancy can be evidenced by a peak in antibody levels (Stenlund *et al.*, 1999; Quintanilla-Gozalo *et al.*, 2000).

Of the different serological tests available, the IFAT and ELISA (see section on serology, above) are the most commonly used (Bjorkman and Uggla, 1999; Atkinson *et al.*, 2000). If post-abortion sera are examined, a restrictive cut-off should be used, while with individual sera from non-aborting cattle, a cut-off value for maximal sensitivity is more appropriate (Schares *et al.*, 1999a; Alvarez-Garcia *et al.*, 2003). Western blotting can be used to confirm inconclusive results.

A comparative study has shown that most tests have a high level of agreement if standardized cut-offs are used, which may be useful in cross-sectional seroprevalence studies (von Blumröder *et al.*, 2004). Individual milk samples can be analysed by ELISA (Bjorkman *et al.*, 1997; Schares *et al.*, 2004a).

Avidity tests could be applied to determine a recent exogenous infection in an individual cow. However, avidity test results are more reliable when used at the herd level. A summary of the diagnostic approach in an individual aborting cow is given in Table 3.1.3.

KEY POINTS

- Serum or milk samples
- Different cut-off points in IFAT or ELISA should be considered, depending on the purpose of testing: diagnosis of abortion or infection. Western blot is the recommended confirmative method to clarify inconclusive results
- A high antibody titre in a post-abortion sample is a strong indicator of a causal relationship between infection and abortion.

HERD LEVEL Serious abortion problems should be examined on a herd level (see Table 3.1.4). Thereby, an estimation of the serological status of the whole herd is essential. A classical approach is to accomplish whole-herd serological screening by means of IFAT or ELISA using a cut-off for maximal sensitivity. Bulk milk antibody testing can be applied to gain a rapid impression of the prevalence of *N. caninum* infection in a herd (Chanlun *et al.*, 2002). With this approach, a minimal prevalence of *N. caninum* infection of 10–15% in the lactating cows in a herd can be measured (Schares *et al.*, 2003). However, a low bulk milk adsorbance does not exclude the presence of infection in the herd.

If the results of a whole-herd screening are available, a sero-epidemiological approach can be applied to estimate the impact of *Neospora* infection in relation to the incidence of abortions (Thurmond and Hietala, 1995; Sager *et al.*, 2001). The rationale is to determine whether the proportion of seropositivity in aborting cows is higher than in non-aborting cows. If *N. caninum* contributes to the abortion problem, the proportion of seropositive aborting cows should be significantly higher than the proportion of seropositive non-aborting cows that are also at risk.

The differences in seropositivity can be assessed by statistical procedures such as Fisher's exact test or the Chi-squared test. If the seropositivity rate is not significantly higher for aborting than for non-aborting cows, there is no evidence for a causal relationship between the infection with *N. caninum* and the abortion problem. If the proportion of seropositivity is higher among aborting than among non-aborting cows, an additional calculation can be done to estimate the extent to which *N. caninum* is involved in the problem.

By means of odds ratio calculation, a magnitude of about 2 would be indicative of an endemic pattern of abortion, whereas higher values would be expected for an epidemic presentation of abortions. If results of a whole-herd screening are not available, a representative number of samples should be submitted for this evaluation: at least ten samples from each risk group (aborting and non-aborting pregnant cows), and also samples from cows that are not pregnant (not at risk), if possible from matched pairs (Kinsel, 2002).

Once a causal relationship between *N. caninum* infection and the abortions in a herd has been established, further analyses should be performed to determine the predominant route of transmission in the herd (i.e. endogenous or exogenous trans-placental transmission). After a serological screening of a whole herd, an analysis of the age distribution of the seropositive animals and a comparison of dam–daughter serostatus can help in interpreting the mode of transmission (Dijkstra *et al.*, 2003).

In the case of endogenous trans-placental (vertical) transmission, seropositive animals are equally distributed across the different age groups, and seropositive daughters have seropositive dams. In the case of (additional) exogenous (horizontal) transmission, an age-related clustering of seropositive animals can be found, together with a lack of association between the serological status of dams and daughters (Dijkstra *et al.*, 2001a).

Such a lack of association between serological status of dams and daughters has been documented by several researchers, particularly in relation to abortion outbreaks (Thurmond *et al.*, 1997; Patitucci *et al.*, 1999; Waldner *et al.*, 1999). Age clusters of *N. caninum*-seropositive cattle may have either seronegative dams or seronegative offspring (Dijkstra *et al.*, 2001a).

The first situation indicates post-natal infection of the daughters, and the second situation indicates post-natal infection of the dams after the birth of the seronegative progeny. Apart from comparison of dam–daughter serology and analysis of the age distribution of seropositive animals, the pattern of the abortions (endemic or epidemic) – and the avidity serological responses in aborting dams – may give indications concerning the predominant route of transmission of *N. caninum* (Jenkins *et al.*, 2000; McAllister *et al.*, 2000; Bjorkman *et al.*, 2003). After primary infection, the *N. caninum*-specific antibody response is characterized by an initially low avidity, which later increases.

In herds with epidemic abortion, the majority of animals will have an IgG response of low avidity, while animals in herds with endemic abortion will predominantly exhibit high avidity responses (Dijkstra *et al.*, 2002a; Schares *et al.*, 2002a). It is a prerequisite for a correct interpretation of avidity values that samples are obtained shortly after abortion.

KEY POINTS

- Diagnosis of infection should be carried out by employing IFAT or ELISA methods with restrictive cut-off points in order to increase sensitivity.
  - Serological screening of a whole herd is a rapid and valid method of estimating herd prevalence and the serological status of individual animals.
- Analysis of the serological status of aborting and non-aborting cattle permits the establishment of statistical associations between seropositivity and abortion.
- Analysis of the age distribution of the seropositive animals and a comparison of dam–daughter serostatus can help to interpret the mode of transmission.
  - Bulk milk testing can also be done by ELISA if the intra-herd prevalence is higher than 10–15%.
- Avidity tests can be used to provide evidence of acute horizontal (post-natal, exogenous) infection.

Dog

Gereon Schares and Bertrand Losson

General considerations

Epidemic bovine abortion due to infection with *N. caninum* is considered to occur after the ingestion of food or water contaminated with *N. caninum* oocysts (McAllister *et al.*, 1996a). In case reports of epidemic bovine abortion low-avidity antibodies have regularly been observed, suggesting recent post-natal infection of the dams (McAllister *et al.*, 2000; Schares *et al.*, 2002a; Sager *et al.*, 2005). Oocysts shed by definitive hosts are regarded as the most likely source of post-natal infections. Dogs and coyotes are known definitive hosts (McAllister *et al.*, 1998; Gondim *et al.*, 2004a), whereas the European fox does not currently seem to be a definitive host of this parasite (Schares *et al.*, 2002b).

Thus in Europe, the dog (*Canis familiaris*) is the only known definitive host for *N. caninum*. Risk factor analyses have indicated that farm dogs (Paré *et al.*, 1998; Bartels *et al.*, 1999; Mainar-Jaime *et al.*, 1999; Ould-Amrouche *et al.*, 1999; Schares *et al.*, 2004b) and dogs in the neighbourhood of farms may pose an infection risk (Schares *et al.*, 2003). Consequently, there is a need to identify dogs or other definitive hosts as sources of post-natal infection in cattle or other intermediate hosts. In the following, information is summarized on diagnostic methods for identifying dogs that are actively infected, or have been infected, with *N. caninum.*

It must be stressed that dogs can also be intermediate hosts of *N. caninum.* However, the diagnosis of dogs as intermediate hosts of *N. caninum* is beyond the scope of this chapter.

As yet, there is no unambiguous proof that contamination of the environment, the fodder or the drinking water with oocysts causes *N. caninum* infections on farms and there is no tool for identifying those definitive hosts retrospectively that have contributed to these oocyst contaminations.

However, if aborting or infected intermediate hosts have low-avidity antibodies against immunodominant *N. caninum* tachyzoite antigens and if one or more of the following observations are made, this increases the evidence for oocyst contamination being responsible for the infection (see Fig. 3.1.14):

- *N. caninum* oocysts are detected in the faeces of a putative definitive host that had access to the animals' environment (e.g. the environment of a cattle herd) or to their sources of fodder and drinking water.
- *N. caninum* oocysts are detected in fodder, drinking water or in environmental samples close to an affected herd or animal.
- Antibodies against a 152 kDa antigen of *N. caninum* tachyzoites are detected in the serum of a putative definitive host.
- Antibodies against the immunodominant antigens of *N. caninum* are detected in the sera of a putative definitive host.

*Diagnostic significance of low-avidity, anti-*N. caninum *IgG response in intermediate hosts*

Low-avidity responses against *N. caninum* tachyzoite antigens in intermediate hosts usually indicate that the animals have become infected recently. The most likely sources are oocyst contaminations. However, at present, the finding of low-avidity antibodies is no proof that the animals became infected by the ingestion of oocysts.

Diagnostic significance of oocyst detection

If *N. caninum* oocysts, either in dog faeces, in fodder, drinking water or in the environment of the infected animals are detected, this is a strong indicator that the animals have become infected by this mode.

Little is known about the shedding of oocysts by naturally infected dogs. In experimentally infected dogs, however, oocyst shedding did not commence earlier than 5 days after the ingestion of *N. caninum*-infected intermediate host tissues. If all published results – with a total of 42 dogs (McAllister *et al.*, 1998;

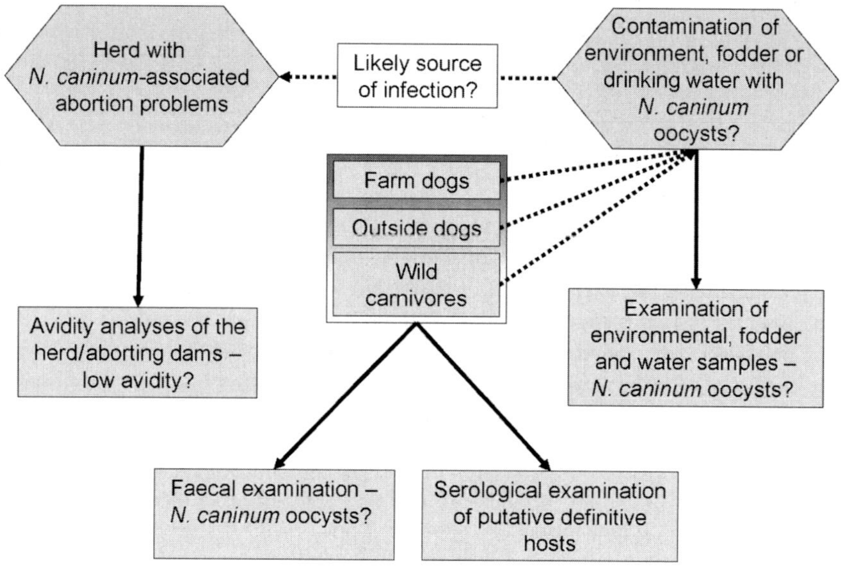

Fig. 3.1.14. Diagnostic tools for obtaining evidence that oocyst contamination caused by a definitive host is responsible for *Neospora caninum* infection in cattle or other intermediate hosts.

Lindsay *et al.*, 1999b; Dijkstra *et al.*, 2001b; Schares *et al.*, 2001a; Gondim *et al.*, 2002) – are taken into account, the observations of the *N. caninum* oocyst shedding peak between days 8 and 13 after oral infection (see Fig. 3.1.15).

Most dogs shed oocysts until day 13, after the ingestion of infectious material. This period is followed by intermittent shedding for up to 30 days post-feeding in many dogs. In all published experiments, the observation periods ended after 15–37 days post-ingestion of infectious material. Consequently, it is unknown whether this intermittent shedding can last for longer under field conditions. There is experimental evidence that dogs that have shed *N. caninum* oocysts in the past might be able to shed (perhaps smaller numbers of oocysts) again after a new infection (Dijkstra *et al.*, 2001b; Gondim *et al.*, 2005).

While epidemics of bovine abortion due to neosporosis are considered to have been caused by oocyst contamination of fodder or drinking water of the dams, shedding of oocysts by the definitive host has not yet been observed on such farms and no oocyst contamination has been demonstrated. The reason for this might be that abortions do not occur immediately after the oocyst-induced infection, but with a delay of several days or weeks, and this is supported by experimental infections with tachyzoites (Dubey *et al.*, 1992; Schares *et al.*, 1999b; Andrianarivo *et al.*, 2000; Williams *et al.*, 2000). Therefore, oocyst shedding by the responsible dog might have already ended when the abortions in the cattle were

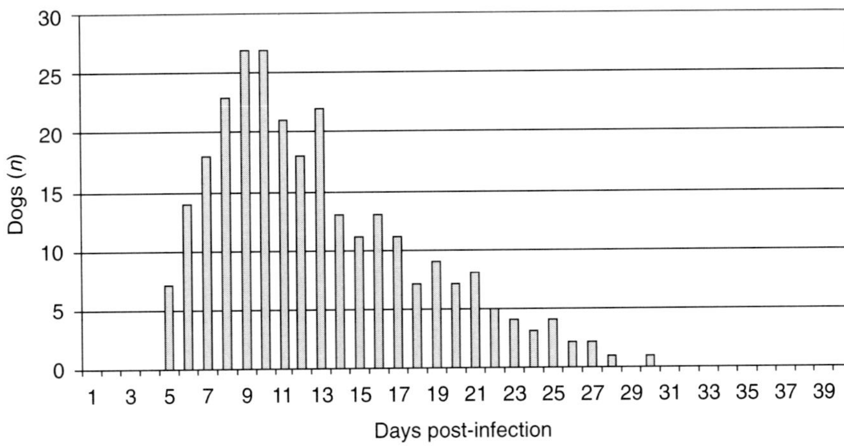

Fig. 3.1.15. Summary of data from the literature showing the shedding of *Neospora caninum* oocysts by 42 experimentally infected dogs. Shedding had started in most dogs by day 8 and ended 13 d post-infection. Some dogs started to shed oocysts earlier (but not before day 5). A significant number of dogs also released oocysts after day 13.

first detected. Consequently, if dog faeces are negative on faecal examination, this does not mean that the respective dog has not shed oocysts before. In addition, the intensity of oocyst shedding might be low, i.e. oocyst shedding might be missed when only a small proportion of dog faeces is examined.

As yet, no methods have been described for the identification of *N. caninum* oocysts in fodder, drinking water or in the environment of infected intermediate hosts. However, methods used to detect *T. gondii* oocysts in soil or water might be applicable after modification (Dumetre and Darde, 2003, 2005). Little is known of the sensitivity of these techniques. Consequently, oocyst contaminations of the fodder, the drinking water or the environment might be too low to be easily detected.

Diagnostic significance of reactions against a 152 kDa antigen of N. caninum *tachyzoites in the sera of putative definitive hosts*

Reactions against a 152 kDa tachyzoite antigen were recognized in the sera of dogs after they had shed oocysts following feeding with infected intermediate host tissues (Schares *et al.*, 2001b). The reactions were observed in dogs between days 35 and 447 after feeding with infected tissue, but not prior to oocyst shedding. Recognition of the 152 kDa tachyzoite antigen may thus be a marker for dogs having shed *N. caninum* oocysts in the past. The 152 kDa tachyzoite antigen was also recognized by those dogs which neither showed positive IFAT reactions nor recognized *N. caninum* immunodominant antigens (IDAs) in the immunoblot. Possibly this 152 kDa is an *N. caninum* antigen not only present in the tachyzoite stage but also in stages involved in the asexual or sexual development of *N. caninum* in the intestine of the definitive host.

However, it has to be stressed that the results regarding the 152 kDa tachyzoite antigen are still preliminary. Further experiments are needed to confirm that this antigen is species specific. In particular, it has to be shown that dogs shedding oocysts of related coccidians or infected with coccidian parasites do not recognize the 152 kDa antigen. Unpublished results on dogs shedding *H. heydorni* after experimental infection showed no reactions against the 152 kDa in these animals (Schares *et al.*, unpublished results). In addition, not all dogs that have shed oocysts in the past recognize this antigen. Consequently, a negative result does not allow any conclusions regarding oocyst shedding at any stage. Since the antibody reactions against the 152 kDa antigen persist for months, positive reactions are not necessarily associated with a recent intermediate host infection. However, when this technique was applied in the analysis of sera from dogs naturally shedding *N. caninum*, all these sera reacted with the 152 kDa antigen (Schares *et al.*, 2005).

Diagnostic significance of reactions against N. caninum *tachyzoites in sera of putative definitive hosts by the IFAT or NAT*

The detection of IFAT- or NAT-positive reactions against *N. caninum* tachyzoites in dog sera is of almost no significance regarding the identification of dogs as a source of oocyst contaminations. Such reactions indicate only that the respective animal is or was an intermediate host of *N. caninum*.

In many studies, the majority of dogs with a history of shedding oocysts showed no seroconversion with respect to *N. caninum* tachyzoite surface antigens examined by IFAT or NAT (see Table 3.1.5). Immunoblot-based tests detecting antibodies against immunodominant antigens (IDAs) were expected to be more sensitive than the IFAT, but were not able to demonstrate seroconversion against immunodominant tachyzoite antigens in dogs that had shed oocysts (Schares *et al.*, 2001b, 2005). This was also confirmed by the examination of sera from naturally infected dogs, shedding *N. caninum* oocysts (Schares *et al.*, 2005).

Thus, during the development of *N. caninum* in the canine intestine, which ultimately leads to oocyst shedding, no or only a few extra-intestinal stages (e.g. tachyzoites) may develop. These apparently fail to cause a detectable antibody response against the specific tachyzoite antigens (i.e. surface antigens) in the majority of dogs.

The observation that some dogs seroconverted after they had shed *N. caninum* oocysts (see Table 3.1.5) is also in agreement with the hypothesis that no extra-intestinal stages occur in dogs during the intestinal development of *N. caninum*. Dogs that shed oocysts may have a chance of ingesting some of those

Table 3.1.5. Antibodies against *N. caninum* tachyzoites in experimentally infected dogs having shed *N. caninum* oocysts.

Reference	Dogs shedding *N. caninum* oocysts (*n*)	Blood sampling post-feeding (d)	Dogs IFAT-positive[a] (*n*)	Dogs NAT-positive[a] (*n*)
McAllister *et al.*, 1998	5	37	4 (≥ 1:50)	nd
Lindsay *et al.*, 1999	2	42	1 (≥ 1:25)	nd
Dijkstra *et al.*, 2001	3	77	0 (≥ 1:20)	nd
Schares *et al.*, 2001	6	30–71	0 (≥ 1:20)	nd
Lindsay *et al.*, 2001	2	36	nd	1 (≥ 1:25)
Gondim *et al.*, 2002	14	25	6 (≥ 1:50)	nd
Total	32	–	11	1

[a] Cut-off applied.
nd, no data.

oocysts that they had shed themselves. If these oocysts are sporulated when ingested, auto-infection may occur.

Infection with oocysts may thus lead to an antibody response against tachyzoite antigens, as in many other intermediate hosts. Consequently, the presence of tachyzoite-specific antibodies in a dog indicates only that the animal has become exposed to tachyzoites. Whether the ingestion of *N. caninum* oocysts or other modes of transmission are responsible for the seroconversion to tachyzoites remains unanswered.

Future trends

Bruno Gottstein and Franz J. Conraths

Introduction

Neospora caninum is regarded as a major cause of infectious bovine abortion worldwide. Diagnosis of a *N. caninum* infection is primarily based on the examination of the aborted fetus and, secondarily, of the dam. In addition, sero-epidemiological analyses in afflicted herds may demonstrate statistical association between seropositivity and abortion.

Examination of fetal material

Examination of fetal material includes the histological examination of fetal tissues for characteristic lesions, the demonstration of specific *N. caninum* antigens or DNA by immunohistochemistry (IHC) or PCR, respectively, and the detection of specific antibodies in fetal fluids. The latter possibility is of restricted interest due to the fact that some fetuses fail to develop a detectable antibody response. Immunoblot (IB), as an alternative method to detect specific antibodies against *N. caninum*, has been introduced into routine laboratory diagnostics and has become an important tool in the elucidation of unclear reactions observed in other serological tests such as ELISA and IFAT.

Serology

Serology has also been adapted to detect antibodies in milk samples and may thus be easily applied for sero-epidemiological surveys. The use of bulk milk samples may allow a fast and inexpensive assessment of the herd status. The avidity ELISA has improved the application spectrum of serology by yielding information on the recency of exposure to infection. Standardization of serology has been approached in Europe by carrying out respective studies (von Blumröder *et al.*, 2004).

PCR

Technologically, pathohistology and immunohistochemistry have been well established and evaluated. Diagnostic findings provide a good insight into both the damage the infection has induced and the intensity of infection in the affected organ. An emergent request for further developments has thus not been claimed in the field. PCR has considerably enlarged the field of application, in that autolysed and mummified samples can also be investigated.

So far, PCR has rather been used to qualitatively discriminate between presence and absence of *N. caninum* DNA, and not to address the question of intensity of infection. While histology and IHC require the presence of approximately 100,000 parasites/g of tissue, PCR already operates with five parasites/g of tissue. Due to the high methodical sensitivity, the diagnostic sensitivity of PCR is thus considerably higher than that of histology or IHC.

Future trends in PCR diagnosis are therefore focusing on the development of quantitative PCRs, which may also reflect the viability status of the parasite. Such approaches have already been described (Liddell *et al.*, 1999; Collantes-Fernandez *et al.*, 2002; Müller *et al.*, 2002), but not yet introduced into routine laboratory diagnosis. An appropriately validated RT-PCR approach could, in future, become the unique tool for investigation of diagnostic tissue samples and thus completely replace histology or IHC. All PCRs used so far yield very sensitive detection of parasite genomes in diagnostic specimens, the diagnosis operating at the species-specific level.

For future studies, many groups wish to include molecular-epidemiological aspects into their diagnosis. Therefore, for example, multiplex PCRs enabling the characterization of putative 'strains', 'geographical isolates' or 'subspecies' of the parasites harbouring respective genetic markers (which first need to be detected and characterized). Moreover, PCR has also been used to detect the parasite *in vivo* in the blood and semen of cattle (Okeoma *et al.*, 2004, 2005; Ferre *et al.*, 2005).

Detection of N. caninum *in wildlife*

Due to the detection of *N. caninum* in wildlife, it seems now likely that a sylvatic cycle exists – at least in North America (Rosypal and Lindsay, 2005), but probably also in Europe and perhaps elsewhere. In North America, the white-tailed deer (*Odocoileus virginianus*) has been identified as a frequently infected natural intermediate host of *N. caninum* (Dubey *et al.*, 1999; Lindsay *et al.*, 2002). A *N. caninum* isolate obtained from deer fed to dogs led to oocyst excretion (Gondim *et al.*, 2004b). Coyotes (*Canis latrans*) have been revealed as natural definitive hosts (Gondim *et al.*, 2004 a, b). Techniques specifically validated for examining wildlife species may be required in future to analyse potential interactions between the sylvatic and the domestic life cycles of *N. caninum* infections.

Future developments in serodiagnosis

Although serodiagnosis has been markedly improved within recent years, new technological developments will have to be included in the field of neosporosis. Mass seroscreening is presently being carried out with ELISA, the most appropriate test for large-scale use.

One of the most promising new tools in this respect is the fluorescence polarization (FP) test. This test has been used as a tool to monitor protein–protein, protein–peptide and other intermolecular interactions. Most fluorophores, including fluorescein, emit light in the same direction in which it is absorbed. When a fluorophore is freely rotating in solution, the light is emitted in all directions as a result of the molecule's rotation during the lifetime of the fluorescence emission; it is non-polarized. If, however, the fluorophore is part of a slowly rotating molecule (one that is large or in a viscous environment), the molecule does not rotate quickly with respect to the lifetime of the fluorescence, and the emission will occur in roughly the same direction as the absorption: it is therefore polarized. This property of fluorescence can therefore be used to distinguish small molecules (e.g. fluorescently labelled peptides) from large ones (e.g. peptide bound to antibody).

Relatively recent advances in instrumentation have allowed the use of this phenomenon in the development of rapid immunoassays for a large number of analytes, including antibodies to infectious agents (Buchli *et al.*, 2004; Gall and Nielsen, 2004; Aoyiagi and Kudo, 2005). These assays can be performed in a matter of minutes (versus hours or days for the other tests) and usually do not require extensive sample preparation. In addition, the materials required for the assay are relatively simple and highly stable, making this technique attractive for field use.

For neosporosis, FP would also hold the advantage of being applicable to various animal species without changing (and respectively evaluating) test reagents. The development of a *Neospora* FP test would require the previous isolation and standardization of small antigenic molecules, most likely based on a proteinic or carbohydrate nature. Recombinant oligopeptides or synthetic polysaccharides are target candidates and, at the protein level, various recombinant antigens have been described so far that now need to be appropriately investigated for an FP approach.

References

Aguado-Martinez, A., Álvarez-Garcia, G., Arnaiz-Seco, I., Innes, E. and Ortega-Mora, L.M. (2005) Use of avidity enzyme-linked immunosorbent assay and avidity Western blot to discriminate between acute and chronic *Neospora caninum* infection in cattle. *Journal of Veterinary Diagnostic Investigation* 17, 442–450.

Álvarez-Garcia, G., Pereira-Bueno, J., Gomez-Bautista, M. and Ortega-Mora, L.M. (2002) Pattern of recognition of *Neospora caninum* tachyzoite antigens by naturally infected pregnant cattle and aborted foetuses. *Veterinary Parasitology* 107, 15–27.

Álvarez-Garcia, G., Collantes-Fernandez, E., Costas, E., Rebordosa, X. and Ortega-Mora, L.M. (2003) Influence of age and purpose for testing on the cut-off selection of serological methods in bovine neosporosis. *Veterinary Research* 34, 341–352.

Anderson, M.L., Blanchard, P.C., Barr, B.C., Dubey, J.P., Hoffman, R.L. and Conrad, P.A. (1991) *Neospora*-like protozoan infection as a major cause of abortion in California dairy cattle. *Journal of the American Veterinary Medical Association* 198, 241–244.

Anderson, M.L., Palmer, C.W., Thurmond, M.C., Picanso, J.P., Blanchard, P.C., Breitmeyer, R.E., Layton, A.W., McAllister, M., Daft, B., Kinde, H., Read, D.H., Dubey, J.P., Conrad, P.A. and Barr, B.C. (1995) Evaluation of abortions in cattle attributable to neosporosis in selected dairy herds in California. *Journal of the American Veterinary Medical Association* 207, 1206–1210.

Anderson, M.L., Reynolds, J.P., Rowe, J.D., Sverlow, K.W., Packham, A.E., Barr, B.C. and Conrad, P.A. (1997) Evidence of vertical transmission of *Neospora* sp. infection in dairy cattle. *Journal of the American Veterinary Medical Association* 210, 1169–1172.

Andrianarivo, A.G., Rowe, J.D., Barr, B.C., Anderson, M.L., Packham, A.E., Sverlow, K.W., Choromanski, L., Loui, C., Grace, A. and Conrad, P.A. (2000) A POLYGEN-adjuvanted killed *Neospora caninum* tachyzoite preparation failed to prevent foetal infection in pregnant cattle following iv/im experimental tachyzoite challenge. *International Journal for Parasitology* 30, 985–990.

Andrianarivo, A.G., Barr, B.C., Anderson, M.L., Rowe, J.D., Packam, A.E., Sverlow, K.W. and Conrad, P.A. (2001) Immune responses in pregnant cattle and bovine fetuses following experimental infections with *Neospora caninum*. *Parasitology Research* 87, 817–825.

Aoyagi, S. and Kudo M. (2005) Development of fluorescence change-based, reagent-less optic immunosensor. *Biosensors and Bioelectronics* 20, 1680–1684.

Atkinson, R., Harper, P.A.W., Reichel, M.P. and Ellis, J.T. (2000) Progress in the serodiagnosis of *Neospora caninum* infections of cattle. *Parasitology Today* 16, 110–114.

Barling, K.S., Sherman, M., Peterson, M.J., Thompson, J.A., McNeill, J.W., Craig, T.M. and Adams, L.G. (2000) Spatial associations among density of cattle, abundance of wild canids, and seroprevalence to *Neospora caninum* in a population of beef calves. *Journal of the American Veterinary Medical Association* 217,1361–1365.

Barling, K.S., McNeill, J.W., Paschal, J.C., McCollum, F.T., Craig, T.M., Adams, L.G. and Thompson, J.A. (2001) Ranch-management factors associated with antibody seropositivity for *Neospora caninum* in consignments of beef calves in Texas, USA. *Preventive Veterinary Medicine* 52, 53–61.

Barr, B.C., Anderson, M.L., Blanchard, P.C., Daft, B.M., Kinde, H. and Conrad, P.A. (1990) Bovine foetal encephalitis and myocarditis associated with protozoal infections. *Veterinary Pathology* 27, 354–361.

Barr, B.C., Anderson, M.L., Dubey, J.P. and Conrad, P.A. (1991) *Neospora*-like protozoal infections associated with bovine abortions. *Veterinary Pathology* 28, 110–116.

Barr, B.C., Anderson, M.L., Woods, L.W., Dubey, J.P. and Conrad, P.A. (1992) *Neospora*-like protozoal infections associated with abortion in goats. *Journal of Veterinary Diagnostic Investigation* 4, 365–367.

Barr, B.C., Conrad, P.A., Breitmeyer, R., Sverlow, K., Anderson, M.L., Reynolds, J., Chauvet, A.E., Dubey, J.P. and Ardans, A.A. (1993) Congenital *Neospora* infection in calves born from cows that had previously aborted *Neospora*-infected fetuses: four cases (1990–1992). *Journal of the American Veterinary Medical Association* 202, 113–117.

Barr, B.C., Anderson, M.L., Sverlow, K.W. and Conrad, P.A. (1995) Diagnosis of bovine foetal *Neospora* infection with an indirect fluorescent antibody test. *Veterinary Record* 137, 61–613.

Barta, J.R. and Dubey, J.P. (1992) Characterization of anti-*Neospora caninum* hyperimmune rabbit serum by Western blot analysis and immunoelectron microscopy. *Parasitology Research* 78, 689–694.

Bartels, C.J., Wouda, W. and Schukken, Y.H. (1999) Risk factors for *Neospora caninum*-associated abortion storms in dairy herds in the Netherlands (1995–1997). *Theriogenology* 52, 247–257.

Bartels, C.J.M., Ruiz-Santa-Quitera, J.A., Arnaiz-Seco, I., Björkman, C., Frössling, J., von Blumröder, D., Conraths, F.J., Schares, G., Van Maanen, C., Wouda, W. and Ortega-Mora, L.M. (2006). Supranational comparison of *Neospora caninum* seroprevalences in cattle in Germany, The Netherlands, Spain and Sweden. *Veterinary Parasitology* 137, 17–27.

Bartley, P.M., Kirvar, E. Wright, S. Swales, C. Esteban-Redondo, I. Buxton, D., Maley, S.W., Schock, A., Rae, A.G., Hamilton, C. and Innes, E.A. (2004) Maternal and fetal immune responses of cattle inoculated with *Neospora caninum* at mid-gestation. *Journal of Comparative Pathology* 130, 81–91.

Basso, W., Venturini, L., Venturini, M.C., Hill, D.E., Kwok, O.C., Shen, S.K. and Dubey, J.P. (2001) First isolation of *Neospora caninum* from the feces of a naturally infected dog. *Journal of Parasitology* 87, 612–618.

Baszler, T.V., Adams, S., Vander-Schalie, J., Mathison, B.A. and Kostovic, M. (2001) Validation of a commercially available monoclonal antibody-based competitive-inhibition enzyme-linked immunosorbent assay for detection of serum antibodies to *Neospora caninum* in cattle. *Journal of Clinical Microbiology* 39, 3851–3857.

Bergeron, N., Fecteau, G., Villeneuve, A., Girard, C. and Paré, J. (2001a) Failure of dogs to shed oocysts after being fed bovine fetuses naturally infected by *Neospora caninum*. *Veterinary Parasitology* 97, 145–152.

Bergeron, N., Girard, C., Pare, J., Fecteau, G., Robinson, J. and Baillargeon, P. (2001b) Rare detection of *Neospora caninum* in placentas from seropositive dams giving birth to full-term calves. *Journal of Veterinary Diagnostic Investigation* 13, 173–175.

Berlin, O.G., Peter, J.B., Gagne, C., Conteas, C.N. and Ash, L.R. (1998) Autofluorescence and the detection of cyclospora oocysts. *Emerging Infectious Diseases* 4, 127–128.

Billman-Jacobe, H., Carrigan, M., Cockram, F., Corner, L.A., Gill, I.J., Hill, J.F., Jessep, T., Milner, A.R. and Wood, P.R. (1992) A comparison of the interferon gamma assay with the absorbed ELISA for the diagnosis of Johne's disease in cattle. *Australian Veterinary Journal* 69, 25–28.

Bjerkås, I., Mohn, S.F. and Presthus, J. (1984) Unidentified cyst-forming-sporozoon causing encephalomyelitis and myositis in dogs. *Zeitschrift für Parasitenkunde.* 70, 271–274.

Bjerkås, I., Jenkins, M.C. and Dubey, J.P. (1994) Identification and characterization of *Neospora caninum* tachyzoite antigens useful for diagnosis of neosporosis. *Clinical and Diagnostic Laboratory Immunology* 1, 214–221.

Björkman, C. and Uggla, A. (1999) Serological diagnosis of *Neospora caninum* infection. *International Journal for Parasitology* 29, 1497–1507.

Björkman, C., Johannsson, O., Stenlund, S., Holmdahl, O.J.M. and Uggla, A. (1996) *Neospora* species infection in a herd of dairy cattle. *Journal of the American Veterinary Medical Association* 208, 1441–1444.

Björkman, C., Holmdahl, O.J.M. and Uggla, A. (1997) An indirect enzyme-linked immunoassay (ELISA) for demonstration of antibodies to *Neospora caninum* in serum and milk of cattle. *Veterinary Parasitology* 68, 251–260.

Björkman, C., Naslund, K., Stenlund, S., Maley, S.W., Buxton, D. and Uggla, A. (1999) An IgG avidity ELISA to discriminate between recent and chronic *Neospora caninum* infection. *Journal of Veterinary Diagnostic Investigation* 11, 41–44.

Björkman, C., Alenius, S., Emanuelsson, U. and Uggla, A. (2000) *Neospora caninum* and bovine virus diarrhoea virus infections in Swedish dairy cows in relation to abortion. *Veterinary Journal* 159, 201–206.

Björkman, C., McAllister, M.M., Frössling, J., Näslund, K., Leung, F. and Uggla, A. (2003) Application of the *Neospora caninum* IgG avidity ELISA in assessment of chronic reproductive losses following an outbreak of neosporosis in a herd of beef cattle. *Journal of Veterinary Diagnostic Investigation* 15, 3–7.

Björkman, C., Álvarez-Garcia, G., Conraths, F.J., Mattsson, J.G., Ortega-Mora, L.M., Sager, H. and Schares, G. (2006) *Neospora caninum* IgG avidity tests: an interlaboratory comparison. *Veterinary Parasitology* 140, 273–280.

Blagburn, B.L., Lindsay, D.S., Swango, L.J., Pidgeon, G.L. and Braund, K.G. (1988) Further characterization of the biology of *Hammondia heydorni*. *Veterinary Parasitology* 27, 193–198.

Boulton, J.G., Gill, P.A., Cook, R.W., Fraser, G.C., Harper, P.A.W. and Dubey, J.P. (1995) Bovine *Neospora* abortion in north-eastern New South Wales. *Australian Veterinary Journal* 72, 119–120.

Buchli, R., VanGundy, R.S., Hickman-Miller, H.D., Giberson, C.F., Bardet, W. and Hildebrand, W.H. (2004) Real-time measurement of *in vitro* peptide binding to soluble HLA-A*0201 by fluorescence polarization. *Biochemistry* 43, 14852–14863.

Buxton, D. (1998) Protozoan infections (*Toxoplasma gondii, Neospora caninum* and *Sarcocystis* spp.) in sheep and goats: recent advances. *Veterinary Research* 29, 289–310.

Buxton, D., Maley, S.W., Thomson, K.M., Trees, A.J. and Innes, E.A. (1997) Experimental infection of non-pregnant and pregnant sheep with *Neospora caninum*. *Journal of Comparative Pathology* 117, 1–16.

Buxton, D., Maley, S.W., Wright, S., Thomson, K.M., Rae, A.G. and Innes, E.A. (1998) The pathogenesis of experimental neosporosis in pregnant sheep. *Journal of Comparative Pathology* 118, 267–279.

Caetano-da-Silva, A., Ferre, I., Collantes-Fernández, E., Navarro, V., Aduriz, G., Ugarte-Garagalza, C. and Ortega-Mora, L.M. (2004) Occasional detection of *Neospora caninum* DNA in frozen extended semen from naturally infected bulls. *Theriogenology* 62, 1329–1336.

Caldow, G.L. (1998) Bovine abortion outbreak associated with *Neospora* and other infectious agents. *Veterinary Record* 142, 118–119.

Canada, N., Carvalheira, J., Meireles, C.S., Correia da Costa, J.M. and Rocha, A. (2004) Prevalence of *Neospora caninum* infection in dairy cows and its consequences for reproductive management. *Theriogenology* 62, 1229–1235.

Chanlun, A., Naslund, K., Aiumlamai, S. and Björkman, C. (2002) Use of bulk milk for detection of *Neospora caninum* infection in dairy herds in Thailand. *Veterinary Parasitology* 110, 35–44.

Cole, R.A., Lindsay, D.S., Dubey, J.P. and Blagburn, B.L. (1993) Detection of *Neospora caninum* in tissue sections using a murine monoclonal antibody. *Journal of Veterinary Diagnostic Investigation* 4, 579–584.

Collantes-Fernandez, E., Zaballos, A., Álvarez-Garcia, G. and Ortega-Mora, L.M. (2002) Quantitative detection of *Neospora caninum* in bovine aborted fetuses and experimentally infected mice by real-time PCR. *Journal of Clinical Microbiology* 40, 1194–1198.

Conrad, P.A., Barr, B.C., Sverlow, K.W., Anderson, M., Daft, B., Kinde, H., Dubey, J.P., Munson, L. and Ardans, A. (1993a) *In vitro* isolation and characterization of a *Neospora* sp. from aborted bovine foetuses. *Parasitology* 106, 239–249.

Conrad, P.A., Sverlow, K., Anderson, M., Rowe, J., BonDurant, R., Tuter, G., Breitmeyer, R., Palmer, C., Thurmond, M., Ardans, A., Dubey, J.P., Duhamel, G. and Barr, B. (1993b) Detection of serum antibody responses in cattle with natural or experimental *Neospora* infections. *Journal of Veterinary Diagnostic Investigation* 5, 572–578.

Corbellini, L.G., Colodel, E.M. and Driemeier, D. (2001) Granulomatous encephalitis in a neurologically impaired goat kid associated with degeneration of *Neospora caninum* tissue cysts. *Journal of Veterinary Diagnostic Investigation* 13, 416–419.

Dannatt, L. (1997) *Neospora caninum* antibody levels in an endemically infected dairy herd. *Cattle Practitioner* 5, 335–337.

Davison, H.C., Trees, A.J., Guy, F., Otter, A., Holt, J.J., Simpson, V.R. and Jeffrey, M. (1997) Isolation of bovine *Neospora* in Britain. *Veterinary Record* 141, 607.

Davison, H.C., Otter, A. and Trees, A.J. (1999) Estimation of vertical and horizontal transmission parameters of *Neospora caninum* infections in dairy cattle. *International Journal for Parasitolology* 29, 1683–1689.

Davison, H.C., Guy, C.S., McGarry, J.W., Guy, F., Williams, D.J.L., Kelly, D.F. and Trees, A.J. (2001) Experimental studies on the transmission of *Neospora caninum* between cattle. *Research in Veterinary Science* 70, 163–168.

De Marez, T., Liddell, S., Dubey, J.P., Jenkins, M.C. and Gasbarre, L. (1999) Oral infection of calves with *Neospora caninum* oocysts from dogs: humoral and cellular immune responses. *International Journal for Parasitology* 29, 1647–1657.

Dijkstra, T., Barkema, H.W., Eysker, M. and Wouda, W. (2001a) Evidence of post-natal transmission of *Neospora caninum* in Dutch dairy herds. *International Journal for Parasitology* 31, 209–215.

Dijkstra, T., Eysker, M., Schares, G., Conraths, F.J., Wouda, W. and Barkema, H.W. (2001b) Dogs shed *Neospora caninum* oocysts after ingestion of naturally infected bovine placenta but not after ingestion of colostrum spiked with *Neospora caninum* tachyzoites. *International Journal for Parasitology* 31, 747–752.

Dijkstra, T., Barkema, H.W., Björkman, C. and Wouda, W. (2002a) A high rate of seroconversion for *Neospora caninum* in a dairy herd without an obvious increased incidence of abortions. *Veterinary Parasitology* 109, 203–211.

Dijkstra, T., Barkema, H.W., Eysker, M., Hesselink, J.W. and Wouda, W. (2002b) Natural transmission routes of *Neospora caninum* between farm dogs and cattle. *Veterinary Parasitology* 105, 99–104.

Dijkstra, T., Barkema, H.W., Hesselink, J.W. and Wouda, W. (2002c) Point source exposure of cattle to *Neospora caninum* consistent with periods of common housing and feeding and related to the introduction of a dog. *Veterinary Parasitology* 105, 89–98.

Dijkstra, T., Barkema, H.W., Eysker, M., Beiboer, M.L. and Wouda, W. (2003) Evaluation of a single serological screening of dairy herds for *Neospora caninum* antibodies. *Veterinary Parasitology* 110, 161–169.

Dubey, J.P. (1999) Recent advances in *Neospora* and neosporosis. *Veterinary Parasitology* 84, 349–367.

Dubey, J.P. (2003a) Neosporosis in cattle. *Journal of Parasitology* 89, S42–S56.

Dubey, J.P. (2003b) Review of *Neospora caninum* and neosporosis in animals. *Korean Journal of Parasitology* 41, 1–16.

Dubey, J.P. and Lindsay, D.S. (1993) Neosporosis. *Parasitology Today* 9, 452–457.

Dubey, J.P. and Lindsay, D.S. (1996) A review of *Neospora caninum* and neosporosis. *Veterinary Parasitology* 67, 1–59.

Dubey, J.P. and Lindsay, D.S. (2000) Gerbils (*Meriones unguiculatus*) are highly susceptible to oral infection with *Neospora caninum* oocysts. *Parasitology Research* 86, 165–168.

Dubey, J.P. and Porterfield, M.L. (1990) *Neospora caninum* (Apicomplexa) in an aborted equine fetus. *Journal of Parasitology* 76, 732–734.

Dubey, J.P., Carpenter, J.L., Speer, C.A., Topper, M.J. and Uggla, A. (1988a) Newly recognized fatal protozoan disease of dogs. *Journal of the American Veterinary Medical Association* 192, 1269–1285.

Dubey, J.P., Hattel, A.L., Lindsay, D.S. and Topper, M.J. (1988b) Neonatal *Neospora caninum* infection in dogs: isolation of the causative agent and experimental transmission. *Journal of the American Veterinary Medical Association* 193, 1259–1263.

Dubey, J.P., Lindsay, D.S., Anderson, M.L., Davis, S.W. and Shen, S.K. (1992) Induced transplacental transmission of *Neospora caninum* in cattle. *Journal of the American Veterinary Medical Association* 201, 709–713.

Dubey, J.P., Lindsay, D.S., Adams, D.S., Gay, J.M., Baszler, T.V., Blagburn, B.L. and Thulliez, P. (1996a) Serologic responses of cattle and other animals infected with *Neospora caninum. American Journal of Veterinary Research* 57, 329–336.

Dubey, J.P., Rigoulet, J., Lagourette, P., George, C., Longeart, L. and LeNet, J.L. (1996b) Fatal transplacental neosporosis in a deer (*Cervus eldi siamensis*). *Journal of Parasitology* 82, 338–339.

Dubey, J.P., Jenkins, M.C., Adams, D.S., McAllister, M.M., Anderson-Sprecher, R.C., Baszler, T.V., Kwok, O.C.H., Lally, N.C., Björkman, C. and Uggla, A. (1997) Antibody responses of cows during an outbreak of neosporosis evaluated by indirect fluorescent antibody test and different enzyme-linked immunosorbent assays. *Journal of Parasitology* 83, 1063–1069.

Dubey, J.P., Hollis, K., Romand, S., Thulliez, P., Kwok, O.C., Hungerford, L., Anchor, C. and Etter, D. (1999) High prevalence of antibodies to *Neospora caninum* in white-tailed deer (*Odocoileus virginianus*). *International Journal of Parasitology* 29, 1709–1711.

Dubey, J.P., Barr, B.C., Barta, J.R., Bjerkås, I., Björkman, C., Blagburn, B.L., Bowman, D.D., Buxton, D., Ellis, J.T., Gottstein, B. *et al.* (2002) Redescription of *Neospora caninum* and its differentiation from related coccidia. *International Journal for Parasitology* 32, 929–946.

Dumetre, A. and Darde, M.L. (2003) How to detect *Toxoplasma gondii* oocysts in environmental samples? *FEMS Microbiology Reviews* 27, 651–661.

Dumetre, A. and Darde, M.L. (2005) Immunomagnetic separation of *Toxoplasma gondii* oocysts using a monoclonal antibody directed against the oocyst wall. *Journal of Microbiological Methods* 61, 209–217.

Dyer, R.M., Jenkins, M.C., Kwok, O.C.H., Douglas, L.W. and Dubey, J.P. (2000) Serologic survey of *Neospora caninum* infection in a closed dairy cattle herd in

Maryland: risk of serologic reactivity by production groups. *Veterinary Parasitology* 90, 171–181.

Eleni, C., Crotti, S., Manuali, E., Costarelli, S., Filippini, G., Moscati, L. and Magnino, S. (2004) Detection of *Neospora caninum* in an aborted goat foetus. *Veterinary Parasitology* 123, 271–274.

Ellis, J.T., Morrison, D.A., Liddell, S., Jenkins, M.C., Mohammed, O.B., Ryce, C. and Dubey, J.P. (1999) The genus *Hammondia* is paraphyletic. *Parasitology* 118, 357–362.

Entrican, G. (2002) Immune regulation during pregnancy and host–pathogen interactions in infectious abortion. *Journal of Comparative Pathology* 126, 79–94.

Entrican, G., McInnes, C.J., Rothel, J.S. and Haig, D.M. (1992). Kinetics of ovine interferon-gamma production: detection of mRNA and characterisation of biological activity. *Veterinary Immunology and Immunopathology* 33, 171–178.

Ferre, I., Aduriz, G., Del-Pozo, I., Regidor-Cerrillo, J., Atxaerandio, R., Collantes-Fernandez, E., Hurtado, A., Ugarte-Garagalza, C. and Ortega-Mora, L.M. (2005) Detection of *Neospora caninum* in the semen and blood of naturally infected bulls. *Theriogenology* 63, 1504–1518.

Fioretti, D.P., Pasquai, P., Diaferia, M., Mangili, V. and Rosignoli, L. (2003) *Neospora caninum* infection and congenital transmission: serological and parasitological study of cows up to the fourth gestation. *Journal of Veterinary Medicine* B 50, 399–404.

Fischer, I., Furrer, K., Audigé, L., Fritsche, A., Giger, T., Gottstein, B. and Sager, H. (2003) The importance of bovine neosporosis for abortion in Switzerland. *Schweizerisches Archiv für Tierheilkunde* 145, 114.

French, N.P., Clancy, D., Davison, H.C. and Trees, A.J. (1999) Mathematical models of *Neospora caninum* infection in dairy cattle: transmission and options for control. *International Journal for Parasitology* 29, 1691–1704.

Frössling, J., Bonnett, B., Lindberg, A. and Björkman, C. (2003) Validation of a *Neospora caninum* iscom ELISA without a gold standard. *Preventive Veterinary Medicine* 57, 141–153.

Gall, D. and Nielsen, K. (2004) Serological diagnosis of bovine brucellosis: a review of test performance and cost comparison. *Scientific and Technical Review (International Office of Epizooties)* 23, 989–1002.

Gardner, I.A. and Greiner, M. (2000) Diagnostic testing in veterinary epidemiologic studies. *Preventive Veterinary Medicine* 45 (1–2), 1–162.

Godfroid, J., Wellemans, G., Van Opdenbosch, E. and Letesson, J.J. (1995) Infectious bovine rhinotracheitis (IBR) diagnosis based on an *in vitro* bovine herpesvirus-1 (BHV-1) antigen-specific IFNg production. In: Schwyzer, M. and Ackermann, M. (eds) *Proceedings of the 3rd Congress of the European Society for Veterinary Virology*, pp. 159–163.

Godfroid, J., Czaplicki, G., Kerkhofs, P., Weynants, V., Wellemans, G., Thiry, E. and Letesson, J.J. (1996) Assessment of the cell-mediated immunity in cattle infection after bovine herpesvirus 4 infection using an *in vitro* antigen-specific interferon assay. *Veterinary Microbiology* 53, 133–141.

Gondim, L.F.P., Gao, L. and McAllister, M.M. (2002) Improved production of *Neospora caninum* oocysts, cyclical oral transmission between dogs and cattle, and *in vitro* isolation from oocysts. *Journal of Parasitology* 88, 1159–1163.

Gondim, L.F.P., McAllister, M.M., Pitt, W.C. and Zemlicka, D.E. (2004a) Coyotes (*Canis latrans*) are definitive hosts of *Neospora caninum. International Journal for Parasitology* 34, 159–161.

Gondim, L.F.P., McAllister, M.M., Mateus-Pinilla, N.E., Pitt, W.L., Mech, L.D. and Nelson, M.E. (2004b) Transmission of *Neospora caninum* between wild and domestic animals. *Journal of Parasitology* 90, 1361–1365.

Gondim, L.F., McAllister, M.M. and Gao, L. (2005) Effects of host maturity and prior exposure history on the production of *Neospora caninum* oocysts by dogs. *Veterinary Parasitology* 134, 1–2.

Gottstein, B., Hentrich, B., Wyss, R., Thür, B., Busato, A., Stärk, K.D.C. and Müller, N. (1998) Molecular and immunodiagnostic investigations on bovine neosporosis in Switzerland. *International Journal for Parasitology* 28, 679–691.

Guarino, A., Fusco, G., Savini, G., Di Francesco, G. and Cringoli, G. (2000) Neosporosis in water buffalo (*Bubalus bubalis*) in Southern Italy. *Veterinary Parasitology* 91, 15–21.

Guy, C.S., Williams, D.J.L., Kelly, D.F., McGarry, J.W., Guy, F., Björkman, C., Smith, R.F. and Trees, A.J. (2001) *Neospora caninum* in persistently infected, pregnant cows: spontaneous transplacental infection is associated with an acute increase in maternal antibody. *Veterinary Record* 149, 443–449.

Hancock, K. and Tsang, V.C.W. (1983) India ink staining of proteins on nitrocellulose paper. *Analytical Biochemistry* 133, 157–162.

Hässig, M., Sager, H., Reitt, K., Ziegler, D., Strabel, D. and Gottstein, B. (2003) *Neospora caninum* in sheep: a herd case report. *Veterinary Parasitology* 117, 213–220.

Helmick, B., Otter, A., McGarry, J. and Buxton, D. (2002) Serological investigation of aborted sheep and pigs for infection by *Neospora caninum. Research in Veterinary Science* 73, 187–189.

Hemphill, A. (1999) The host–parasite relationship in neosporosis. *Advances in Parasitology* 43, 47–104.

Hemphill, A. and Gottstein, B. (2000) A European perspective on *Neospora caninum. International Journal for Parasitology* 30, 877–924.

Heydorn, A.O. (1973) Zum lebenszyklus der kleinen form von *Isospora bigemina* des hundes. *Berliner und Münchener Tierärztliche Wochenschrift* 86, 323–329.

Hill, D.E., Liddell, S., Jenkins, M.C. and Dubey, J.P. (2001) Specific detection of *Neospora caninum* oocysts in fecal samples from experimentally infected dogs using the polymerase chain reaction. *Journal of Parasitology* 87, 395–398.

Holmdahl, O.J.M. and Mattsson, J.G. (1996) Rapid and sensitive identification of *Neospora caninum* by *in vitro* amplification of the internal transcribed spacer 1. *Parasitology* 112, 177–182.

Hurtado, A., Aduriz, G., Moreno, B., Barandika, J. and Garcia-Perez, A.L. (2001) Single tube-nested PCR for the detection of *Toxoplasma gondii* in fetal tissues from naturally aborted ewes. *Veterinary Parasitology* 102, 17–27.

Innes, E.A., Panton, W.R.M., Marks, J., Trees, A.J., Holmdahl, J. and Buxton, D. (1995) Interferon gamma inhibits the intracellular multiplication of *Neospora caninum* as shown by the incorporation of ^{3}H Uracil. *Journal of Comparative Pathology* 115, 95–100.

Innes, E.A., Wright, S.E., Maley, S., Rae, A., Schock, A., Kirvar, E., Bartley, P., Hamilton, C., Carey, I.M. and Buxton, D. (2001) Protection against vertical transmission in bovine neosporosis. *International Journal for Parasitology* 31, 1523–1534.

Innes, E.A., Andrianarivo, A.G., Björkman, C., Williams, D. and Conrad, P.A. (2002) Immune responses of *Neospora caninum* and prospects for vaccination. *Trends in Parasitology* 18, 497–504.

Jardine, J.E. (1996) The ultrastructure of bradyzoites and tissue cysts of *Neospora caninum* in dogs: absence of distinguishing morphological features between parasites of canine and bovine origin. *Veterinary Parasitology* 62, 231–240.

Jenkins, M.C., Caver, J.A., Bjorkman, C., Anderson, T.C., Romand, S., Vinyard, B., Uggla, A., Thulliez, P. and Dubey, J.P. (2000) Serological investigation of an outbreak of *Neospora caninum*-associated abortion in a dairy herd in southeastern United States. *Veterinary Parasitology* 94, 17–26.

Jenkins, M., Baszler, T., Björkman, C., Schares, G. and Williams, D. (2002) Diagnosis and seroepidemiology of *Neospora caninum*-associated bovine abortion. *International Journal for Parasitology* 32, 631–636.

Kinsel, M.L. (2002) An epidemiological approach to investigating abortion problems in dairy herds. *Compendium on Continuing Education for the Practising Veterinarian* 24, S34–S39.

Kobayashi, Y., Yamada, M., Omata, Y., Koyama, T., Saito, A., Matsuda, T., Okuyama, K., Fujimoto, S., Furuoka, H. and Matsui, T. (2001) Naturally occurring *Neospora caninum* infection in an adult sheep and her twin fetuses. *Journal of Parasitology* 87, 434–436.

Koyama, T., Kobayashi, Y., Omata, Y., Yamada, M., Furuoka, H., Maeda, R., Matsui, T., Saito, A. and Mikami, T. (2001) Isolation of *Neospora caninum* from the brain of a pregnant sheep. *Journal of Parasitology* 87, 1486–1488.

Kritzner, S., Sager, H., Blum, J., Krebber, R., Greif, G. and Gottstein, B. (2002) An explorative study to assess the efficacy of Toltrazuril-sulfone (Ponazuril) in calves experimentally infected with *Neospora caninum. Annals of Clinical Microbiology and Antimicrobials* 1, 4.

Liddell, S., Jenkins, M.C. and Dubey, J.P. (1999) A competitive PCR assay for quantitative detection of *Neospora caninum. International Journal for Parasitology* 29, 1583–1587.

Lindsay, D.S. and Dubey, J.P. (1989) Immunohistochemical diagnosis of *Neospora caninum* in tissue sections. *American Journal Veterinary Research* 11, 1981–1983.

Lindsay, D.S., Upton, S.J. and Dubey, J.P. (1999a) A structural study of the *Neospora caninum* oocyst. *International Journal for Parasitology* 29, 1521–1523.

Lindsay, D.S., Dubey, J.P. and Duncan, R.B. (1999b) Confirmation that the dog is a definitive host for *Neospora caninum. Veterinary Parasitology* 82, 327–333.

Lindsay, D.S., Ritter, D.M. and Brake, D. (2001) Oocyst excretion in dogs fed mouse brains containing tissue cysts of a cloned line of *Neospora caninum. Journal of Parasitology* 87, 909–911.

Lindsay, D.S., Little, S.E. and Davidson, W.R. (2002) Prevalence of antibodies to *Neospora caninum* in white-tailed deer, *Odocoileus virginianus*, from the south-eastern United States. *Journal of Parasitology* 88, 415–417.

Longo, M.C., Berninger, M.S. and Hartley, J.L. (1990) Use of uracil DNA glycosidase to control carry-over contamination in polymerase chain reaction. *Gene* 93, 125–128.

Lunden, A., Marks, J., Maley, S. and Innes, E.A. (1998) Cellular immune responses in cattle experimentally infected with *Neospora caninum*. *Parasite Immunology* 20, 519–526.

Macaldowie, C., Maley, S.W., Wright, S., Bartley, P., Esteban-Redondo, I., Buxton, D. and Innes, E.A. (2004) Placental pathology associated with foetal death in cattle experimentally infected with *Neospora caninum* by two different challenge routes in early pregnancy. *Journal of Comparative Pathology* 131, 142–156.

Mainar-Jaime, R.C., Thurmond, M.C., Berzal-Herranz, B. and Hietala, S.K. (1999) Seroprevalence of *Neospora caninum* and abortion in dairy cows in northern Spain. *Veterinary Record* 145, 72–75.

Mainar-Jaime, R.C., Berzal-Herranz, B., Arias, P. and Rojo-Vazquez, F.A. (2001) Epidemiological pattern and risk factors associated with bovine viral diarrhoea virus (BVDV) infection in a non-vaccinated dairy-cattle population from the Asturias region of Spain. *Preventive Veterinary Medicine* 52, 63–73.

Maley, S.W., Buxton, D., Thomson, K.M., Schriefer, C.E. and Innes, E.A. (2001) Serological analysis of calves experimentally infected with *Neospora caninum*: a 1-year study. *Veterinary Parasitology* 96, 1–9.

Maley, S.W., Buxton, D., Rae, A.G., Wright, S.E., Schock, A., Bartley, P.M., Esteban-Redondo, I., Swales, C., Hamilton, C.M., Sales, J. and Innes, E.A. (2003) The pathogenesis of neosporosis in pregnant cattle: inoculation at mid-gestation. *Journal of Comparative Pathology* 129, 186–195.

Marks, J., Lunden, A., Harkins, D. and Innes, E.A. (1998) Identification of *Neospora* antigens recognised by CD4+ T-cells and immune sera from experimentally infected cattle. *Parasite Immunology* 20, 303–309.

Marsh, A.E., Barr, B.C., Packham, A.E. and Conrad, P.A. (1998) Description of a new *Neospora* species (Protozoa: Apicomplexa: Sarcocystidae). *Journal of Parasitology* 84, 983–991.

McAllister, M.M., Huffman, E.M., Hietala, S.K., Conrad, P.A., Anderson, M.L. and Salman, M.D. (1996a) Evidence suggesting a point source exposure in an outbreak of bovine abortion due to neosporosis. *Journal of Veterinary Diagnostic Investigation* 8, 355–357.

McAllister, M.M., McGuire, A.M., Jolley, W.R., Lindsay, D.S., Trees, A.J. and Stobart, R.H. (1996b) Experimental neosporosis in pregnant ewes and their offspring. *Veterinary Pathology* 33, 647–655.

McAllister, M.M., Dubey, J.P., Lindsay, D.S., Jolley, W.R., Wills, R.A. and McGuire, A.M. (1998) Dogs are definitive hosts of *Neospora caninum*. *International Journal for Parasitology* 28, 1473–1478.

McAllister, M.M., Björkman, C., Anderson-Sprecher, R. and Rogers, D.G. (2000) Evidence of point-source exposure to *Neospora caninum* and protective immunity in a herd of beef cows. *Journal of the American Veterinary Medical Association* 217, 881–887.

McGarry, J.W., Stockton, C.M., Williams, D.J.L. and Trees, A.J. (2003) Protracted shedding of oocysts of *Neospora caninum* by a naturally infected foxhound. *Journal of Parasitology* 89, 628–630.

Moen, A.R., Wouda, W., Mul, M.F., Graat, E.A. and van Werven, T. (1998) Increased risk of abortion following *Neospora caninum* abortion outbreaks: a retrospective and prospective cohort study in four dairy herds. *Theriogenology* 49, 1301–1309.

Müller, N., Zimmermann, V., Hentrich, B. and Gottstein, B. (1996) Diagnosis of *Neospora caninum* and *Toxoplasma gondii* infection by PCR and DNA hybridization immunoassay. *Journal of Clinical Microbiology* 34, 2850–2852.

Müller, N., Vonlaufen, N., Gianinazzi, C., Leib, S.L. and Hemphill, A. (2002) Application of real-time fluorescent PCR for quantitative assessment of *Neospora caninum* infections in organotypic slice cultures of rat central nervous system tissue. *Journal of Clinical Microbiology* 40, 252–255.

Okeoma, C.M., Williamson, N.B., Pomroy, W.E., Stowell, K.M. and Gillespie, L. (2004) The use of PCR to detect *Neospora caninum* DNA in the blood of naturally infected cows. *Veterinary Parasitology* 122, 307–315.

Okeoma, C.M., Stowell, K.M., Williamson, N.B. and Pomroy, W.E. (2005) *Neospora caninum*: quantification of DNA in the blood of naturally infected aborted and pregnant cows using real-time PCR. *Experimental Parasitology* 110, 48–55.

Ortega-Mora, L.M., Ferre, I., del Pozo, I., Caetano da Silva, A., Collantes-Fernández, E., Regidor-Cerrillo, J., Ugarte-Garagalza, C. and Aduriz, G. (2003) Detection of *Neospora caninum* in semen of bulls. *Veterinary Parasitology* 117, 301–308.

Osburn, B.I. (1988) Ontogeny of immune responses in cattle. In: Morrison, I. (ed.) *The Ruminant Immune System in Health and Disease*, 2nd edn. Cambridge University Press, Cambridge, UK, pp. 252–260.

O'Toole, D. and Jeffrey, M. (1987) Congenital sporozoan encephalomyelitis in a calf. *Veterinary Record* 121, 563–566.

Otranto, D., Llazari, A., Testini, G., Traversa, D., Frangipane di Regalbono, A., Badan, M. and Capelli, G. (2003) Seroprevalence and associated risk factors of neosporosis in beef and dairy cattle in Italy. *Veterinary Parasitology* 118, 7–18.

Otter, A., Jeffrey, M., Griffiths, I.B. and Dubey, J.P. (1995) A survey of the incidence of *Neospora caninum* infection in aborted and stillborn bovine fetuses in England and Wales. *Veterinary Record* 136, 602–606.

Ould-Amrouche, A., Klein, F., Osdoit, C., Mohammed, H.O., Touratier, A., Sanaa, M. and Mialot, J.P. (1999) Estimation of *Neospora caninum* seroprevalence in dairy cattle from Normandy, France. *Veterinary Research* 30, 531–538.

Owen, M.R. and Trees, A.J. (1999) Genotyping of *Toxoplasma gondii* associated with abortion in sheep. *Journal of Parasitology* 85, 382–384.

Packham, A.E., Sverlow, K.W., Conrad, P.A., Loomis, E.F., Rowe, J.D., Anderson, M.L., Marsh, A.E., Cray, C. and Barr, B.C. (1998) A modified agglutination test for *Neospora caninum*: development, optimization, and comparison to the indirect fluorescent-antibody test and enzyme-linked immunosorbent assay. *Clinical and Diagnostic Laboratory Immunology* 5, 467–473.

Paré, J., Hietala, S.K. and Thurmond, M.C. (1995a) An enzyme-linked immunosorbent assay (ELISA) for serological diagnosis of *Neospora* sp. infection in cattle. *Journal of Veterinary Diagnostic Investigation* 7, 352–359.

Paré, J., Hietala, S.K. and Thurmond, M.C. (1995b) Interpretation of an indirect fluorescent antibody test for diagnosis of *Neospora* sp. infection in cattle. *Journal of Veterinary Diagnostic Investigation* 7, 273–275.

Paré, J., Thurmond, M.C. and Hietala, S.K. (1996) Congenital *Neospora caninum* infection in dairy cattle and associated calfhood mortality. *Canadian Journal of Veterinary Research* 60, 133–139.

Paré, J., Thurmond, M.C. and Hietala, S.K. (1997) *Neospora caninum* antibodies in cows during pregnancy as a predictor of congenital infection and abortion. *Journal of Parasitology* 83, 82–87.

Paré, J., Fecteau, G., Fortin, M., Marsolais, G. (1998) Seroepidemiologic study of *Neospora caninum* in dairy herds. *Journal of the American Veterinary Medical Association* 213, 1595–1598.

Parish, S.M., Maag-Miller, L., Besser, T.E., Weidner, J.P., McElwain, T., Knowles, D.P. and Leathers, C.W. (1987) Myelitis associated with protozoal infection in newborn calves. *Journal of the American Veterinary Medical Association* 191, 1599–1600.

Patitucci, A.N., Charleston, W.A.G., Alley, M.R., O'Connor, R.J. and Pomroy, W.E. (1999) Serological study of a dairy herd with a recent history of *Neospora* abortion. *New Zealand Veterinary Journal* 47, 28–30.

Pereira-Bueno, J., Quintanilla-Gozalo, A., Perez-Perez, V., Espi-Felgueroso, A., Álvarez, G., Collantes-Fernandez, E. and Ortega-Mora, L.M. (2003) Evaluation by different diagnostic techniques of bovine abortion associated with *Neospora caninum* in Spain. *Veterinary Parasitology* 111, 143–152.

Peters, M., Lütkefels, E., Heckeroth, A.R. and Schares, G. (2001a) Immunohistochemical and ultrastructural evidence for *Neospora caninum* tissue cysts in skeletal muscles of naturally infected dogs and cattle. *International Journal for Parasitology* 31, 1144–1148.

Peters, M., Wohlsein, P., Knieriem, A. and Schares, G. (2001b) *Neospora caninum* infection associated with stillbirths in captive antelopes (*Tragelaphus imberbis*). *Veterinary Parasitology* 97, 153–157.

Quintanilla-Gozalo, A., Pereira-Bueno, J., Seijas-Carballedo, A., Costas, E. and Ortega-Mora, L.M. (2000) Observational studies in *Neospora caninum*-infected dairy cattle: relationship between infection/abortion and gestational antibody fluctuations. *International Journal for Parasitology* 30, 900–906.

Reichel, M.P. and Pfeiffer, D.U. (2002) An analysis of the performance characteristics of serological tests for the diagnosis of *Neospora caninum* infection in cattle. *Veterinary Parasitolology* 107, 197–207.

Rinaldi, L., Fusco, G., Musella, V., Veneziano, V., Guarino, A., Taddei, R. and Cringoli, G. (2005) *Neospora caninum* in pastured cattle: determination of climatic, environmental, farm management and individual risk factors using remote sensing and geographic information systems. *Veterinary Parasitology* 128, 219–230.

Rodrigues, A.A., Gennari, S.M., Aguiar, D.M., Sreekumar, C., Hill, D.E., Miska, K.B., Vianna, M.C. and Dubey, J.P. (2004) Shedding of *Neospora caninum* oocysts by dogs fed tissues from naturally infected water buffaloes (*Bubalus bubalis*) from Brazil. *Veterinary Parasitology* 124, 139–150.

Rodriguez, I., Choromanski, L., Rodgers, S.J. and Weinstock, D. (2002) Survey of *Neospora caninum* antibodies in dairy and beef cattle from five regions of the United States. *Veterinary Therapy* 3, 396–401.

Romand, S., Thulliez, P. and Dubey, J.P. (1998) Direct agglutination test for serologic diagnosis of *Neospora caninum* infection. *Parasitology Research* 84, 50–53.

Romero, J.J. and Frankena, K. (2003) The effect of the dam–calf relationship on serostatus to *Neospora caninum* on 20 Costa Rican dairy farms. *Veterinary Parasitology* 114, 159–171.

Romero, J.J., Perez, E., Dolz, G. and Frankena, K. (2002) Factors associated with *Neospora caninum* serostatus in cattle of 20 specialised Costa Rican dairy herds. *Preventive Veterinary Medicine* 53, 263–273.

Rosypal, A.C. and Lindsay, D.S. (2005) The sylvatic cycle of *Neospora caninum*: where do we go from here? *Trends in Parasitology* 21, 439–440.

Ruttkowski, B., Joachim, A. and Daugschies, A. (2001) PCR-based differentiation of three porcine *Eimeria* species and *Isospora suis*. *Veterinary Parasitology* 95, 17–23.

Sager, H., Fischer, I., Furrer, K., Strasser, M., Waldvogel, A., Boerlin, P., Audige, L. and Gottstein, B. (2001) A Swiss case-control study to assess *Neospora caninum*-associated bovine abortions by PCR, histopathology and serology. *Veterinary Parasitology* 102, 1–15.

Sager, H., Gloor, M., Björkman, C., Kritzner, S. and Gottstein, B. (2003) Assessment of antibody avidity in aborting cattle by a somatic *Neospora caninum* tachyzoite antigen IgG avidity ELISA. *Veterinary Parasitology* 112, 1–10.

Sager, H., Hussy, D., Kuffer, A., Schreve, F. and Gottstein, B. (2005) Mise en évidence d'un cas de «abortion storm» (transmission transplacentaire exogène de *Neospora caninum*) dans une exploitation de vaches laitières: une première en Suisse. *Schweizerisches Archiv für Tierheilkunde* 147, 113–120.

Schares, G., Peters, M., Wurm, R., Bärwald, A. and Conraths, F.J. (1998) The efficiency of vertical transmission of *Neospora caninum* in dairy cattle analysed by serological techniques. *Veterinary Parasitology* 80, 87–98.

Schares, G., Conraths, F.J. and Reichel, M.P. (1999a) Bovine neosporosis: comparison of serological methods using outbreak sera from a dairy herd in New Zealand. *International Journal for Parasitology* 29, 1659–1667.

Schares, G., Rauser, M., Zimmer, K., Peters, M., Wurm, R., Dubey, J.P., de Graaf, D.C., Edelhofer, R., Mertens, C., Hess, G. and Conraths, F.J. (1999b) Serological differences in *Neospora caninum*-associated epidemic and endemic abortions. *Journal of Parasitology* 85, 688–694.

Schares, G., Rauser, M., Söndgen, P., Rehberg, P., Bärwald, A., Dubey, J.P., Edelhofer, R. and Conraths, F.J. (2000) Use of purified tachyzoite surface antigen p38 in an ELISA to diagnose bovine neosporosis. *International Journal for Parasitology* 30, 1123–1130.

Schares, G., Heydorn, A.O., Cüppers, A., Conraths, F.J. and Mehlhorn, H. (2001a) *Hammondia heydorni*-like oocysts shed by a naturally infected dog and *Neospora caninum* NC-1 cannot be distinguished. *Parasitology Research* 87, 808–816.

Schares, G., Heydorn, A.O., Cuppers, A., Conraths, F.J. and Mehlhorn, H. (2001b) Cyclic transmission of *Neospora caninum*: serological findings in dogs shedding oocysts. *Parasitology Research* 87, 873–877.

Schares, G., Bärwald, A., Staubach, C., Söndgen, P., Rauser, M., Schröder, R., Peters, M., Wurm, R., Selhorst, T. and Conraths, F.J. (2002a) p38-avidity-ELISA: examination of herds experiencing epidemic or endemic *Neospora caninum*-associated bovine abortion. *Veterinary Parasitology* 106, 293–305.

Schares, G., Heydorn, A.O., Cuppers, A., Mehlhorn, H., Geue, L., Peters, M. and Conraths, F.J. (2002b) In contrast to dogs, red foxes (*Vulpes vulpes*) did not shed *Neospora caninum* upon feeding of intermediate host tissues. *Parasitology Research* 88, 44–52.

Schares, G., Bärwald, A., Staubach, C., Ziller, M., Klöß, D., Wurm, R., Rauser, M., Labohm, R., Dräger, K., Fasen, W., Hess, R.G. and Conraths, F.J. (2003) Regional distribution of bovine *Neospora caninum* infection in the German state of Rhineland-Palatinate modelled by logistic regression. *International Journal for Parasitology* 33, 1631–1640.

Schares, G., Bärwald, A., Staubach, C., Wurm, R., Rauser, M., Conraths, F.J. and Schroeder, C. (2004a) Adaptation of a commercial ELISA for the detection of antibodies against *Neospora caninum* in bovine milk. *Veterinary Parasitology* 120, 55–63.

Schares, G., Bärwald, A., Staubach, C., Ziller, M., Klöss, D., Schröder, R., Labohm, R., Dräger, K., Fasen, W., Hess, R.G. and Conraths, F.J. (2004b) Potential risk factors for bovine *Neospora caninum* infection in Germany are not under the control of the farmers. *Parasitology* 129, 301–305.

Schares, G., Pantchev, N., Barutzki, D., Heydorn, A.O., Bauer, C. and Conraths, F.J. (2005) Oocysts of *Neospora caninum, Hammondia heydorni, Toxoplasma gondii* and *Hammondia hammondi* in faeces collected from dogs in Germany. *International Journal for Parasitology* 35, 1525–1537.

Serrano-Martinez, E., Collantes-Fernandez, E., Rodriguez-Bertos, A., Casas-Astos, E., Álvarez-Garcia, G., Chavez-Velasquez, A. and Ortega-Mora, L.M. (2004) *Neospora* species-associated abortion in alpacas (*Vicugna pacos*) and llamas (*Llama glama*). *Veterinary Record* 155, 748–749.

Slapeta, J.R., Koudela, B., Votypka, J., Modry, D., Horejs, R. and Lukes, J. (2002a) Coprodiagnosis of *Hammondia heydorni* in dogs by PCR-based amplification of ITS 1 rRNA: differentiation from morphologically indistinguishable oocysts of *Neospora caninum*. *Veterinary Journal* 163, 147–154.

Slapeta, J.R., Modry, D., Kyselova, I., Horejs, R., Lukes, J. and Koudela, B. (2002b) Dog shedding oocysts of *Neospora caninum*: PCR diagnosis and molecular phylogenetic approach. *Veterinary Parasitology* 109, 157–167.

Söndgen, P., Peters, M., Bärwald, A., Wurm, R., Holling, F., Conraths, F.J. and Schares, G. (2001) Bovine neosporosis: immunoblot improves foetal serology. *Veterinary Parasitology* 102, 279–290.

Staska, L.M., McGuire, T.C., Davies, C.J., Lewin, H.A. and Bazler, T.V. (2003) *Neospora caninum*-infected cattle develop parasite-specific CD4+ cytotoxic T-lymphocytes. *Infection and Immunity* 71, 3272–3279.

Stenlund, S., Kindahl, H., Magnusson, U., Uggla, A. and Björkman, C. (1999) Serum antibody profile and reproductive performance during two consecutive pregnancies of cows naturally infected with *Neospora caninum*. *Veterinary Parasitology* 85, 227–234.

Thilsted, J.P. and Dubey, J.P. (1989) Neosporosis-like abortions in a herd of dairy cattle. *Journal of Veterinary Diagnostic Investigation* 1, 205–209.

Thurmond, M. and Hietala, S. (1995) Strategies to control *Neospora* infection in cattle. *The Bovine Practitioner* 4, 29–32.

Thurmond, M.C. and Hietala, S.K. (1997) Effect of congenitally acquired *Neospora caninum* infection on risk of abortion and subsequent abortions in dairy cattle. *American Journal of Veterinary Research* 58, 1381–1385.

Thurmond, M.C., Hietala, S.K. and Blanchard, P.C. (1997) Herd-based diagnosis of *Neospora caninum*-induced endemic and epidemic abortion in cows and evidence for congenital and post-natal transmission. *Journal of Veterinary Diagnostic Investigation* 9, 44–49.

Trees, A.J. and Williams, D.J. (2002) Neosporosis in the United Kingdom. *International Journal for Parasitology* 30, 891–893.

Trees, A.J. and Williams, D.J. (2005) Endogenous and exogenous transplacental infection in *Neospora caninum* and *Toxoplasma gondii*. *Trends in Parasitology* 21, 558–561.

Trees, A.J., Guy, F., Low, J.C., Roberts, L., Buxton, D. and Dubey, J.P. (1994) Serological evidence implicating *Neospora* species as a cause of abortion in British cattle. *Veterinary Record* 134, 405–407.

Trees, A.J., Davison, H.C., Innes, E.A. and Wastling, J.M. (1999) Towards evaluating the economic impact of bovine neosporosis. *International Journal for Parasitology* 29, 1195–1200.

Trees, A., McAllister, M.M., Guy, C.S., McGarry, J., Smith, R. and Williams, D. (2002) *Neospora caninum*: oocyst challenge of pregnant cows. *Veterinary Parasitology* 109, 147–154.

Uggla, A., Stenlund, S., Holmdahl, O.J.M., Jakubek, E.-B., Thebo, P., Kindahl, H. and Björkman, C. (1998) Oral *Neospora caninum* inoculation of neonatal calves. *International Journal for Parasitology* 28, 1467–1472.

van Maanen, C., Wouda, W., Schares, G., von Blumröder, G., Conraths, F.J., Norton, R., Williams, D.J.L., Esteban-Redondo, I., Innes, E.A., Mattsson, J.G., Björkman, C., Fernández-García, A., Ortega-Mora, L.M., Müller, N., Sager, H. and Hemphill, A. (2004) An interlaboratory comparison of immunohistochemistry and PCR methods for detection of *Neospora caninum* in bovine foetal tissues. *Veterinary Parasitology* 126, 351–364.

Vianna, M.C.B., Sreekumar, C., Miska, K.B., Hill, D.E and Dubey, J.P. (2005) Isolation of *Neospora caninum* from naturally infected white-tailed deer (*Odocoileus virginianus*). *Veterinary Parasitology* 129, 253–257.

von Blumröder, D., Schares, G., Norton, R., Williams, D.J.L., Esteban-Redondo, I., Wright, S., Björkman, C., Frossling, J., Risco-Castillo, V., Fernandez-Garcia, A., Ortega-Mora, L.M., Sager, H., Hemphill, A., Van Maanen, C., Wouda, W. and Conraths, F.J. (2004) Comparison and standardisation of serological methods for the diagnosis of *Neospora caninum* infection in bovines. *Veterinary Parasitology* 120, 11–22.

Waldner, C.L., Janzen, E.D., Henderson, J. and Haines, D.M. (1999) Outbreak of abortion associated with *Neospora caninum* infection in a beef herd. *Journal of the American Veterinary Medical Association* 215, 1485–1489.

Weynants, V., Godfroid, J., Limbourg, B., Saegerman, C. and Letesson, J. (1995) Specific bovine brucellosis diagnosed on *in vitro* antigen specific gamma interferon production. *Journal of Clinical Microbiology* 33, 706–712.

Williams, D.J.L., Davison, H.C., Helmick, B., McGarry, J., Guy, F., Otter, A. and Trees, A.J. (1999) Evaluation of a commercial ELISA for detecting serum antibody to *Neospora caninum* in cattle. *Veterinary Record* 145, 571–575.

Williams, D., Guy, C., McGarry, J., Guy, F., Tasker, L., Smith, R., MacEachern, K., Cripps, P.J., Kelly, D. and Trees, A.J. (2000) *Neospora caninum*-associated abortion in cattle: the time of experimentally induced parasitaemia during gestation determines foetal survival. *Parasitology* 121, 347–358.

Williams, J.H., Espie, I., van Wilpe, E. and Matthee, A. (2002) Neosporosis in a white rhinoceros (*Ceratotherium simum*) calf. *Journal of the South African Veterinary Association* 73, 38–43.

Wood, P.R., Corner, L.A., Rothel, J.S., Baldock, C., Jones, S.L., Cousins, D.B., McCormick, B.S., Francis, B.R., Creeper, J. and Tweddle, N.E. (1991) Field comparison of the interferon gamma assay and the intradermal tuberculin test for the diagnosis of bovine tuberculosis. *Australian Veterinary Journal* 68, 286–290.

Woods, L.W., Anderson, M.L., Swift, P.K. and Sverlow, K.W. (1994) Systemic neosporosis in a California black-tailed deer (*Odocoileus hemionus columbianus*). *Journal of Veterinary Diagnostic Investigation* 6, 508–510.

Wouda, W., Dubey, J.P. and Jenkins, M.C. (1997a) Serological diagnosis of bovine fetal neosporosis. *Journal of Parasitology* 83, 545–547.

Wouda, W., Moen, A.R., Visser, I.J.R. and van Knapen, F. (1997b) Bovine fetal neosporosis: a comparison of epizootic and sporadic abortion cases and different age classes with regard to lesion severity and immunohistochemical identification of organisms in brain, heart and liver. *Journal of Veterinary Diagnostic Investigation* 9, 180–185.

Wouda, W., Brinkhof, J., Van Maanen, C., De Gee, A.L.W. and Moen, A.R. (1998a) Serodiagnosis of neosporosis in individual cows and dairy herds: a comparative study of three enzyme-linked immunosorbent assays. *Clinical and Diagnostic Laboratory Immunology* 5, 711–716.

Wouda, W., Moen, A.R. and Schukken, Y.H. (1998b) Abortion risk in progeny of cows after a *Neospora caninum* epidemic. *Theriogenology* 49, 1311–1316.

Wouda, W., Bartels, C.J.M. and Moen, A.R. (1999a) Characteristics of *Neospora caninum*-associated abortion storms in dairy herds in the Netherlands (1995–1997). *Theriogenology* 52, 233–245.

Wouda, W., Dijkstra, T., Kramer, A.M.H., van Maanen, C. and Brinkhof, J.M.A. (1999b) Seroepidemiological evidence for a relationship between *Neospora caninum* infections in dogs and cattle. *International Journal for Parasitology* 29, 1677–1682.

Wouda, W., Bartels, C.J.M. and Dijkstra, T. (2000) The epidemiology of bovine neosporosis with emphasis on risk factors. In: Hemphill, A. and Gottstein, B. (eds) *A European Perspective on* Neospora caninum. *International Journal for Parasitology* 30, 877–924.

Yaeger, M.J., Shawd-Wessels, S. and Leslie-Steen, P. (1994) *Neospora* abortion storm in a midwestern dairy. *Journal of Veterinary Diagnostic Investigation* 6, 506–508.

Yamane, I., Shibahara, T., Kokuho, T., Shimura, K., Hamaoka, T., Haritani, M., Conrad, P.A., Park, C.H., Sawada, M. and Umemura, T. (1998) An improved isolation technique for bovine *Neospora* species. *Journal of Veterinary Diagnostic Investigation* 10, 364–368.

Zhao, X., Duszynski, D.W. and Loker, E.S. (2001) A simple method of DNA extraction for *Eimeria* species. *Journal of Microbiological Methods* 44, 131–137.

3.2 Toxoplasmosis

General Considerations

David Buxton and Bertrand Losson

Toxoplasma gondii is an obligate intracellular protozoan parasite that may cause abortion in sheep, goats and pigs. The parasite also has zoonotic potential.

Sheep

In 2003, over 86 million sheep were farmed in the 15 countries of the EU, of which an estimated 63.5 million were recorded as being females of breeding age (Eurostat Statistics in Focus, 20/2005, http://epp.eurostat.ec.europa.eu), with three-quarters located in the UK, Spain, Italy and France (Eurostat Statistics in Focus, 20/2005, http://epp.eurostat.ec.europa.eu) (see Table 3.2.1). *Toxoplasma* abortion and fetal loss may occur wherever sheep are kept, but precise figures do not exist.

Perhaps the best estimates were made by Blewett and Trees (1987), who concluded that the average incidence of clinical toxoplasmosis in the UK is between 1 and 2% per annum. This figure is an underestimate, as it does not include the loss of fetuses early in gestation that are resorbed, these ewes presenting as cases of infertility. If the incidence of toxoplasmosis is assumed to be similar throughout the sheep-rearing areas of the EU, then in 2003 between 635,000 and 1.27 million lambs would have been lost (assuming a conservative average lambing of one lamb per ewe per year). However, this is a risky assumption and the incidence of clinical toxoplasmosis will vary from region to region.

Table 3.2.1. Distribution of sheep and goats and their output of milk in the EU15 countries (Eurostat Statistics in Focus, 20/2005, http://epp.eurostat.ec.europa.eu).

Country	Sheep (%)	Sheep milk production per annum (l, %)	Goats (%)	Goat milk production per annum (l, %)
UK	28.5	–	–	–
Spain	26.5	18.8	26.2	27.8
Italy	9.3	38.0	8.6	6.9
France	10.2	11.2	10.6	32.8
Greece	10.6	30.0	45.0	27.8
Ireland	5.2	–	–	–
Portugal	4.1	4.4	4.8	1.9
Others	5.6	–	4.8	–
Total (million)	86	2236	11.6	1614

Goats

The 11.4 million goats in the EU15 (Eurostat Statistics in Focus, 20/2005, http://epp.eurostat.ec.europa.eu) (see Table 3.2.1) are equally susceptible to toxoplasmosis as are sheep, and the incidence of clinical disease will be influenced in similar ways to those for sheep. Goats and sheep are also kept for milk production (see Table 3.2.1), particularly in Greece, France, Italy and Spain. Both liquid milk and cheese represent an important source of income for many farmers, with cheese production providing significant added value (see Table 3.2.2). Toxoplasmosis and any other cause of abortion may severely reduce herd and flock milk production, thus adversely affecting farm income.

Table 3.2.2. Sheep and goat production of cheese in the EU15 countries in 2004, expressed as million kg (no significant production in other countries) (Eurostat Statistics in Focus, 25/2005, http://epp.eurostat.ec.europa.eu).

	Sheep	Goats
France	17.2	76.0
Italy	54.5	2.7
Netherlands	0.0	6.0
Portugal	1.9	1.0
Spain	26.0	11.9

Cattle

In cattle, clinical toxoplasmosis is not considered to be a significant disease, as the parasite seems to be cleared very effectively (Dubey, 1986a), probably because cattle develop a more effective immune response to the parasite than do sheep and goats (Esteban-Redondo and Innes, 1997). However, infection was detected by means of the polymerase chain reaction (PCR) in four of 86 aborted bovine fetuses examined in Switzerland (Gottstein *et al.*, 1998), suggesting that *Toxoplasma* abortion may occur infrequently in cattle. In this case, it needs to be differentiated from bovine neosporosis amongst other causes.

Biology, Transmission and Clinical Signs

David Buxton and Bertrand Losson

Life cycle

The life cycle of *Toxoplasma* can be divided into two parts: (i) an asexual cycle with little host specificity; and (ii) a sexual cycle, confined to the entero-epithelial cells of cats and other Felidae (Miller *et al.*, 1972), resulting in the production of oocysts (Dubey and Beattie, 1988) (see Fig. 3.2.1).

Asexual cycle

In the asexual cycle, two developmental stages are involved, the tachyzoite and the bradyzoite. Each crescent-shaped tachyzoite (about 5 by 1.5 μm) can actively

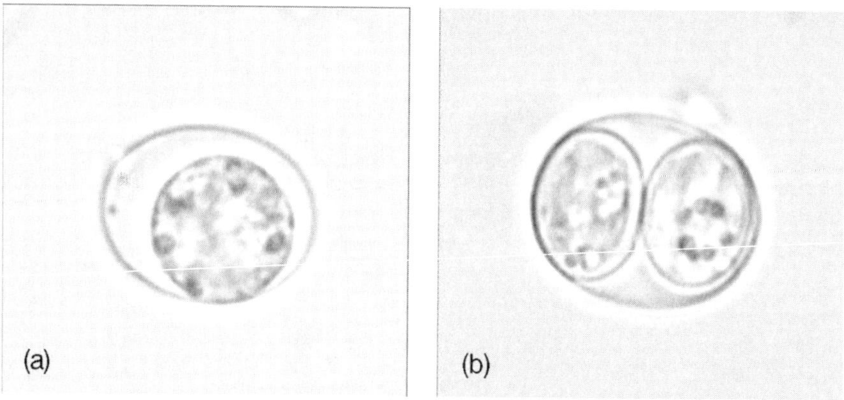

Fig. 3.2.1. *Toxoplasma gondii* oocysts isolated from cat faeces: (a) unsporulated, (b) sporulated.

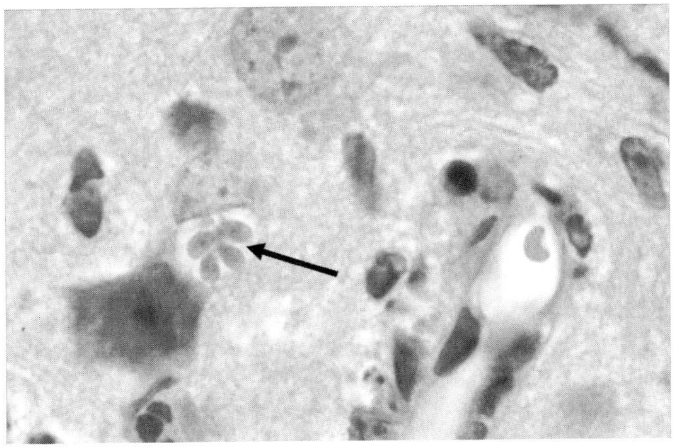

Fig. 3.2.2. *Toxoplasma gondii* tachyzoites in a parasitophorous vacuole (arrow) in the brain of an experimentally infected mouse (haematoxylin and eosin stain).

penetrate a host cell, where it becomes surrounded by a parasitophorous vacuole (see Fig. 3.2.2), in which it multiplies by endodyogeny (two daughter cells form within the mother cell). Multiplication continues until the host cell ruptures, when the organisms are released to parasitize further cells. This process proceeds until the host dies or, more usually, develops immunity to the parasite. In the latter case, extracellular organisms are eliminated, intracellular multiplication slows and the tachyzoites transform into slowly multiplying bradyzoites in tissue cysts, to establish a persistent infection (see Fig. 3.2.3).

A small cyst contains only a few bradyzoites, but a large one may contain many hundreds. When a cyst ruptures, the bradyzoites are released and, transformed into tachyzoites, enter other cells to complete the asexual cycle

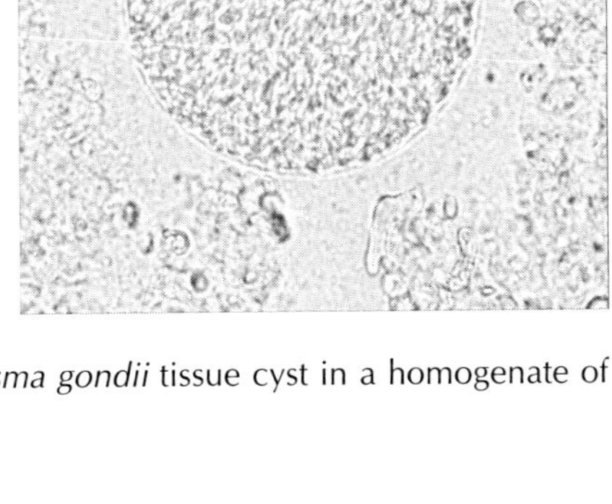

Fig. 3.2.3. *Toxoplasma gondii* tissue cyst in a homogenate of mouse brain.

(Dubey and Beattie, 1988). These microscopic tissue cysts are present most frequently in brain and muscle and represent the quiescent stage of the parasite within the host. In sheep, goats, pigs, man and many other animals it is thought likely that tissue cysts remain viable for the rest of the life of the individual (Petersen and Dubey, 2000).

Sexual cycle

Initiation of the sexual cycle occurs when a non-immune cat ingests food contaminated by oocysts or containing tachyzoites or tissue cysts. In the case of the latter, the cyst wall is dissolved by proteolytic enzymes in the stomach and small intestine and the released bradyzoites penetrate the epithelial cells of the small intestine. While the parasite spreads to brain and muscles where tissue cysts will develop (asexual cycle), simultaneously toxoplasms also undergo gametogeny (sexual cycle) in entero-epithelial cells. Here, in the small intestine (most commonly the ileum) gametocytes develop, over 3–15 d post-infection. Microgametes form and are released to penetrate mature macrogametes, triggering the formation of an oocyst wall around each fertilized gamete.

Each oocysts, almost filled by the sporont, is discharged into the intestinal lumen to pass out in the faeces. Sporulation occurs within 1–5 d (depending on aeration and temperature) to produce two ellipsoidal sporocysts, each containing four sporozoites within each oocyst. Sporulated oocysts measure 11 × 13 μm in diameter (Dubey and Beattie, 1988; Petersen and Dubey, 2000).

Thus, the cat is capable of shedding millions of oocysts in its faeces for around 10 d, starting 4 d after first ingesting tissue cysts. After this, it will not normally excrete the parasite again, although stress can trigger the recrudescence of infection (Dubey and Frenkel, 1974). Unrelated illness may therefore lead to the re-excretion of oocysts in smaller numbers and for a shorter time than in a primary infection. Cats may also become infected by ingesting oocysts or tachyzoites, but in this case they tend to produce oocysts after 19 or 20 d for only a day or two, and in relatively small numbers and, even then, about half of these animals will not excrete oocysts (Dubey and Beattie, 1988) (for further details see section on cats, below).

Oocysts are highly resistant and survive for long periods (> 500 d) at room temperature in moist conditions (they are destroyed by a 5% ammonium solution within minutes, 95% ethanol within an hour and 10% formalin within 24 h, but none of these methods could be used to treat food or water) (Dubey and Beattie, 1988). In contrast, tachyzoites and bradyzoites do not survive long outside the host and are readily killed by freezing and thawing, desiccation, standard disinfectants and even water.

Sources of infection for cats

Kittens are normally born free of infection with *T. gondii* and subsequent excretion of oocysts follows establishment of a primary infection. Thus, young cats are more likely to be found to be shedding oocysts, although it has been estimated that less than 1% of all cats may be excreting at any one time (Dubey and Beattie, 1988). Numbers of young cats are influenced by the numbers of breeding adults, bearing in mind that female feral cats can produce two to three litters a year, each of up to eight kittens, and may rear their young communally (Macdonald, 1980).

In rural areas, male cats may have territories of 60–80 ha (250–200 acres), while females usually only occupy a one-tenth of this area (Macdonald, 1980), and in an urban situation these territories are considerably smaller (Tabor, 1980). The area occupied by feral cats will depend upon the supply of food, which includes mice, voles, shrews, rats, rabbits and small birds (Macdonald, 1980). Such animals persistently infected with *T. gondii* are an important source of infection (Jackson and Hutchison, 1989; Peach *et al.*, 1989).

Also mice (Eichenwald, 1948; Beverley, 1959; De Roever-Bonnet, 1969; Owen and Trees, 1998), but probably not rats (Dubey *et al.*, 1997), are particularly important because they can transmit *Toxoplasma* parasite to their developing young before birth without causing overt clinical disease or fetopathy. In this way, a reservoir of *T. gondii* tissue cyst infection for cats can exist in a particular population of mice for a long time.

Clinical symptoms

Toxoplasma does not usually cause clinical illness in cattle, camelids or deer but it can cause fatal disease in New World monkeys, marsupials and hares (*Lepus europeus* and *L. timidus*) (Petersen and Dubey, 2000). Toxoplasmosis arguably has its greatest impact in sheep in which, following a primary infection during pregnancy, it may cause fetal death and abortion. It can also cause abortion in goats and pigs but no significant disease is produced in cattle and horses (Lind and Buxton, 2000). Recrudescence of toxoplasmosis during pregnancy has been observed in goats (Dubey, 1982; Dubey *et al.*, 1985) and, more recently, it has been suggested as occurring in sheep (Duncanson *et al.*, 2001).

Clinical ovine (or caprine) toxoplasmosis occurs following a primary *Toxoplasma* infection in a pregnant ewe (or doe) and is usually initiated when the mother is infected in the middle part of pregnancy. In a typical case, the mother produces a stillborn lamb (kid) that may be accompanied by a weakly sibling or a 'mummified' fetus. This usually occurs a few days earlier than the predicted end of pregnancy (Buxton, 2000), but otherwise the ewe/doe remains clinically normal. White necrotic foci (sometimes mineralized) up to 3 mm in diameter may be visible in the placental cotyledons, but the intercotyledonary membranes appear normal (see Fig. 3.2.4).

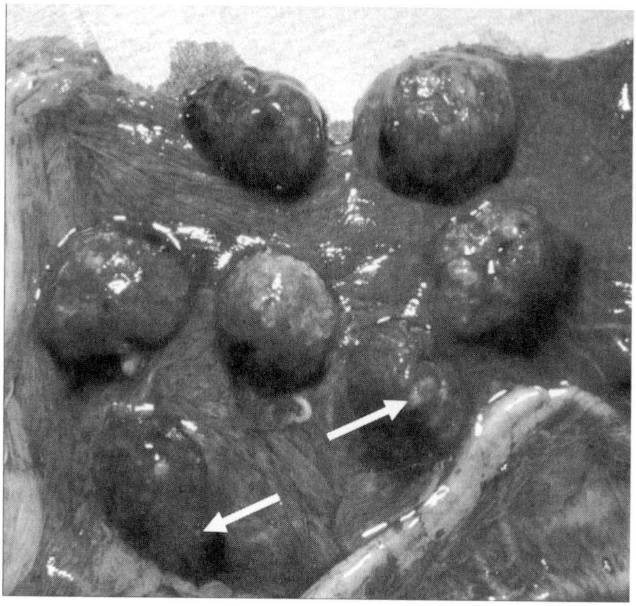

Fig. 3.2.4. Placenta from a sheep that had aborted due to infection with *Toxoplasma gondii*. Focal necrosis, visible as white spots of varying size (arrows), is present in the cotyledons; the intercotyledonary membranes, while congested, are normal.

Infection initiated early in pregnancy, when the fetus has only a rudimentary immune system, results in fetal death and resorption. In this case, the mother may present as barren. When a significant number of animals are affected it may appear to be a flock/herd infertility problem so that, for example, the ram may be wrongly put under suspicion in an affected flock of sheep.

Mothers that become infected in late pregnancy would be expected to remain clinically normal and to produce clinically normal, but infected, offspring (Watson and Beverley, 1971; Hartley and Moyle, 1974; Blewett *et al.*, 1982). However, clinically normal – but infected – lambs have been reported as being born to a small number of persistently infected ewes (Hartley, 1966).

Transmission

In a primary infection of a pregnant ewe (or goat), infection of the fetus occurs following the ingestion of sporulated oocysts in contaminated food or water (exogenous trans-placental transmission; Trees and Williams, 2005). Environmental contamination with oocysts from cat faeces has long been considered the most important source of infection for sheep (Plant *et al.*, 1974; Blewett and Watson, 1984; Faull *et al.*, 1986; Owen, 1996), and farmers have been advised to control disease both by preventing the soiling of feed and water by cats and by implementing vaccination (Buxton, 1990; Buxton and Innes, 1995) (see Fig. 3.2.5).

Sheep previously exposed to *T. gondii* will develop an effective immune response to the parasite and would not be expected to subsequently suffer a clinical infection (McColgan *et al.*, 1988), and while the situation is essentially similar for goats (Obendorf *et al.*, 1990), repeat abortion due to *Toxoplasma* has

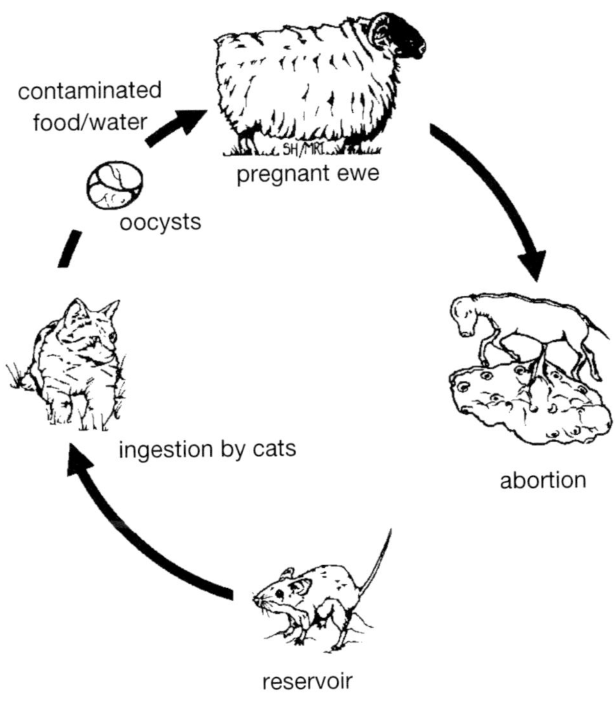

Fig. 3.2.5. *Toxoplasma gondii* oocysts are produced after a naïve cat ingests *T. gondii* tissue cysts. Tissue cysts occur in the brain and muscle of persistently infected birds and small rodents, but mice are a particularly important source of infection, as pregnant female mice can pass the parasite trans-placentally to their developing fetuses. The female offspring are therefore born infected and remain so for life and will, in turn, pass the parasite on trans-placentally to their offspring. In this way, *T. gondii* can be maintained in a population of mice over time. Oocysts deposited in the environment in cat faeces may remain viable and infectious for very many months. Pregnant sheep becoming infected for the first time, by ingesting food or water contaminated with oocysts, may go on and abort. The weight of evidence suggests that the ewe and her offspring will not transmit infection to any significant extent in subsequent pregnancies, although recent research challenges this conclusion (see text).

been recorded (Dubey, 1982). Recrudescence of a persistent infection during pregnancy that leads to infection of the placenta and fetus has been termed endogenous trans-placental transmission (Trees and Williams, 2005).

In sheep, endogenous trans-placental transmission is reported as occurring infrequently and at a very low level. For example, Munday (1972) analysed 178 precolostral lamb sera from 135 ewes persistently infected with *T. gondii*, collected over a 4-year period, but failed to demonstrate any serological evidence of congenital transmission, although one placenta from a seropositive ewe was shown to contain *T. gondii*.

However, recent reports have suggested that endogenous trans-placental transmission of *T. gondii* in sheep may occur more readily than previously thought (Duncanson *et al.*, 2001; Williams *et al.*, 2005). These authors reported transmission of *T. gondii* in 41–69% of pregnancies and an abortion rate that ranged from 4.5–20.6% of the potential lamb crop. They also described a strong association between congenital infection with *T. gondii* (the detection of parasite DNA by PCR) and fetal mortality (90% of dead lambs were positive for *T. gondii* by PCR compared with 46.4% of live lambs) (Williams *et al.*, 2005).

However, the apparent causal association between *T. gondii* infection and lamb mortality/abortion in the flocks described by Duncanson *et al.* (2001) and Williams *et al.* (2005) is difficult to assess, as these authors relied solely on the PCR, with no pathological or serological data presented. A subsequent smaller, but more thorough, study by Rodger *et al.* (2006) failed to detect significant endogenous transmission.

To date, the majority of evidence from the field and scientific literature supports the hypothesis that oocysts shed by cats are the most epidemiologically significant source of *T. gondii* infection for sheep. Outbreaks of disease have been associated with the contamination of feed with cat faeces (Plant *et al.*, 1974), epidemiological studies have shown an association between the presence of cats on farms, the contamination of pasture with *Toxoplasma* infectivity and the occurrence of toxoplasmosis or evidence of exposure to *T. gondii* in sheep (Faull *et al.*, 1986; Skjerve *et al.*, 1998).

In addition, serological surveys have shown an increasing seroprevalence to *T. gondii* with age in sheep (Waldeland, 1977; Blewett, 1983; Lundén *et al.*, 1994; Owen, 1996) and in reindeer (Oksanen *et al.*, 1997). Oocyst-contaminated drinking water has caused *T. gondii* infection in humans (Bowie *et al.*, 1997) and recent evidence shows that while pigs maintained exclusively inside are unlikely to encounter *Toxoplasma*, if they are maintained in outdoor housing systems then they run the risk of becoming infected (Kijlstra *et al.*, 2004).

In the light of this evidence it seems probable that endogenous trans-placental transmission by persistently infected ewes is of relatively minor significance, while environmental contamination with *T. gondii* oocysts is of considerable importance as a source of infection for sheep and other animals.

Because clinical toxoplasmosis occurs following primary infection during pregnancy, its incidence is influenced by the number of susceptible sheep and the degree of contamination of food and water by viable oocysts, which itself is dependent both on the cat population and environmental conditions. A lower incidence than occurs in, for example, the UK will be recorded in a region if infection, seroconversion and immunity are established in a high proportion of sheep before they reach breeding age. This may be the result of significant contamination of food and/or water with oocysts and/or environmental conditions that favour their prolonged viability (relatively little desiccation and frost). A low incidence of clinical toxoplasmosis would be expected in a region where viable oocysts are scarce.

Thus few cats, extensive grazing practices, no supplementary feeding and relatively harsher weather conditions, with significant frosts and/or desiccation, will lead to low infection rates. While such conditions would favour a relatively high proportion of susceptible sheep, they would be exposed to minimal challenge by oocysts and so the incidence would normally remain low.

Abortion in sheep and goats, due to *T. gondii*, must be differentiated from other causes, including infections with *Chlamydophila abortus* (*Chlamydia psittaci*) (enzootic abortion of ewes; ovine chlamydial abortion), *Coxiella burnetii* (Q fever), *Brucella melitensis*, *Campylobacter fetus fetus*, *Salmonella* spp., Border disease virus and certain arthropod-borne viral diseases (see Chapter 4, this volume). In addition, pregnant sheep are very susceptible to experimental infection with *N. caninum* and rare outbreaks have been recorded under natural conditions (Haessig *et al.*, 2003) (see Chapter 4, this volume).

Zoonotic risks

Toxoplasma gondii readily infects human beings and, while infection is relatively common (approximately 30% of the population depending on age and environment), clinical illness is relatively uncommon (Hall *et al.*, 2000). People particularly at risk include pregnant women, as the parasite can pose a serious threat to the unborn child if the mother becomes infected for the first time while pregnant (Thulliez, 2000). Also, those who are immunosuppressed, such as tissue transplant patients, victims of AIDS, people suffering from certain types of cancer and those undergoing certain forms of cancer therapy, are at risk of developing acute lethal infection if left untreated (Mariuz and Steigbigel, 2000; Wreghitt and Joynson, 2000).

The very young and very old may also be more susceptible. On occasions, people with no apparent immune system deficiency may develop an illness characterized by general malaise, fever and lymphadenopathy (Ho-Yen, 2000). The two most likely sources of human infection are: (i) ingestion of raw or lightly cooked meat containing live *T. gondii* tissue cysts; or (ii) ingestion of oocysts derived from cat faeces, such as may be encountered in gardens and children's sand pits (Hall *et al.*, 2000).

Diagnostic Techniques

Histology

Laura Kramer and David Buxton

The birth of stillborn or weak lambs (or kids), sometimes accompanied by a mummified fetus, is suggestive of ovine (or caprine) toxoplasmosis. Typically, gross lesions are apparent in the placenta and present as multiple white foci of necrosis, 2–3 mm in diameter, within the cotyledons (Buxton, 2000). A more certain diagnosis of *T. gondii*-induced abortion would ideally include histopathology and immunohistochemistry of the placenta and fetal brain, liver, lung and heart. Preparation of samples for examination under the microscope should be carried out according to universally accepted protocols, involving the fixation of tissues.

Procedure

1. Fix tissues in 10% formol saline
2. Dehydrate through graded alcohols
3. Clear in xylene (or an equivalent)
4. Embed in paraffin wax
5. Cut 4 μm sections with a microtome
6. Mount sections on glass histology slides
7. Stain with haematoxylin and eosin (H & E)
8. Mount under a coverslip
9. Histopathological examination by a pathologist

PLACENTA In affected placental cotyledons from sheep and goats, the white spots visible macroscopically are large foci of coagulative necrosis, which may have become mineralized with time. Associated inflammation is characteristically slight and non-suppurative. Well-preserved samples of placental cotyledons may show moderate oedema of the mesenchyme of the fetal villi, with a diffuse hypercellularity due to the presence of large mononuclear cells.

Sometimes, small numbers of intracellular and extracellular toxoplasms are visible, usually on the periphery of a necrotic area or in a villus that is in the early stages of infection. The organisms appear ovoid, 1–4 μm long, with nuclei that are moderately basophilic and located centrally or towards the blunt end.

BRAIN In the fetal brain, primary and secondary lesions may develop (Buxton *et al.*, 1982; Buxton and Finlayson, 1986), with the former manifest as inflammatory foci, composed of microglia and other mononuclear inflammatory cells, and typically with a necrotic and sometimes mineralized centre (see Fig. 3.2.6). They represent a fetal immune response following direct damage by local parasite multiplication and are often associated with a mild focal lymphoid meningitis. Toxoplasms are only rarely found, usually at the periphery of these lesions. Secondary lesions, manifest as focal leukomalacia, are more commonly encountered (see Fig. 3.2.7). They are considered to be due to fetal anoxia in late gestation caused by the progressive multifocal necrosis in the placentome preventing sufficient oxygen transfer from mother to fetus (Buxton *et al.*, 1982). Leukomalacia occurs most commonly in the cerebral white matter cores, but sometimes also in the cerebellar white matter.

Examined under a low-power objective, they can be differentiated from the surrounding white matter as areas with an altered staining intensity in which there is also a loss of supporting oligodendroglia. There may also be associated small haemorrhages. Examined with a higher power objective, the affected white matter appears less refractile than normal white matter, the axons are swollen and there may be ghosts of glial nuclei.

When both inflammation and focal leukomalacia are seen together, diagnosis is reasonably certain. While the former is characteristic of protozoal infection, the latter may also occur, albeit less commonly, with other infections such as ovine chlamydial abortion (Buxton *et al.*, 2002) or tick-borne fever (Chianini *et al.*, 2004).

LIVER, LUNG AND HEART Focal necrosis with associated lymphoid inflammation may be seen but, as other pathogens may cause similar lesions, to confirm a

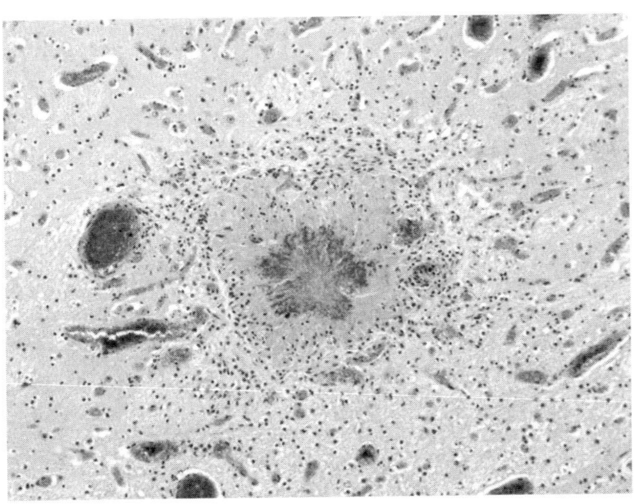

Fig. 3.2.6. Focal necrosis and surrounding inflammation in the brain of a stillborn lamb. The lesion is the result of primary damage by the *Toxoplasma gondii*. Note the mineralization at the centre of the lesion (haematoxylin and eosin stain).

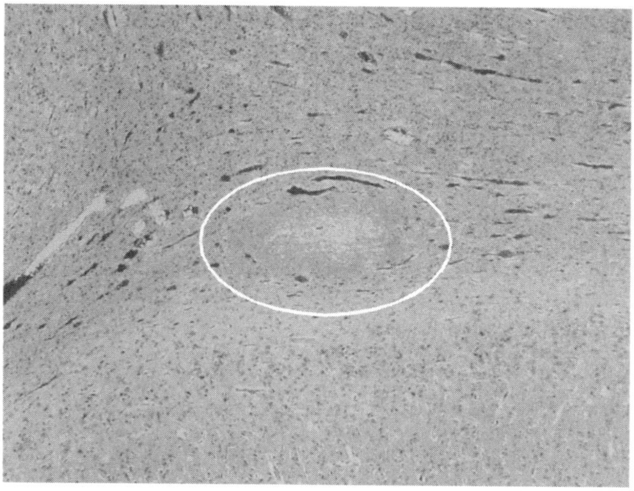

Fig. 3.2.7. Focal leukomalacia in the cerebral white matter of a stillborn lamb following infection with *Toxoplasma gondii*. The lesion is the result of fetal anoxia arising from advanced placental lesions compromising the efficiency of the placenta (haematoxylin and eosin stain).

diagnosis of toxoplasmosis the parasite should be demonstrated by immunohistochemistry.

Immunohistochemistry

Laura Kramer and David Buxton

A definitive diagnosis of toxoplasmosis depends on the identification of *T. gondii* in fetal tissues associated with the lesions described above. This is best achieved by immunohistochemistry, in which specific antibody binds to toxoplasma antigen in the tissue section and is then visualized by applying reagent(s) that locate on the bound antibody and produce a colour reaction. The method labels intact *T. gondii* or antigenic debris in tissue sections from normally fixed paraffin wax-embedded tissues.

Several immunohistochemical procedures are suitable, including: (i) the peroxidase–antiperoxidase (PAP) technique (Uggla *et al.*, 1987); (ii) the Vectastain avidin biotin complex method (Vector Laboratories, California, USA); and (iii) the DAKO EnVision+ System (DAKO Corporation, California, USA). Polyclonal antibodies (e.g. antiserum #210-70-TOXO, VMRD, Pullman, Washington, USA) against *T. gondii* are more commonly used as a primary antibody in diagnosis, as they will recognize a number of different parasite antigens, but certain monoclonal antibodies, such as against the P30 protein (e.g.

clone NCL-TG, Novocastra Ltd., Newcastle-upon-Tyne, UK) may also be appropriate. The procedure for the PAP-technique is briefly outlined below.

Reagents, materials and equipment

- Tris buffer solution (TBS) (pH 7.6)
- Methanol 100%
- Ethanol 100%
- Ethanol 96%
- Ethanol 70%
- Xylene (or an equivalent)
- H_2O_2 30%
- Haematoxylin
- Anti-*Toxoplasma* antiserum (polyclonal or monoclonal) prepared in (non-ungulate) animal A
- Anti-animal A-IgG antibodies prepared in (non-ungulate) animal B
- Peroxidase–antiperoxidase complex
- Diamino benzidine (DAB) solution (DAKO)
- Demineralized water
- Histological slides
- Coverslips
- Usual equipment of an (immuno)histological laboratory

Procedure

Note: In each laboratory, optimization of antiserum dilutions should be performed, using a positive control sample, before application of the test for routine use.

In each test session, duplicate negative control sections (not treated with specific antiserum) should be prepared together with the test sections, as well as positive control sections from a confirmed positive case.

1. Prepare unstained paraffin wax sections (as detailed in the previous section).
2. Dewax sections via xylene, reverse graded alcohols and water.
3. Incubate for 30 min in 1% H_2O_2 in methanol to inhibit endogenous peroxidase.
4. Wash with water and TBS.
5. Incubate with anti-toxoplasma serum of animal A (for negative control use TBS instead).
6. Wash with TBS.
7. Incubate with anti-animal A-IgG antibodies for 1 h.
8. Wash with TBS.
9. Incubate with peroxidase–antiperoxidase complex for 1 h.
10. Wash with TBS.
11. Incubate with diamino benzidine (DAB) solution for up to 5 min.

12. Wash with TBS.
13. Stain with haematoxylin.
14. Dehydrate via graded alcohols.
15. Clear with xylene.
16. Mount under a coverslip.
17. Histopathological examination by a pathologist.

An advantage of immunohistochemistry, compared with the PCR, is that the presence of parasites, or parasite antigen, can be related to the lesions. However, immunohistochemistry has a lower sensitivity than the PCR. Also, due to methodological differences, the experience of the operator, the time devoted to screening and the sensitivity of the procedure may vary between laboratories.

Although likely to be rare when examining ovine or caprine tissues, it is important to be aware that there may be some cross-reactivity with *N. caninum* (Van Maanen *et al.*, 2004), necessitating that the specificity of *Toxoplasma* antisera be checked against tissues containing *Neospora*.

Other fetal tissues that may be useful include heart, lung and liver, in which typical lesions appear as foci of coagulative necrosis. Associated inflammation may not be present in younger fetuses, but those infected in the latter part of gestation may have lymphoblastic cells associated with the necrotic foci. Intracellular and extracellular *T. gondii* tachyzoites, when present, are usually found at the periphery of microscopic lesions.

Polymerase chain reaction

Ana Hurtado and Gorka Aduriz

PCR assays targeting different genes have been developed for the detection of *T. gondii*, with the B1 repetitive sequence (35 copies), the P30 (SAG1) surface antigen (single copy) and the ribosomal RNA (110 copies) being the most widely used targets (Burg *et al.*, 1989; Savva *et al.*, 1990; Wastling *et al.*, 1993; Tenter *et al.*, 1994). Since the sensitivity of the PCR is dependent on the copy number of the target sequence, the ribosomal RNA, both the small subunit rRNA gene (Tenter *et al.*, 1994) and the more variable internal transcribed spacer (ITS1) sequence (Hurtado *et al.*, 2001), have been shown to be more sensitive.

Although two consecutive PCR runs (the nested PCR) are generally used in order to improve the sensitivity of the assay, the nested procedure increases the chance of carry-over contamination, as well as being more costly and time-consuming. Therefore, a procedure intended to minimize the risk of contamination without loss of detection power, single-tube nested PCR (Hurtado *et al.*, 2001), is presented here. The B1 gene is also widely used as a target in many *Toxoplasma* PCR detection methods, mainly as a nested protocol based on that described by Burg *et al.* (1989), as described below.

Recently, real-time PCR has emerged as a powerful diagnosic tool in which the accumulation of the amplicon is monitored in real time by the labelling of primers, probe or amplicons with fluorogenic molecules. This technique allows quantitative detection of the pathogen, since quantification is determined during the log-linear phase of amplification when the transcript number truly reflects the starting copy number.

In addition, real-time PCR eliminates post-amplification processing of the products, reducing the chances of carry-over contamination and speeding up the process. However, although several protocols have been developed for different *T. gondii* genes (Costa *et al.*, 2000; Lin *et al.*, 2000; Jauregui *et al.*, 2001; Reischl *et al.*, 2003), the technique is still expensive and it becomes cost-effective only when used in diagnostic laboratories with a considerable throughput.

PCR allows direct detection of the parasite DNA extracted from different types of fresh biological samples, such as blood and fetal tissues from aborted ewes, including the placenta, central nervous system, heart, skeletal muscle and fetal fluid. Formalin-fixed paraffin wax-embedded tissues can also be a very useful diagnostic source of tissue for PCR. While sensitivity might be reduced, autolysis is stopped and contamination may be less of a problem. Thereby, the technique can be applied to the diagnosis of *T. gondii*-associated abortion in sheep.

Reagents, materials and equipment

- Sterile scalpel blade
- TE (10 mM Tris and 1 mM EDTA, pH 8.0)
- Proteinase K
- 1% SDS
- Phenol-chloroform-isoamyl alcohol
- $MgCl_2$
- Deoxynucleotides
- Primers
- *Thermus aquaticus* (*Taq*) polymerase and reaction buffer
- Agarose, molecular biology grade
- Electrophoresis buffer (1 × TBE)
- Ethidium bromide 0.5 μg/ml
- Stomacher
- Thermocycler
- Electrophoresis tank and power supply
- UV light transilluminator
- Fluorescence-labelled hybridization probe
- Real-time PCR Master Mix.

Procedure

DNA EXTRACTION

1. Finely chop approximately 1 g of tissue with a sterile scalpel blade.
2. Resuspend in 7 ml of TE and homogenize with a Stomacher.
3. Digest a 400 μl aliquot of each of these homogenates with proteinase K (200 μg/ml) and 1% SDS.
4. Extract the DNA with phenol-chloroform-isoamyl alcohol (Sambrook *et al.*, 1989) (for fetal fluid samples, 200 μl are processed as above).
5. For every ten samples extracted, one control (400 μl of TE without tissue sample) is included to detect any possible contamination occurring during the extraction process.

PCR AMPLIFICATION

Procedure 1: single-tube nested PCR amplification of the ITS1 region

Single-tube nested PCR amplification, targeting the 18S-5.8S rRNA (ITS1) region, is carried out using two pairs of primers: the external pair (with a higher 'melting' temperature, T_m) at a lower concentration and the internal pair (with a lower T_m) at a higher concentration, to prevent the first amplification round interfering with the second (Tang *et al.*, 1997; Ellis *et al.*, 1999).

The external primers are NN1 (5'-CCTTTGAATCCCAAGC AAAACATGAG-3') and NN2 (5'-GCGAGCCAAGACATCCATTGCTGA-3'), which hybridize to a region of the ITS1 common to both *T. gondii* and *N. caninum*. The internal primers are Tg-NP1 (5'-GTGATAGTATCGA AAGGTAT-3') and Tg-NP2 (5'-ACTCTCTCTCAAATGTTCCT-3'), which amplify a region of 227 bp of the ITS1 of *T. gondii*.

1. The PCR is performed in 25 μl volumes including the following:

- 250–500 ng of total genomic DNA
- 10 mM Tris-HCl (pH 8.3)
- 50 mM KCl
- 1.5 mM $MgCl_2$
- 0.01% (w/v) gelatin
- 0.2 mM of each deoxynucleotide
- 0.01 μM of each external primer
- 0.4 μM of each internal primer
- 1 U of *Thermus aquaticus* (*Taq*) polymerase.

2. Subject the mixture to the following cycling conditions using a RoboCycler 40 thermocycler (Stratagene, La Jolla, California, USA):

- 94°C for 3 min, followed by treatment at 94°C for 30s, then at 65°C for 45s and at 72°C for 1 min
- Repeat this three-part cycle for a total of 15 times

- Treat at 94°C for 20s, followed by treatment at 53°C for 30s and then at 72°C for 30s
- Repeat this three-part treatment for a total of 35 cycles
- Treat at 72°C for 5 min.

(NB: positive controls (purified RH strain *T. gondii* DNA) and negative controls (no template DNA) are included in each PCR run).

3. The PCR products are analysed by electrophoresis through 1.5% (w/v) agarose gels in TBE buffer.
4. The bands are visualized by staining with ethidium bromide and excitation under UV light on a transilluminator.

The PCR amplifies a single band of the expected size (227 bp) from the *T. gondii* ITS1 region, whose identity can be confirmed by sequencing. No amplicons are generated from the related protozoans *Sarcocystis gigantea*, *Neospora caninum* or DNA extracted from non-infected sheep. The detection limit of the PCR assay determined by serial dilutions of pure *T. gondii* genomic DNA is 0.1 pg, which corresponds to one *T. gondii* organism (Cornelissen *et al.*, 1984).

The ability of the assay to detect *T. gondii* in the background of ovine tissue DNA has been assessed by PCR amplification of extracts from *T. gondii*-negative ovine tissues spiked with different amounts of RH strain *T. gondii* DNA. Thus, between 250 and 500 ng of host DNA has been found to be the best amount to avoid interference with the amplification of *T. gondii* and still detect as little as 1 pg of *T. gondii* DNA.

Procedure 2: nested PCR amplification of the B1 gene

Two pairs of oligonucleotide primers directed against the B1 gene of *T. gondii* (Burg *et al.*, 1989) are used to perform a nested PCR. Primary amplification is performed with primers 1 (5'GGAACTGCATCCGTTCATGAG-3') and 4 (5'-TCTTTAAAGCGTTCGTGGTC-3') to give a 193 bp product, and a secondary amplification using nested primers 2 (5'-TGCATAGGTTGC AGTCACTG-3') and 3 (5'-GGCGACCAATCTGCGAATACACC-3') is then carried out to give a 94 bp product.

1. PCR mixture contains the following:

- 10 mM Tris (pH 8.3)
- 2.5 mM $MgCl_2$
- 40 mM KCl
- 0.01% gelatine
- 0.1 mM dNTPs
- 0.2 μM of each primer
- 2.5 units of *Taq* polymerase.

2. Subject the mixture to the following cycling conditions: 94°C for 3 min, followed by treatment at 94°C for 1m, 50°C for 1m and 72°C for 1 min, and repeat this cycle for a total of 40 times.

3. One µl of each reaction is then used in a second reaction with the nested primers.
4. Subject the mixture to the same cycling conditions as above.
5. The PCR products are analysed by electrophoresis through 2% (w/v) agarose gels in TBE buffer.
6. The bands are visualized by staining with ethidium bromide and excitation under UV light on a transilluminator.

(NB: negative control samples from first-round amplification (omitted template DNA, which was substituted with sterile water) and an additional second-round negative control of sterile water are included in the nested reactions.

Procedure 3: real-time quantitative PCR amplification of the B1 gene

The B1 gene is widely used as a target in many real-time PCR detection methods for *Toxoplasma*. Here, we present the procedure described by Lin and colleagues (2000), which targets the same region of the B1 gene as Procedure 2 described above and claims a detection limit of 0.05 tachyzoite. A 98 bp fragment is amplified using the forward primer TOXO-F (5'-TCCCCTCTGCTGGCGAAAAGT-3'), the reverse primer TOXO-R (5'-AGCGTTCGTGGTCAACTATCGATTG-3') and a fluorescence-labelled hybridization probe (6FAM-TCTGTGCAACTTTGGTGTATTCGCAG-TAMRA) which, when cleaved, releases fluorescence proportionally to the amount of PCR product generated.

1. PCR mixture contains the following:
- 2 × PCR universal Master Mix (Applied Biosystems, Foster City, California, USA)
- 0.5µM of each primer
- 0.2µM of the TaqMan probe in a final volume of 50 µl.

2. Subject the mixture to the following cycling conditions using the GenAmp Sequence Detection System (Applied Biosystems): 95°C for 10 min, followed by 40 cycles of 95°C each for 15 s and 60°C for 1min.
3. Determine the cycle threshold value (Ct) at which the fluorescence of a given sample exceeds the baseline signal.

For a qualitative (presence/absence) interpretation of results, signals above the defined threshold are considered positive and those below are considered negative.

For a quantitative expression of the results, Ct values have to be plotted against the standard curve obtained from serial dilutions of known concentration. The standard curve is generated by plotting the log concentrations of known standards (serial dilutions) against their Ct values (NB: for quantification analysis, it is recommended to carry out experiments in duplicates or triplicates).

Serology

Gorka Aduriz, Stephen Maley, Ana L. Garcia-Pérez and Juana Pereira-Bueno

Introduction

There are several serological tests available for the detection of *T. gondii* antibodies in both humans and animals. In one type of test, the observer judges the given colour of tachyzoites under a microscope, such as with the Sabin-Feldman dye test (DT) and the indirect fluorescent antibody test (IFAT). Another depends upon the principle of agglutination of *Toxoplasma* tachyzoites, red blood cells (rbc) or latex particles, as in the modified agglutination test (MAT), the indirect haemagglutination test (IHAT) and the latex agglutination test (LAT), respectively.

With the enzyme linked immunosorbent assay – or ELISA – the degree of colour change defines the quantity of specific antibody in a given solution. The DT, IFAT, MAT and ELISA may all be used for diagnosing toxoplasmosis in sheep and goats and are outlined below with the IFAT and ELISA described in more detail.

THE DT TEST (SABIN AND FELDMAN, 1948) This the so-called 'gold standard' serological test for *Toxoplasma* antibody is not species-specific. Live *Toxoplasma* tachyzoites are incubated with an 'accessory factor' (complement contained in serum from an uninfected individual) and the test serum at 37°C for 1 h before methylene blue is added. Specific antibody induces membrane permeability in the parasite, so that the cytoplasm is able to leak out and the tachyzoite does not incorporate the dye and therefore appears colourless.

Tachyzoites not exposed to specific antibody (i.e. a negative serum sample) take up the dye and appear blue. The DT is both highly sensitive and specific but it is potentially hazardous, as live parasite is used. The method is expensive (the production of a sufficient number of live parasites), requires a degree of technical expertise and is now rarely used in clinical laboratories for the diagnosis of toxoplasmosis in animals.

THE IFAT TEST (MUNDAY AND CORBOULD, 1971) This a simple and widely used species-specific method. The test can be employed to assess specific anti-*Toxoplasma* IgG or IgM (the latter an indication of recent infection) in many species. Whole, formalinized (see below) *Toxoplasma* tachyzoites are incubated with diluted test serum and, subsequently, the appropriate fluorescent anti-species serum is added and the result then viewed with a fluorescence microscope.

Fluorescent-labelled antibodies (IgG- or IgM- conjugated with fluorescein isothiocyanate (FITC)) for a variety of animal species are commercially available (as are kits) and the method is relatively inexpensive. However, the test is time-consuming, with large numbers of sera, and requires a fluorescence microscope. The results involve subjective determination of the titre and endpoint and may therefore give rise to some variation. The IFAT results compare well with those of DT and conventional ELISAs.

THE MAT TEST (DESMONTS AND REMINGTON, 1980) This test uses whole *T. gondii* tachyzoites for the antigen and measures only agglutinating IgG antibodies. This test is both sensitive and specific and can be used with human or animal sera. Formalinized *Toxoplasma* tachyzoites (see below) are added to U-shaped-well microtitre plates and dilutions of test sera are then applied. Test serum is diluted in a buffer containing 2-mercaptoethanol (2ME) to inactivate IgM antibodies or, alternatively, the 2ME can be incorporated in the antigen (Dubey *et al.*, 1986). Positive samples will produce agglutination that can be graded, while negative samples will produce a 'button' of precipitated tachyzoites at the bottom of the well. Qualitative results with the MAT agree well with results of the DT and the IFAT.

The test has certain advantages: it is safe to use, is simple and easy to perform and it is not species-specific. While it requires relatively large amounts of antigen, kits are commercially available. The method, being simple to perform and read, can be used to diagnose toxoplasmosis both on an individual basis as well as on a large number of samples. The method of growth and harvesting of parasites is given below.

ENZYME LINKED IMMUNOSORBENT ASSAY (ELISA) This is probably the most widely used test for screening toxoplasmosis in sheep and goats. Like the IFAT, the ELISA is species-specific and can distinguish between immunoglobulin classes (IgG or IgM). The original ELISA (Voller *et al.*, 1976) uses a soluble antigen preparation, made from RH-strain *Toxoplasma* tachyzoites (as described below), layered into wells in a microtitre plate.

Ovine (or caprine) sera are added, followed by an anti-ovine-IgG (this antibody would be expected to recognize caprine IgG equally well) labelled with an enzyme such as horseradish peroxidase. In this way, any conjugate that binds to a well in turn induces a degree of colour change in an added substrate, which directly relates to the amount of bound antibody. The colour change can then be read with a spectrophotometer at the absorbance specific to the substrate used.

The assay is simple and easy to perform with the chosen anti-species conjugate, although it requires a spectrophotometer. Defined anti-species conjugates, substrates and whole kits are commercially available. This method is best automated and can readily test a large number of samples, and so the ELISA is very appropriate for laboratories that perform large-scale surveys. The reading of the results is objective and the test is both highly sensitive and specific (some false positives have been reported) when compared to the IFAT (used as a reference test). It is appropriate for routine serology and for epidemiological surveys of *Toxoplasma* infection.

Recently, a kinetics ELISA (KELA) as a diagnostic test for exposure of sheep to *T. gondii* was developed (Werre *et al.*, 2002). Although the principle of the KELA and the conventional ELISA is the same, the KELA system measures the rate of reaction between bound enzyme and the substrate solution that leads to development of colour. The protocol for this test includes reading three optical densities (OD) at 45 s intervals (using the KELA data management program) and the results are reported in terms of slopes. The correlation between the ELISA and the KELA is very high and, therefore, the two tests are very good diagnostic tools, differing only in their convenience of application.

To improve the specificity of the traditional *Toxoplasma* ELISA, other assays based on recombinant antigens (H4 and H11) (Johnson and Illana, 1991) and affinity-purified *Toxoplasma*-specific antigens (*T. gondii* P30 surface antigen – also called SAG1) (Lekutis *et al.*, 2001) have been developed for their potential use in sheep (Tenter *et al.*, 1992; Sager *et al.*, 2003). Compared with ELISAs based on tachyzoite crude antigens or soluble extracts, the sensitivity rates of the recombinant antigen ELISAs are lower, with a range of 34–79% and with H4 appearing to be a better antigen than H11.

However, these ELISAs are highly specific and may be a cost-effective alternative (the production of recombinant antigens is relatively inexpensive) when evaluating the presence of the infection in an area (Tenter *et al.*, 1992). In contrast to the recombinant antigen ELISAs, the *Toxoplasma* P30 ELISA has proved to be highly specific and sensitive (Sager *et al.*, 2003).

The detection of *Toxoplasma*-specific IgG and IgM antibodies is an imprecise way of attempting to discriminate between acute and chronic toxoplasmosis, and so has led to the development of avidity assays. These are based on the fact that as the immune response matures, after infection is established, so antibodies of increasing avidity (functional affinity) for the antigen develop. This avidity can be measured and used to indicate active or recent *T. gondii* infection.

The *Toxoplasma*-P30-ELISA has been adapted to detect avidity of IgG to P30 *T. gondii* in sheep (Sager *et al.*, 2003). The avidity ELISA involves analysing each serum in two fourfold serial dilutions (starting at 1:25) and washing, after the serum has been incubated with the antigen, and is performed with PBS containing 0.3% Tween 20 (PBS-T) and with PBS-T containing 6M urea for the two dilution series, respectively. Low-affinity antibodies are eluted with urea, and the antibodies measured with and without urea treatment are then used to calculate an IgG avidity value, expressed as the percentage avidity: (endpoint titre with urea/endpoint titre without urea) × 100 (Jenum *et al.*, 1997).

Avidities of up to 25% are considered to be low, those between 26% and 35% as intermediate and those above 35% as being high. This test works well as a diagnostic tool over a relatively short time after initial infection but, once avidity has matured and reached a high value (chronic stage), it is less useful.

IMMUNOBLOTTING (IB-TG) (WASTLING *ET AL.*, 1994) This is not considered to be a routine tool for screening in small ruminants, but is useful for verifying reactivity to *T. gondii* antigens (such as TgSAG-1, the specific tachyzoite antigen most obviously recognized by sera from infected animals) in sera that produce divergent results with other serological tests (e.g. IFAT and ELISA). The antigens from a whole extract of *T. gondii* tachyzoites are submitted to electrophoresis in 12% polyacrylamide gel (SDS-PAGE) and the proteins separated are transferred to membranes. After blocking, membranes are incubated with test serum and processed in an ELISA to detect bound antibody. The recognition of antigens by problematic sera is considered as a confirmation of infection.

General procedure

PREPARATION OF ANTISERA AND ANTIGENS Antisera to *T. gondii* and conjugated antisera for use in the IFAT and ELISA, to allow screening of a variety of animal species, may be obtained commercially. International standards for animal sera are not available.

Below are protocols for the preparation of tachyzoite antigen for use in the IFAT and ELISA. Tachyzoites may be grown in mice or in tissue culture and retained as whole parasite for the IFAT, or prepared as soluble antigen for the ELISA.

PRODUCTION OF TACHYZOITES IN MICE

1. Inject each of six mice intraperitoneally with 0.2 ml of a 1 × 10^7/ml suspension of *T. gondii* tachyzoites of the RH strain, using a 1 ml syringe and a 26 G needle, collected fresh from a previous mouse passage or from tissue culture.

2. Three days later, kill mice by CO_2 inhalation (avoid cervical dislocation as this may cause contamination of peritoneal fluid with blood).

3. Pin the mouse out on its back on a clean cork mat, or equivalent. Reflect the abdominal skin and, with aseptic precautions, ensuring that the abdominal wall is left intact, remove any peritoneal fluid by means of a 21 g needle attached to a 1 ml syringe and gently eject the harvested exudate into an equal volume of PBS. The optimum time to collect tachyzoites is 72 h after initial inoculation, when there will be sufficient organisms but before there is significant contamination by host cells. It is also important not to delay harvesting peritoneal fluid past 3 days for welfare reasons. However, if tachyzoites for mouse inoculation are taken from a frozen stabilate (see below), they may be less virulent and it may be necessary to harvest mice 4 or 5 d after initial inoculation and pass the parasite once more before using it as an antigen in the above tests.

4. Centrifuge the fluid at 500 *g* for 5 min, aspirate the supernatant and resuspend it in Hanks' balanced salt solution (HBSS). Alternate between PBS and HBSS washes by centrifugation.

5. Calculate the concentration of tachyzoites and contaminating host cells with an improved Neubauer counting chamber (count numbers of tachyzoites at 1:1000 dilution and cellular contamination at 1:10).

6. Carry out further washes (see above) as required to reduce cellular contamination to < 0.5% host mononuclear cells and < 0.25% for rbc.

7. Resuspend the tachyzoites in PBS to give a final concentration of 1×10^7/ ml.
8. Tachyzoites may be maintained by continual passage in this manner without the addition of penicillin/streptomycin by observing strict aseptic procedures.

PRODUCTION OF TACHYZOITES IN TISSUE CULTURE

1. *Toxoplasma gondii* can be grown and maintained in tissue culture by twice weekly passage in VERO cells.
2. Cells and the parasite are grown in Iscove's modified Dulbecco's medium (IMDM) (Gibco BRL, Paisley, UK) supplemented with 50 IU/ml penicillin, 50 μg/ml streptomycin and 2% fetal bovine serum.
3. Tachyzoites are harvested from tissue culture flasks by scraping off the cell monolayer using a sterile cell scraper.
4. Using 25 cm^2 vented tissue culture flasks that have each been seeded with 1×10^5 VERO cells, add tachyzoites at the rate of two tachyzoites per monolayer cell and incubate at 37°C in a 5% CO_2 humidified chamber. Harvest after 3–4 d.

TO PREPARE ALIQUOTS OF A FROZEN STABILATE OF TACHYZOITES

1. Produce tachyzoites in mice or tissue culture as described.
2. Centrifuge at 500 *g* for 5 min and resuspend in Iscove's modified Dulbecco's medium (IMDM) approximately three times.
3. Add the following solutions to give the following final concentrations:

- dimethyl sulphoxide (DMSO) (10%)
- normal horse serum (free of antibody to *T. gondii*) (20%)
- resuspended tachyzoites (70%) to give a final concentration of 1×10^8 tachyzoites/ml.

4. Allow the preparation to stand on the bench for 1 h.
5. Dispense 1 ml aliquots into screw-topped tubes appropriate for liquid nitrogen storage.
6. Put the tubes into a small container, wrap in thick insulating material and place in −70°C freezer to allow the tachyzoites to freeze gradually.
7. Next day, transfer to liquid nitrogen, keeping well insulated while transferring.
8. This stabilate may then be used for mouse inoculation or tissue culture growth of the parasite. When removing from storage, thaw the sample rapidly in warm water.

PREPARATION OF WHOLE TACHYZOITES FOR USE IN THE IFAT

1. Produce a suspension of 4×10^7/ml suspension of RH strain *T. gondii* tachyzoites in PBS.
2. Add formaldehyde (40%) to give a final concentration of 0.2% v/v.
3. Incubate at 4°C overnight and divide into aliquots in suitable sealed tubes and store at −20°C until required.

PRODUCTION OF SOLUBLE ANTIGEN FOR ELISA

1. Produce a suspension of RH strain *T. gondii* tachyzoites in PBS.

2. Centrifuge at 2000 *g* for 15 min, retain the pellet and resuspend it in nine times its volume of distilled water.
3. Rupture the tachyzoites by freezing and thawing three times.
4. The antigen preparation is then sonicated for 20 s at 4°C, at an amplitude of 20 microns.
5. Remove any cellular debris by centrifugation at 10,000 *g* for 30 min at 4°C.
6. Retain the supernatant and store at –20°C until required (a protein estimation might be expected to give a value of between 2 and 4 μg/ml).

Protocol for the IFAT

The following is a protocol for carrying out an IFAT, for anti-*Toxoplasma* IgG antibodies in sheep serum. It requires no modification for the testing of goat serum, as anti-sheep IgG will also bind to goat IgG. For measuring IgM antibodies specific to *T. gondii*, an anti-sheep specific IgM conjugated with FITC would be used.

1. Clean the required number of tissue culture 15-well multitest slides (ICN Biomedicals Inc, Aurora, Ohio, USA) and allow to dry.
2. Layer 5 μl of a whole tachyzoite preparation on to each well and allow to dry.
3. Fix in methanol for 10 min.
4. Wash twice in PBS, 0.3M, pH 7.4, each for 10 min.
5. Add 5μl of the given test sheep serum (diluted in PBS) to each well (prepare serial dilutions of the test sera, e.g. 1:16, 1:32, etc. up to 1:1024). Incubate for 30 min at room temperature. Ensure that, for each batch of sera tested, both a positive and a negative control serum are set up on a slide along with a 'PBS-only' sample (see Fig. 3.2.8).
6. Wash slides for 10 min, twice, in PBS.
7. Add 5μl of rabbit-anti-sheep IgG conjugated to FITC, diluted 1:500 in 0.2% filtered Evans blue dye in PBS, to each well and incubate for 30 min at room temperature.
8. Wash the slides three times for 10 min each in PBS.
9. Mount the slides under coverslips with buffered glycerol (nine parts PBS:one part glycerol) or Citifluor (Citifluor Ltd., London, UK).
10. Examine using a fluorescence microscope, fitted with × 20 and × 40 objective lenses.

With a negative test serum, the whole parasite will appear red due to the autofluorescence of the Evans blue dye. They may also present with a green fluorescent cap at the parasite pole (non-specific polar fluorescence). With a positive test serum, the parasites will fluoresce red and at least 80% of them within a given well will be surrounded by an unbroken band of green fluorescence.

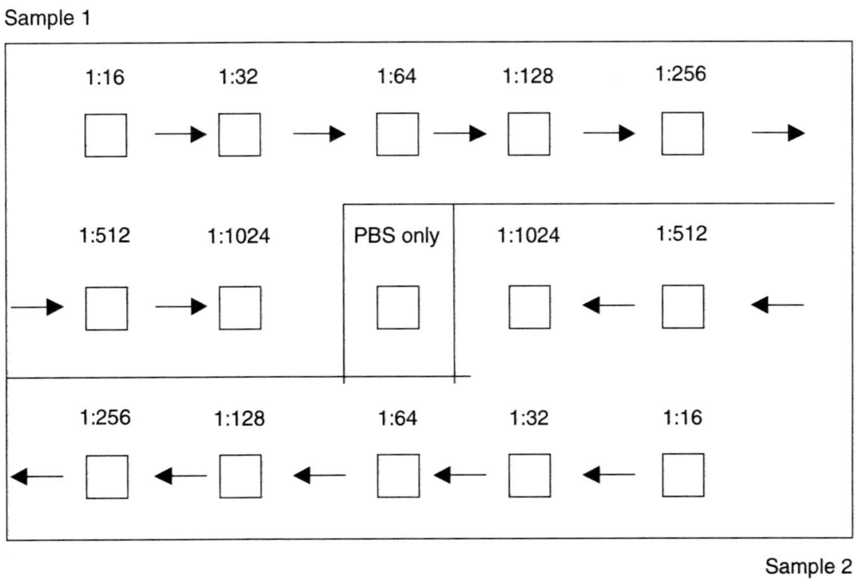

Fig. 3.2.8. Suggested layout of slides for carrying out the IFAT.

In an adult sheep/goat, a positive titre could be defined as ≥ 1:64 and a negative titre as < 1:64. For precolostral lamb/kid sera and fetal sera, a positive titre could be defined as ≥ 1:32.

Protocol for the ELISA

The following is a protocol for carrying out a conventional *Toxoplasma* ELISA based on soluble extract antigen (prepared as described above), for the detection of anti-*Toxoplasma* IgG antibodies in sheep serum. It requires no modification for testing goat serum, as anti-sheep IgG will also bind to goat IgG. For measuring IgM antibodies specific to *T. gondii*, use peroxidase-labelled anti-sheep IgM. The indicator system for detection of specifically bound antibody is conversion of a chromogen to a coloured product in the presence of a specific bound enzyme and substrate. The resulting colour development is read spectrophotometrically.

ELISA PROCEDURE

1. Add 100 μl/well of *T. gondii* soluble antigen preparation (diluted to around 1 μg protein per well) diluted in carbonate–bicarbonate buffer (pH 9.6) to each well of columns 2–12 (leaving Column 1 as a blank) of the ELISA plate (high-binding microtiter plate) and leave covered overnight at 4°C. To reduce non-specific binding 200 μl/well of PBS containing 0.05% Tween 20 (PBS-T) and 3% bovine serum albumin (BSA) is added to each well and incubated for 2 h at 37°C.
2. Wash the plates with three 5 min washes, with 200 μl/well of PBS-T.

3. Add 100 μl of serum (control or test sera) diluted 1:100 in PBS-T to each well (of columns 2 to 12) and incubate at 37°C for 1 h. Ensure that wells without test serum are included ('no test serum').
4. Wash three times as in '2' above.
5. Add 100 μl of conjugate e.g. rabbit anti-sheep IgG peroxidase conjugate (Sigma Chemical Co, Poole, Dorset, UK) diluted appropriately, e.g. 1:2000, in PBS-T to each well (of Columns 2–12) and then incubate for 1 h at 37°C.
6. Wash three times as in '2' above.
7. Add to each well (Columns 1–12) 100 μl of substrate solution consisting of chromogen and substrate (9 ml of distilled water, 1 ml of 1 M sodium acetate, 0.1 ml of tetramethyl benzidine (TMB) in dimethylsulphoxide (DMSO) and 20 μl of 30% H_2O_2). Incubate at room temperature in the dark for 20 min.
8. Stop the reaction 20 min after adding the substrate by the addition of 100 μl of 1M H_2SO_4 (stopping solution) to each well.
9. Read the resulting colour development (OD) in an automatic ELISA plate reader at a wavelength of 450 nm (read at a wavelength of 492 nm if orthophenylenediamine (OPD) is the chromogen in the substrate solution).

N.B. each plate should contain known positive and negative control sera and a set of wells incubated without test serum. All samples are in duplicate and the mean of the two absorbance values calculated.

EXPRESSION OF RESULTS ELISA test results can be expressed in various ways; however, for the diagnosis of ovine and caprine toxoplasmosis, they are more frequently expressed as optical density (OD) values.

Optical densities

Test sera with an absorbance value (OD) exceeding the mean absorbance + three standard deviations (SD) of sheep sera, negative in conventional serological tests (e.g. the IFAT), are considered to be positive (Lundén *et al.*, 1992). The cut-off is determined as the mean absorbance value of negative controls plus three standard deviations (Lundén *et al.*, 1992). Currently, the threshold value for a positive result is established according to this criterion when positive control sera of known titre are not available.

International units (IU)

The OD observed on each test serum is expressed in IU. Titration in IU is carried out by comparison with a standard curve created from serial dilutions of a positive control serum of defined titre, following the protocol described by Calamel and Lambert (1988).

Based on our experience with the assay, a fetal serum with a titre ≥ 50 IU is regarded as positive, and in adult sheep an antibody value ≥ 200 IU is considered to indicate a specific response to infection with *T. gondii*.

Percentage (% OD)

Antibody is expressed as a percentage of a standard, high-titre positive serum (sample of *Toxoplasma* serum-positive which consistently produced an OD in the region of 1.0) with the equation:

[(Test serum OD – no test serum OD)/ (Standard positive serum OD – no test serum OD)] × 100

In ewes, OD values of over 25% are regarded as specific for *T. gondii*, and with lambs an OD of 15% or more is considered to be positive (Buxton *et al.*, 1988).

Oocyst detection

Smaragda Sotiraki and Steve Wright

A variety of diagnostic techniques are available for the identification of coccidian oocysts in faeces, usually involving either sedimentation or flotation.

Flotation

The following method of saturated salt flotation has been used to concentrate *Toxoplasma* oocysts from cat faeces over a number of years by one of the authors (SW).

EQUIPMENT

- Balance
- Disposable faeces spoon
- Tea strainer (or sieve of approximately 1 mm^2 mesh)
- 250 ml beaker
- 30 ml plastic 'universal' tube (or conical centrifuge tube)
- Centrifuge
- Pasteur pipette and rubber teat
- McMaster slide
- Saturated salt solution – add salt in excess to water, stirring several times daily for 3 d; the specific gravity should be 1.2.

PROCEDURE

1. Weigh approximately 1 g of cat faeces into a 30 ml 'universal' tube.

2. Add 10 ml tap water to the tube, cap and vortex vigorously to produce an emulsion of faeces. If necessary, the faeces may be mashed up using a spatula to aid dispersion.

3. Strain faecal suspension through a tea strainer into the beaker – this removes larger debris, particularly hair, which may impede the method.
4. Decant suspension into the 30 ml tube, rinse beaker with 10 ml tap water and add to the tube.
5. Centrifuge at 1000 *g* for 10 min.
6. Discard supernatant and re-suspend pellet in 10 ml saturated salt solution.
7. Centrifuge at 1000 *g* for 10 min.
8. Remove the meniscus (approximately 1 ml) using a Pasteur pipette and decant into a McMaster slide.
9. Examine by light microscopy at a magnification of × 200–400.

(Alternatively, a few drop of the meniscus can be placed on a microscope slide, coverslipped and examined as above. Oocysts will be more easily seen using this method – the illumination is much better than when using a McMaster slide) (see Fig. 3.2.9a).

The alternative to Sheather's sugar flotation method (Colville, 1991; Dijkstra *et al.*, 2001; Gondim *et al.*, 2002) is essentially the same as that given above except that sugar solution is used rather than a saturated salt solution:

1. Dissolve 454 g of table sugar in 355 ml of warm water.
2. Stir and allow cooling.
3. You may add 6 ml of formalin (or 6.5 g of phenol) as a preservative.
4. Store refrigerated.
5. Proceed as above and examine by light microscopy.

Oocysts are shed unsporulated, and require several days outwith the host at moderate ambient temperatures to develop and become infectious. Unsporulated

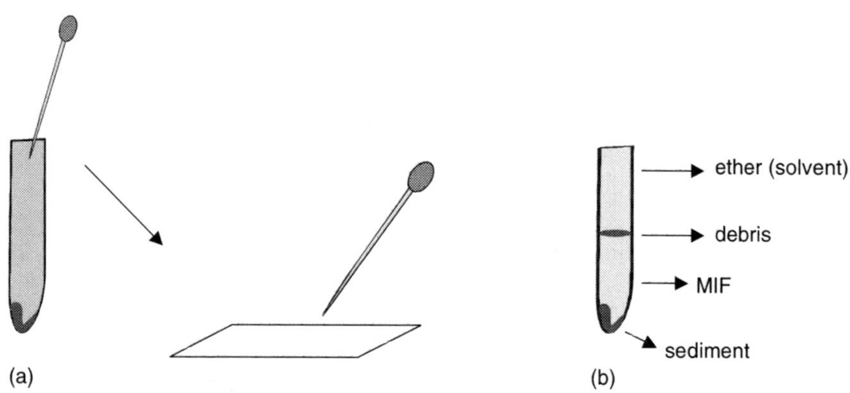

Fig. 3.2.9. (a) Examination of cat faeces for *Toxoplasma gondii* oocysts. After flotation of oocysts transfer a sample of the meniscus to a McMaster slide and examine under a light microscope; (b) sedimentation (MIF) technique for the detection of *Toxoplasma gondii* oocysts. The figure illustrates the four layers after centrifugation.

oocysts are subspherical to spherical and measure approximately 10 ×12 μm. Sporulated oocysts are subspherical to ellipsoid, with dimensions of 11 × 13 μm, containing two ellipsoidal sporocysts, each of which contains four sporozoites (Dubey and Beattie, 1988).

Oocyst shedding usually commences 5–7 days after the consumption of tissue cysts, and may continue for several weeks, though the majority of oocysts will be shed during the first week – this is the best time to harvest oocysts from donor animals. An oral dose of 100 mature tissue cysts to a cat seronegative for *T. gondii* has been found to give reliable, reproducible infection.

Infected cats may become mildly febrile during the early stages of infection, and may experience either mild diarrhoea or become constipated – if the latter, laxatives should be administered. Oocysts recovered from faeces delayed in the gut by constipation seldom sporulate.

Quantitative isolation of Toxoplasma oocysts

If large quantities of oocysts are required for research purposes, the diagnostic technique using saturated salt solution can be scaled up. Faeces and water, in a ratio of 1:10, are homogenized together in a laboratory blender to produce a faecal suspension. Cat hair and particulates are strained out using a 1 mm mesh sieve, stirring the contents of the sieve with a wooden spatula (faeces spoon) to facilitate flow.

1. Place 50 ml of strained faecal suspension in a 250 ml centrifuge bottle, and centrifuge at 1000 *g* for 20 min.
2. Discard supernatant and re-suspend the pellet in 100 ml saturated salt solution, then centrifuge at 1000 *g* for 20 min.
3. Aspirate the meniscus using a pipette (or a vacuum line if possible), place in a 250 ml centrifuge bottle and dilute at least 1:5 with tap water, before centrifuging at 1000 *g* for 20 min.
4. Discard the supernatant and re-suspend the pellet in a small volume of water. If substantial debris is still present, the oocysts can be further cleaned by re-suspending them in 50% tap water/50% saturated salt and centrifuging at 1000 *g* for 20 min.
5. Aspirate the meniscus and dilute at least 1:5 with tap water. Centrifuge at 1000 *g* for 20 min.
6. Discard the supernatant, re-suspend the pellet in 2% sulphuric acid and leave at room temperature for several days to allow sporulation of the oocysts.

It should be noted that the use of 50% saturated salt will incur a reduction of oocyst yield, but will considerably improve the purity of the oocyst preparation.

Other concentration techniques, such as Sheather's sucrose flotation and formalin/ether extraction, could be employed, but saturated zinc sulphate should be avoided – the fat content of cat faeces mitigates against its usefulness. Formalin/ether or formalin/ethyl acetate (see MIF sedimentation below) are useful techniques for defatting faeces, however, if maximal oocyst recovery is required; either one is best used on a dilute faecal suspension, as the formation of a plug of debris at the formalin/ether interface can enmesh significant numbers of oocysts.

Sedimentation technique (MIF)

For further details on this procedure, see Sapero and Lawless (1953); Melvin and Brooke (1974); Truant *et al.* (1981).

EQUIPMENT

- Balance
- Small beaker or plastic container
- Measuring cylinder (or pipette)
- Stirring device (wooden stick)
- Tea strainer
- Centrifuge tube (glass or polypropylene, *not* polystyrene)
- Tube rack
- Centrifuge
- MIF solution (thimerosal 0.02%, formaldehyde 2.3%, glycerol 1.0%)
- Diethyl ether
- Lugol's iodine solution
- Microscope slides and coverslips
- Microscope with × 100–400 magnification.

NB: ethyl acetate may be used as a less explosive alternative to ether.

PROCEDURE

1. Weigh out 1.0 g faeces in a beaker.
2. Add 5 ml of MIF solution (or alternatively 5% acetic acid solution) and stir thoroughly to disperse the faeces.
3. Filter faecal suspension through a tea strainer and decant filtrate into a test tube. Add an equal volume of ether. Shake the mixture vigorously (preferably using a vortex mixer) and centrifuge for 2 min at 500 *g*.
4. This results in four layers: solvent, a plug of debris, formalin (MIF) and sediment. Decant the top three layers after loosening the plug of debris with a wooden applicator.
5. A few drops of the sediment are transferred to a slide, coverslipped and examined under the microscope. Staining of the sediment with a few drops of lugol's iodine solution may improve detection of oocysts (see Fig. 3.2.9b).

NB: there are ready-to-use kits available (Meridian Pra-Pak MIF Stool Collection Kit, Meridian Diagnostics Inc., Cincinnati, USA; Biorepair GmbH, Sinsheim, Germany).

DISADVANTAGES Toxicity of the chemicals used.

Autofluorescence

Autofluorescence describes the ability of some objects to emit light after ultraviolet excitation. It has been observed that some protozoa exhibit this phenomenon: *Cyclospora* (Berlin *et al.*, 1994, 1998; Dytrych and Cooke, 1995) *Cryptosporidium* and *Isospora* (Varea *et al.*, 1998; Daugschies *et al.*, 2001) and, recently, *Toxoplasma* (Lindquist *et al.*, 2003). Intense blue fluorescence of the oocyst wall has been described, using a fluorescence microscope with excitation filter 330–380 nm, barrier filter 420 nm, at × 100–400 magnification.

Sporonts and sporocyst walls may also fluoresce, permitting the observation of the oocyst contents. However, none of the authors quoted actually used autofluorescence for detection of parasites in faecal smears – all used concentrated/cleaned faecal preparations, though some authors claimed the intense fluorescence would enable oocysts to be easily distinguished from 'complex microscopic sample backgrounds'. Mountants should be compatible with fluorescence – for example, buffered glycerol.

Diagnostic Applications and Recommended Diagnostic Schemes

Sheep and goats

David Buxton and Gorka Aduriz

Fetal diagnosis

CLINICAL SIGNS Abortion is the chief clinical manifestation of toxoplasmosis in adult sheep and goats and is the result of naïve animals being exposed to infection in the middle part of pregnancy. Exposure to a primary infection early in pregnancy will result in death of the fetus and its resorption, with the ewe or doe perhaps being seen to return to service (see opening part of section on Toxoplasmosis, above). This may be incorrectly thought to be an infertility problem, possibly involving the ram/billy goat. Male sheep and goats would not be expected to develop clinical signs of toxoplasmosis.

Disseminated acute toxoplasmosis has only been recorded in other species that, for evolutionary reasons, are naturally ill-equipped to cope with the parasite (e.g. marsupials, New World monkeys) or are suffering significant immunosuppression due to other causes (Innes, 1997). In newly born lambs and kids, clinical symptoms of toxoplasmosis are those of a small, weak flaccid animal unable to stand or suck. Otherwise, they appear to have developed normally. They would not normally be expected to display fits or tetanic spasms. The weakness appears to relate to the focal leukomalacia that is generally located in the cerebral white matter cores and, as a result, affected lambs/kids will often die within 48 h despite rigorous nursing support (Buxton, 1998).

DETECTION OF LESIONS AND PARASITES IN FETAL TISSUES

Samples

- Placenta: at post-mortem examination, wash the placental membranes under a gentle stream of tap water and carefully inspect the cotyledons for evidence of the characteristic small white foci of necrosis. Examine also the intercotyledonary membranes. In cases of toxoplasmosis, these should appear normal, being thin and moderately translucent, with the normal vasculature visible. Mummified membranes have an opaque, leathery-brown appearance. Any visible inflammation of the intercotyledonary membranes with thickening and a red or creamy coloured appearance is not characteristic of toxoplasmosis and suggests an alternative (or additional) cause, such as *Chlamydophila abortus*, *Brucella melitensis* or *Coxiella burnetii* (see Chapter 4, this volume). Place selected samples of at least two placental cotyledons, with a fringe of intercotyledonary membranes, into 10% formal saline for processing to histopathology.
- Fetus: collect serum for serological testing. In a fresh carcase, the cardiac blood clot may be withdrawn into a syringe, the clot allowed to retract and the serum separated by centrifugation. In many cases, fetal fluid present in the thoracic or abdominal cavities can be withdrawn by syringe and centrifuged. Supernatant/serum should either be tested immediately for the presence of antibody or it should be stored at –20°C for serological testing at a later time. Remove the fetal brain and sample aseptically for examination by PCR and for attempted isolation of *T. gondii*, if appropriate, and place the remainder of the brain in 10% formal saline along with samples of liver, lung and heart for processing for histopathology. Severe autolysis can be a problem, but histological lesions in the liver and brainstem may still be visible in such cases and the heart is relatively resistant to autolysis.

Histopathological examination

- Placenta: focal necrosis in the placental cotyledons, with or without a degree of lymphoid inflammation, suggests a diagnosis of toxoplasmosis. Although rare, *N. caninum* can infect pregnant sheep and goats and cause similar lesions. As noted above, the intercotyledonary membranes would be expected to look normal. While the presence of inflammatory lesions in the latter suggests another aetiology, care must be taken before one or other cause is ruled out, as it is possible to see mixed infections in some cases.
- Brain: in the brain, in toxoplasmosis, focal leukomalacia is the most commonly encountered lesion. However, as this is considered to represent anoxic damage, it is secondary to placental pathology and, while most commonly associated with toxoplasmosis, it may be triggered by the action

of other pathogens. Less commonly, but more specific to ovine and caprine toxoplasmosis, one observes focal microgliosis with or without central necrosis of the neurophil. These lesions are scattered throughout the brain, with a distribution suggesting blood-borne spread of the parasite. While highly suggestive of toxoplasmosis, they may also occur in ovine/caprine neosporosis.

- Other tissues: lesions may also be seen in lung, liver and heart.

DETECTION OF THE PARASITE IN FETAL TISSUES

Immunohistochemistry, PCR and isolation

- Immunohistochemistry: while histology allows a 'probable' diagnosis, to be more certain additional specific evidence of *T. gondii* is required, such as the immunohistochemical demonstration of the parasite in tissue sections. However, both bradyzoites in tissue cysts and tachyzoites can be very scarce in placenta, brain and other tissues.
- PCR: the demonstration of parasite DNA in tissues by PCR, in conjunction with lesions, permits a certain diagnosis but a positive PCR test alone only confirms the presence of parasite DNA – it does not confirm viable parasite, lesions or disease.
- Isolation: isolation of *Toxoplasma* by inoculation of mice is a relatively straightforward method for confirming a diagnosis of toxoplasmosis but, with the other tests available, its field use is likely to be limited to occasions when an unequivocal diagnosis is required in a new or unusual situation. The most reliable method of isolation of *T. gondii* from aborted fetuses and fetal membranes is the inoculation of laboratory mice. The best tissues are fetal brain and placental cotyledons, and optimum results are obtained with fresh samples free from contamination. As a rule, fetal brain is usually less contaminated than placental cotyledons and is easier to homogenize and inject into mice.

The following method presumes fetal brain is being used. Do not freeze samples at any stage, as this kills the parasite.

1. With aseptic precautions, remove 2–5 g of fetal brain from the aborted fetus.
2. Homogenize the tissue in an equal volume of sterile PBS, 0.3 M pH 7.4, with added antibiotics (100 IU/ml penicillin and 745 IU/ml streptomycin) and homogenize by passing it through a 16 g needle ten times by means of a syringe (macerate placental cotyledon in a 'stomacher' (Seward Ltd, London) or other suitable macerating equipment).
3. Inoculate each of three *Toxoplasma*-free mice intraperitoneally (ip) with 0.5 ml of the homogenate (isolation by inoculation of Vero cells in tissue culture may also be attempted, although success will usually be greater with mouse inoculation when samples have a degree of bacterial contamination).

4. Six to 8 weeks after inoculation, kill the mice and remove brains. Blood should also be recovered from the mice at this stage and the serum separated and stored at –20°C.
5. Homogenize each mouse brain with an equal volume of sterile PBS by passing through a 16 g needle ten times by means of a syringe.
6. Spread one drop (5 μl) of a given suspension on each of five slides.
7. Dry and stain with Giemsa, dehydrate and mount under a coverslip.
8. Examine slides under a microscope. Tissue cysts appear as circular structures measuring 5–50 μm filled with blue-staining, crescent-shaped bradyzoites.

An alternative method for examining mouse brain is to take a portion of brain approximately the size of a match head, place it on a slide and gently squash it flat with a coverslip. Unstained tissue cysts will be visible as circular (spherical) bodies full of refractile crescentic bradyzoites (see Fig. 3.2.3). Failure to demonstrate cysts does not rule out a positive diagnosis. Serum from the mice may be analysed for the presence of antibodies to *Toxoplasma* (e.g. using an IFAT) and, if this analysis is also negative, infection with *Toxoplasma* can be ruled out. The positive isolation of *T. gondii* indicates a viable infection, but without supporting evidence of pathological lesions does not, on its own, indicate disease.

DETECTION OF SPECIFIC ANTIBODIES AGAINST *T. GONDII* IN FETAL FLUIDS The detection of specific antibodies to *T. gondii* indicates fetal exposure to the parasite during pregnancy. As the fetal immune system is only sufficiently mature to respond to 'foreign' antigens by mid-gestation (Nettleton, 2000), the presence of detectable fetal antibody infers infection after this time. In a *Toxoplasma* infection of the fetus in the first half of pregnancy, there will be little inhibition of parasite multiplication, so that fetal death with resorption or mummification is the norm.

If the fetus is sufficiently mature, it will initially produce IgM followed by an IgG response (Buxton and Finlayson, 1986). Ideally, diagnostic serology would seek to detect both separately but economics often preclude this. Testing for one or the other will mean that some cases are missed; however, a test (such as the IFAT) set up to detect Ig would be expected to detect both IgM and IgG and will provide the best chance of establishing a fetal humoral response. The threshold for establishing a positive titre can be set lower in the fetus than in the adult. Hence, an IFAT titre of 1/32 or an ELISA %OD of 15 can be taken as positive (compared with 1/64 and 25% in the adult).

KEY POINTS

- Collect placenta and fetus for examination.
- Collect fetal fluid for serology.
- Sample placenta and brain for histopathological examination.
- Use of the PCR will confirm the presence of *T. gondii* DNA but on its own is insufficient to allow a positive diagnosis.
- Freezing tissues will kill the parasite and preclude isolation.

- Isolation of *T. gondii* also indicates only infection rather than disease.
- Be aware that more than one agent can cause abortion in either an individual sheep/goat or in a flock/herd.

Post-natal detection of infection

Gorka Aduriz and David Buxton

INTRODUCTION Although the majority of *Toxoplasma* infections in sheep and goats are subclinical, in an important minority clinical toxoplasmosis presents as abortion and neonatal mortality and follows primary infection in pregnancy. The ewe or the doe (nanny goat) would not normally be expected to show any signs of illness during the infection.

CLINICAL SIGNS As already stated above, in typical cases of ovine and caprine toxoplasmosis the mother produces a stillborn lamb/kid that may be accompanied by a weakly sibling or a 'mummified' fetus. This usually occurs a few days earlier than the predicted end of pregnancy (Buxton, 2000), but otherwise the ewe/doe remains clinically normal. In placental cotyledons, white foci around 2–3 mm in diameter are usually visible, while the intercotyledonary membranes appear normal.

Infection in early pregnancy, when the fetus has only a rudimentary immune system, results in fetal death and resorption. In this case, the mother may present as barren. When a significant number of animals are so affected it may appear to be a flock/herd infertility problem so that, for example, the ram may be wrongly put under suspicion in an affected flock of sheep. Mothers that become infected in late pregnancy would be expected to remain clinically normal and to produce clinically normal, but infected, offspring.

ADULT SHEEP/GOAT Different techniques have been used to detect *T. gondii* infection in the post-natal period. In live animals, serology is the ideal diagnostic tool for the detection of exposure to the parasite. Following a primary infection with *T. gondii*, specific antibody titres rise in the next 2–3 weeks and may remain elevated for several years (Blewett *et al.*, 1983). In the initial stages of *T. gondii* infection, IgM is more readily detected during the first month. This is followed by a switch to IgG production during the second week of infection, becoming the predominant immunoglobulin during the second month of infection (Trees *et al.*, 1989).

Thus, the presence of antibodies is an indication of exposure to infection, and the presence of high antibody titres in adult animals is suggestive of recent exposure to *T. gondii*, although it is not necessarily diagnostic of recent infection because titres may remain high for several months.

As *Toxoplasma*-infected animals are generally asymptomatic, many serological methods have been evaluated over the years, both in terms of sensitivity and specificity, for use in serological surveys. Serology has also been used to assess the prevalence of infection in meat-producing animals and it has been suggested as an appropriate tool for the certification of meat as being free of *Toxoplasma* (van Knapen *et al.*, 1982; Uggla and Hjort, 1984).

The routinely available serological tests have been described above. The DT remains the reference test for the presence of *Toxoplasma* antibodies in serum and, while it is well suited for use in sheep and goats (Uggla and Buxton, 1990), it is time-consuming and involves the use of live organisms, and is therefore not used as a routine diagnostic test for animal sera.

The agglutination tests – IHAT, latex agglutination test LAT and MAT– do not require species-specific antisera or conjugates and have become popular for serodiagnosis of ovine toxoplasmosis. The IFAT correlates well with the DT and, in addition, it is easier to perform, less hazardous to the operator and is well suited and therefore extensively used for the serological diagnosis of both ovine and caprine toxoplasmosis (Uggla and Buxton, 1990). For serological surveys, the ELISA is highly sensitive and specific (O'Donoghue *et al.*, 1987; Van der Puije *et al.*, 2000; Figueiredo *et al.*, 2001) and suitable for the analysis of large numbers of samples.

Seroconversion in an animal (a fourfold rise in titre) is indicative of a recently contracted active infection. However, unless a test specifically designed to detect acute infection is used, such as one that detects IgM to the parasite or an 'avidity ELISA', a single positive serum sample indicates only infection of the host some time in the past, although as a general rule an IFAT titre of 1/1000 or greater suggests a recent infection (Buxton, 2000). However, a high maternal IFAT antibody titre can not be used on its own to indicate that a given abortion was due to *T. gondii,* as a high serum titre may occur in normally lambing ewes when primary infection has occurred later in gestation. If specific antibodies to the parasite are not found, *Toxoplasma* may be ruled out as the cause of abortion.

NEWBORN LAMB/KID The finding of significant *Toxoplasma* antibodies in the serum of lambs and kids that have not sucked is a good indication of congenital (intrauterine, trans-placental) infection, because maternal antibody cannot cross the materno–fetal placental barrier, even when severely damaged. With postcolostral samples, it is necessary to demonstrate IgM antibody against *Toxoplasma,* as IgG antibody could represent absorbed maternal colostral antibody (Buxton, 2000). Passively acquired *T. gondii* antibodies from the colostrum have usually disappeared from the lambs' (kids') circulation by 3 months of age. Thus, the detection of an appreciable amount of specific antibody in an older animal (weaned lamb or kid) must be regarded as being the result of an actively acquired *T. gondii* infection (Dubey *et al.*, 1987).

Detection of *T. gondii* in tissues from lambs/kids

Various techniques are available for the detection of the parasite within the tissues of lambs and kids. Histopathology may allow the detection of lesions suggesting a *Toxoplasma* infection and, in some cases, immunohistochemistry will reveal the parasite, but the results depend on the presence of adequate lesions and sufficient numbers of the parasites (Esteban-Redondo and Innes, 1998).

When organisms resembling toxoplasms are seen, it is important to be aware that there are morphological similarities between *T. gondii*, *Sarcocystis* spp. and *N. caninum*. More specific tests may therefore be required to confirm the presence of *T. gondii*, such as immunohistochemistry, PCR or the inoculation of susceptible laboratory animals. Immunohistochemical techniques allow the visualization of the parasite and the antigenic residues in tissue sections, and have the advantage of being able to detect *Toxoplasma* antigen, even in decomposed tissues, as well as distinguishing it from other parasites (see above; Jeffrey *et al.*, 1988).

Isolation of *Toxoplasma* is usually carried out by inoculation of samples (placental, fetal tissues, etc.) into mice. Depending on the virulence of the strain, mice may develop either an acute infection, with parasite-rich ascitic fluid, or a chronic infection characterized by the presence of *T. gondii* tissue cysts in the brain (Derouin *et al.*, 1987). This is the most direct and established method to demonstrate *Toxoplasma* infection, but it requires the use of experimental animals and is also slow and expensive.

Use of the PCR technique is preferable to the use of live animals and it is also very sensitive. It is, however, still relatively costly and is not always easily available for routine purposes. The appropriate PCR assays for targeting the different genes have been described above.

Cat

Bertrand Losson

Introduction

Felids, both domestic and feral, are the only known definitive hosts for *T. gondii* and are therefore the main source of infection. Three primary routes of transmission are known: congenital, carnivorism and faecal–oral. All are present in the cat, although their relative importance differs. Ingestion through carnivorism is the most common and most efficient source of infection in the cat.

After a primary ingestion of tissue cysts, the released bradyzoites invade the epithelial cells of the small intestine to perform several shizogonies and a sexual phase, which leads to the emission of unsporulated oocysts. In this case, practically 100% of cats pass very high numbers of oocysts, which can generally be detected from 3 days post-infection. Oocyst shedding is normally completed by days 10–21 post-infection (Dubey, 1986b; Lappin, 1999).

Life cycle

During a primary infection, most bradyzoites enter the gut epithelial cells of the cat. However, some bradyzoites penetrate the *lamina propria* and multiply as tachyzoites. Tachyzoites are then transported via lymph and blood and multiply in many cell types. During this phase, tachyzoites can also cross the placenta and be responsible for congenital toxoplasmosis in kittens (Dubey and Carpenter, 1993a).

With the onset of an immune response, the parasite replication is eventually controlled and tissue cysts are formed in different organs (brain, muscles, heart and other viscera). Tissue cysts remain viable for a very long time, if not for life. It is postulated that they rupture from time to time but that the released bradyzoites are rapidly destroyed by host immune cells. In immunocompromised cats (due, for instance, to infection with feline immunodeficiency virus, FIV), *T. gondii* may renew multiplication and this may lead to excretion of oocysts for a second time and to clinical toxoplasmosis.

As cats are not coprophagous, the oral ingestion of oocysts is considered to be a minor route of infection. Furthermore, this mode of infection is inefficient as, under laboratory conditions, less than 50% of cats that are fed oocysts go on to shed oocysts (Dubey and Frenkel, 1976; Lappin, 1999). The prepatent period is also much longer (18–44 d) and the excretion of oocysts lasts for a maximum of 10 d (Dubey and Lappin, 1998).

The ante-mortem diagnosis of toxoplasmosis in the cat can rely on clinical considerations and on the results obtained from different diagnostic tests.

Clinical signs

Clinical toxoplasmosis in cats may be related to the intestinal phase of replication of the parasite, leading to oocyst shedding or to the multiplication of tachyzoites in extra-intestinal tissues.

Although cats can develop self-limiting diarrhoea after experimental infection, the intestinal phase of feline toxoplasmosis is considered to be relatively unimportant clinically (Lappin, 1999). In fact, most of the time, cats passing very high numbers of *T. gondii* oocysts in their stools are asymptomatic.

Extra-intestinal toxoplasmosis, a rare disease, can lead to either: (i) a severe acute condition due to the overwhelming replication of tachyzoites in different tissues; or (ii) a more chronic sublethal form of the disease.

Kittens acquiring toxoplasmosis trans-placentally or via the mother's milk may develop severe illness (Dubey and Carpenter, 1993a), with anorexia, lethargy, hypothermia, dyspnoea and sudden death all common observations.

Adult cats may show a variety of clinical signs. In one study, the main clinical findings in 100 clinically affected cats included fever, hyperpnoea, dyspnoea, anorexia, lethargy, icterus, CNS signs and signs of ocular inflammation. In a majority of animals, toxoplasmosis was considered to be generalized, whereas in some the lungs, the abdominal viscera – including the liver – were the main targets (Dubey and Carpenter, 1993b).

In the absence of treatment, acute toxoplasmosis is usually fatal. A more chronic sublethal course has been observed in some animals. In this case, the primary clinical sign is uveitis (anterior and posterior), with fever, hyperaesthesia, anorexia, weight loss and ataxia also observed (Lappin *et al.*, 1989).

Laboratory diagnosis of toxoplasmosis in the cat is difficult, although many different tests may be used to assist the clinician. Haematology, biochemistry and urinalysis are often unrewarding, as the findings are not pathognomonic for toxoplasmosis. However, non-regenerative anaemia, lymphocytosis, monocytosis, neutropenia, eosinophilia, hyperbilirubinaemia, hyperproteinaemia, proteinuria, bilirubinuria and elevated transaminase values (creatine kinase, alanine aminotransferase, alkaline phosphatase) are commonly observed.

Thoracic radiographs may reveal a diffuse interstitial and alveolar pneumonia and compensatory emphysema. Abdominal radiographic findings are non-specific, but may include a homogeneous increase in density (Rusbridge, 2001).

The demonstration of tachyzoites in an aspirate from an animal with appropriate clinical signs confirms a diagnosis of toxoplasmosis. The tachyzoites are readily stained by the Giemsa method or an equivalent alternative.

Faecal examination aims at demonstrating *T. gondii* oocysts in feline faeces following flotation, using solutions with high specific gravity (see section above). The detection of oocysts in the faeces may also be possible by autofluorescence, as recently described by Lindquist *et al.* (2003) (see above).

T. gondii oocysts (12×10 μm) must be differentiated from other feline non-pathogenic coccidia such as *Hammondia hammondi*. Microscopically this is not possible, and bioassays in mice must be used (Dubey *et al.*, 1977). The PCR, tissue culture and animal inoculation can be used to detect the organism in blood, aqueous humour or cerebrospinal fluid (CSF). Biopsies can be examined through histology, immunohistochemistry, the PCR, cell culture or animal inoculation. All of these techniques have been described in preceding sections.

Taking into account the short patent period following infection with tissue cysts, most cats with clinical toxoplasmosis have completed the period of oocyst shedding, and therefore the chances of detection of *T. gondii* oocysts by coproscopy are slim. Cats that have excreted oocysts usually do not shed oocysts again, and even when they do so the number of oocysts is negligible (Dubey, 1986b).

Serology is a good diagnostic aid, and many tests have been developed to assess the antibody and/or immune status of a particular animal. Specific antibody detection can be achieved using the ELISA, IFAT, DT, agglutination test or Western blot immunoassay (all described above). Some tests may be adapted to detect IgM, IgG and IgA antibody responses.

Cats usually do not develop antibodies during the period of oocyst shedding (Dubey and Frenkel, 1972), and consequently serology of an individual animal does not guide the clinician as to whether a particular cat might be shedding oocysts. However, a serologically positive cat has probably shed oocysts in the past. Tests based on antigen and immune complex detection are not commercially available and will not be discussed further.

IgM responses develop within 2–4 weeks after oral infection (by tissue cysts), with titres in most cats returning to baseline levels by 16 weeks. IgG titres appear later and can remain elevated for very long periods (probably for life) (Dubey *et al.*, 1995). It is relevant that Lappin *et al.* (1989) concluded that IgM antibodies appear to correlate better with clinical disease than IgG after they observed that they could detect IgM titres in the greater majority of cats with clinical toxoplasmosis, but could demonstrate IgG titres in only 60% of individuals. However, it is relevant that an increasing IgG titre may indicate recent or active disease (this necessitates the collection of two blood samples 2–3 weeks apart).

In ocular toxoplasmosis, the detection of specific IgM antibody in the aqueous humour correlates well with clinical toxoplasmosis (Dubey and Lappin, 1998). Recently, a PCR (see above) was used for the detection of *T. gondii* within the aqueous humour (Lappin *et al.*, 1996; Burney *et al.*, 1998). However, *T. gondii* DNA may also be detected occasionally in the aqueous humour of cats without uveitis (Burney *et al.*, 1998).

In conclusion, the ante-mortem diagnosis of feline clinical toxoplasmosis is difficult. Post-mortem, the presence of tachyzoites in different tissues is usually easy to demonstrate, unlike the situation with live animals. Serology is an important guide, but since *T. gondii*-specific antibodies can be detected in the serum, CSF and aqueous humour of clinically normal animals, the results of such tests must be interpreted with caution.

A presumptive ante-mortem diagnosis of clinical feline toxoplasmosis may be based on a combination of the following criteria:

- demonstration of specific antibodies in serum, CSF or aqueous humour;
- demonstration of a positive IgM titre or a recent marked increase in IgG titre;
- presence of local antibody production or *T. gondii* DNA in the aqueous humour or CSF;
- observation of clinical signs compatible with toxoplasmosis;
- exclusion of other common diseases;
- clinical improvement following a specific therapy (Lappin, 1999).

Future Trends

Bertrand Losson and David Buxton

Since the first description of *T. gondii* by Nicolle and Manceaux in 1908 (Nicolle and Manceaux, 1908), a considerable amount of knowledge has accumulated about this parasitic protozoan, such that after *Plasmodium* spp., *T. gondii* is probably the next most studied apicomplexan. More particularly, many data are available about its host–parasite relationship, immunobiology and antigenic structure and, as a result, many valuable diagnostic tools for use in farm animal veterinary reproductive medicine are now available, so that informed use of histopathology, serology and the PCR may be combined to provide excellent investigative tools.

Significant progress is being made on a number of fronts. Molecular biological techniques such as the PCR tests – designed to detect either the B1 gene or the P30 (SAG1) gene – are finding their way into diagnostic laboratories (see above), and further advances are constantly being made. Real-time PCR is a powerful diagnostic tool that permits the quantification of parasite DNA to be assessed. Currently it is expensive, and cost-effective only in laboratories with very high throughputs, but technological progress will almost certainly ensure that it becomes less expensive and more widely available.

Serological tests are also becoming more sophisticated, with techniques for gauging the duration of an infection not only by differentiating between IgM and IgG but also by determining the avidity of IgG (see above). The kinetics of ELISA has also been discussed and the use of recombinant antigens in traditional ELISA all point to progress in the development of improved serological diagnostic methods.

From a practical point of view, the diagnosis of ovine/caprine congenital toxoplasmosis, although very reliable, is still a fairly slow process. Histopathological and serological examinations, whether combined with the PCR or not, may take 2–3 d at best, as well as being relatively expensive.

With the possibility of producing large amounts of well-characterized recombinant antigen(s) of diagnostic value and at relatively low cost, the development of a rapid test for on-farm use by the veterinary surgeon would represent a major advantage. Immunoreactive strips are already available for use in both human medicine (enteric and respiratory viruses, bacteria and protozoa) as well as companion and some farm animal conditions (FIV infection in cats for example).

Ideally, this kind of product would be used on precolostral blood samples (such as cord blood) or fetal fluids. Furthermore, a one-step combined detection of several abortifacients (such as *Brucella*, *Coxiella*, *Neospora* and *Salmonella*) could represent a further improvement of the technique. In cats, this approach could also be applied in assessing the serological status of either clinically affected or healthy individuals.

While improvements to current serological and molecular diagnostic tests will be forthcoming, it is important that, with their convenience and low cost, they are not relied on to the exclusion of clinical and post-mortem observations made by competent operatives, as an inaccurate or narrow diagnosis could be very costly.

References

Berlin, O.G.W., Novak, S.M., Porschem, R.K., Long, E.G., Stelma, G.N. and Schafer, F.W. (1994) Recovery of *Cyclospora* organisms from patients with prolonged diarrhoea. *Clinical and Infectious Diseases* 18, 606–609.

Berlin, O.G.W., Peter, J.B., Gagne, C., Conteas, C.N. and Ash, L.R. (1998) Autofluorescence and the detection of *Cyclospora* oocysts. *Emerging Infectious Diseases* 4, 127–128.

Beverley, J.K.A. (1959) Congenital transmission of toxoplasmosis through successive generations of mice. *Nature* 183, 1348–1349.

Blewett, D.A. (1983) The epidemiology of ovine toxoplasmosis. I. The interpretation of data for the prevalence of antibody in sheep and other host species. *British Veterinary Journal* 139, 537–545.

Blewett, D.A. and Trees, A.J. (1987) The epidemiology of ovine toxoplasmosis with especial respect to control. *British Veterinary Journal* 143, 128–135.

Blewett, D.A., Miller, J.K. and Buxton, D. (1982) Response of immune and susceptible ewes to infection with *Toxoplasma gondii*. *Veterinary Record* 111, 175–177.

Blewett, D.A. and Watson, W.A. (1984) The epidemiology of ovine toxoplasmosis. III. Observations on outbreaks of clinical toxoplasmosis in relation to possible mechanisms of transmission. *British Veterinary Journal* 140, 54–63.

Blewett, D.A., Bryson, C.E. and Miller, J.K. (1983) Studies of antibody titres in experimental ovine toxoplasmosis. *Research in Veterinary Science* 34, 163–166.

Bowie, W.R., King, A.S., Werker, D.H., Isaac-Renton, J.L., Bell, A., Eng, S.B. and Marion, S.A. (1997) Outbreak of toxoplasmosis associated with municipal drinking water. *Lancet* 350, 173–177.

Burg, J.L., Grover, C.M., Pouletty, P. and Boothroyd, J.C. (1989) Direct and sensitive detection of a pathogenic protozoan, *Toxoplasma gondii*, by polymerase chain reaction. *Journal of Clinical Microbiology* 27, 1787–1792.

Burney, D.P., Chavkin, M.J., Dow, S.W., Potter, T.A. and Lappin, M.R. (1998) Polymerase chain reaction for the detection of *Toxoplasma gondii* within aqueous humor of experimentally inoculated cats. *Veterinary Parasitology* 79, 181–186.

Buxton, D. (1990) Ovine toxoplasmosis: a review. *Journal of the Royal Society of Medicine* 83, 509–511.

Buxton, D. (1998) Protozoan infections (*Toxoplasma gondii*, *Neospora caninum* and *Sarcocystis* spp.) in sheep and goats: recent advances. *Veterinary Research* 29, 289–310.

Buxton, D. (2000) Toxoplasmosis and neosporosis. In: Martin, W.B. and Aitken, I.D. (eds) *Diseases of Sheep*, 3rd edn. Blackwell Science, Oxford, UK, pp. 86–94.

Buxton, D. and Finlayson, J. (1986) Experimental infection of pregnant sheep with *Toxoplasma gondii*: pathological and immunological observations on the placenta and foetus. *Journal of Comparative Pathology* 96, 319–333.

Buxton, D. and Innes, E.A. (1995) A commercial vaccine for toxoplasmosis. *Parasitology* 110, 11–16.

Buxton, D., Gilmour, J.S., Angus, K.W., Blewett, D.A. and Miller, J.K. (1982) Perinatal changes in lambs infected with *Toxoplasma gondii. Research in Veterinary Science* 32, 170–176.

Buxton, D., Donald, K.M. and Finlayson, J. (1987) Monensin and the control of experimental ovine toxoplasmosis: a systemic effect. *Veterinary Record* 20, 618–619.

Buxton, D., Blewett, D.A., Trees, A.J., McColgan, C. and Finlayson, J. (1988) Further studies in the use of monensin in the control of experimental ovine toxoplasmosis. *Journal of Comparative Pathology* 98, 225–236.

Buxton, D., Brebner, J., Wright, S., Maley, S.W., Thomson, K.M. and Millard, K. (1996) Decoquinate and the control of experimental ovine toxoplasmosis. *Veterinary Record* 138, 434–436.

Buxton, D., Anderson, I.E., Longbottom, D., Livingstone, M., Wattegedera, S. and Entrican, G. (2002) Ovine chlamydial abortion: characterization of the inflammatory immune response in placental tissues. *Journal of Comparative Pathology* 127, 133–141.

Calamel, M. and Lambert, M. (1988) *ELISA Standardised Technique*. Laboratoire National de Pathologie des Petits Ruminants at des Abeilles, Nice, France.

Chianini, F., Adams, C. and Buxton, D. (2004) Tick-borne fever: neuropathological changes in ovine foetuses. *Veterinary Record* 155, 805–806.

Colville, J. (1991) *Diagnostic Parasitology for Veterinary Technicians.* AmericanVeterinary Publications Inc., California, pp. 19–26.

Cornelissen, A.W., Overdulve, J.P. and van der Ploeg, M. (1984) Determination of nuclear DNA of five eucoccidian parasites, *Isospora (Toxoplasma) gondii, Sarcocystis cruzi, Eimeria tenella, E. acervulina* and *Plasmodium berghei*, with special reference to gamontogenesis and meiosis in *I. (T.) gondii. Parasitology* 88, 531–553.

Costa, J.M., Pautas, C., Ernault, P., Foulet, F., Cordonnier, C. and Bretagne, S. (2000) Real-time PCR for diagnosis and follow-up of *Toxoplasma* reactivation after allogeneic stem cell transplantation using fluorescence resonance energy transfer hybridization probes. *Journal of Clinical Microbiology* 38, 2929–2932.

Daugschies, A., Bialek, R., Joachim, A. and Mundt, H.C. (2001) Autofluorescence microscopy for the detection of nematode eggs and protozoa, in particular *Isospora suis* in swine faeces. *Parasitology Research* 87, 409–412.

De Roever-Bonnet, H. (1969) Congenital *Toxoplasma* infections in mice and hamsters infected with avirulent and virulent strains. *Tropical Geographical Medicine* 21, 443–450.

Derouin, F., Mazeron, M.C. and Garin, Y.J.K. (1987) Comparative study of tissue culture and mouse inoculation methods for demonstration of *Toxoplasma gondii. Journal of Clinical Microbiology* 25, 1597–1600.

Desmonts, G. and Remington, J.S. (1980) Direct agglutination test for diagnosis of *Toxoplasma* infection: method for increasing sensitivity and specificity. *Journal of Clinical Microbiology* 11, 562–568.

Dijkstra, T., Eyster, M., Schares, G., Conraths, F.J., Wouda, W. and Barkema, H.W. (2001) Dogs shed *Neospora caninum* oocysts after ingestion of naturally infected bovine placenta but not after ingestion of colostrums spiked with *Neospora caninum* tachyzoites. *International Journal for Parasitology* 31, 747–752.

Dubey, J.P. (1982) Repeat transplacental transfer of *Toxoplasma gondii* in goats. *Journal of the American Veterinary Association* 180, 1220–1221.

Dubey, J.P. (1986a) A review of toxoplasmosis in cattle. *Veterinary Parasitology* 22, 177–202.

Dubey, J.P. (1986b) Toxoplasmosis in cats. *Feline Practice* 16, 12–26.

Dubey, J.P. and Beattie, C.P. (1988) *Toxoplasmosis of Animals and Man.* CRC Press, Boca Raton, Florida.

Dubey, J.P. and Carpenter, L. (1993a) Neonatal toxoplasmosis in littermate cats. *Journal of the American Veterinary Medical Association* 203, 1546–1549.

Dubey, J.P. and Carpenter, L. (1993b) Histologically confirmed clinical toxoplasmosis in cats: 100 cases (1952–1990). *Journal of the American Veterinary Medical Association* 203, 1556–1566.

Dubey, J.P. and Frenkel, J.K. (1972) Cyst-induced toxoplasmosis in cats. *Journal of Protozoology* 19, 155–177.

Dubey, J.P. and Frenkel, J.K. (1974) Immunity to feline toxoplasmosis; modification by administration of corticosteroids. *Veterinary Pathology* 11, 350–379.

Dubey, J.P. and Frenkel, J.K. (1976) Feline toxoplasmosis from acutely infected mice and the development of *Toxoplasma* cysts. *Journal of Protozoology* 23, 537–546.

Dubey, J.P. and Lappin, M. (1998) Toxoplasmosis and neosporosis. In: Greene, E. (ed.) *Infectious Diseases of the Dog and Cat.* W.B. Saunders, Philadelphia, pp. 493–503.

Dubey, J.P., Christie, E. and Pappas, P.W. (1977) Characterization of *Toxoplasma gondii* from the faeces of naturally infected cats. *Journal of Infectious Diseases* 3, 432–435.

Dubey, J.P., Desmonts, G., Antunes, F. and McDonald, C. (1985) Serologic diagnosis of toxoplasmosis in experimentally infected pregnant goats and transplacentally infected kids. *American Journal of Veterinary Research* 46, 1137–1140.

Dubey, J.P., Miller, S., Desmonts, G., Thulliez, P. and Anderson, W.R. (1986) *Toxoplasma gondii*-induced abortion in dairy goats. *Journal of the American Veterinary Medical Association* 188, 159–162.

Dubey, J.P., Emond, J.P., Desmonts, G. and Anderson, W.R. (1987) Serodiagnosis of postnatally and prenatally induced toxoplasmosis in sheep. *American Journal of Veterinary Research* 48, 1239–1243.

Dubey, J.P., Lappin, M.R. and Thulliez, P. (1995) Long-term antibody responses of cats fed *Toxoplasma gondii* tissue cysts. *Journal of Parasitology* 81, 887–893.

Dubey, J.P., Shen, S.K., Kwok, O.C.H. and Thulliez, P. (1997) Toxoplasmosis in rats (*Rattus norvegicus*): congenital transmission to first and second generation offspring and isolation of *Toxoplasma gondii* from seronegative rats. *Parasitology* 115, 9–14.

Duncanson, P., Terry, R.S., Smith, J.E. and Hide, G. (2001) High levels of congenital transmission of *Toxoplasma gondii* in a commercial sheep flock. *International Journal for Parasitology* 31, 1699–1703.

Dytrych, J.K. and Cooke, R.P.D. (1995) Autofluorescence of *Cyclospora. British Journal of Biomedical Science* 52, 76.

Eichenwald, H. (1948) Experimental toxoplasmosis. I. Transmission of the infection *in utero* and through the milk of lactating female mice. *American Journal of Diseases of Children* 76, 307–315.

Ellis, J.T., McMillan, D., Ryce, C., Payne, S., Atkinson, R. and Harper, P.A. (1999) Development of a single tube nested polymerase chain reaction assay for the detection of *Neospora caninum* DNA. *International Journal for Parasitology* 29, 1589–1596.

Esteban-Redondo, I. and Innes, E.A. (1997) *Toxoplasma gondii* infection in sheep and cattle. *Comparative Immunology and Microbiology of Infectious Diseases* 20, 191–196.

Esteban-Redondo, I. and Innes, E.A. (1998) Detection of *Toxoplasma gondii* in tissues of sheep orally challenged with different doses of oocysts. *International Journal for Parasitology* 28, 1459–1466.

Faull, W.B., Clarkson, M.J. and Winter, A.C. (1986) Toxoplasmosis in a flock of sheep: some investigations into its source and control. *Veterinary Record* 119, 491–493.

Figueiredo, J.F., Silva, D.A., Cabral, D.D. and Mineo, J.R. (2001) Seroprevalence of *Toxoplasma gondii* infection in goats by the indirect haemagglutination, immunofluorescence and immunoenzymatic tests in the region of Uberlandia, Brazil. *Memórias do Instituto Oswaldo Cruz* 96, 687–692.

Gondim, L.F.P., Gao, L. and McAllister, M.M. (2002) Improved production of *Neospora caninum* oocysts, cyclical oral transmission between dogs and cattle, and *in vitro* isolation from oocysts. *Journal of Parasitology* 88, 1159–1163.

Gottstein, B., Hentrich, B., Wyss, R., Thür, B., Busato, A., Stärk, K.D.C. and Müller, N. (1998) Molecular and immunodiagnostic investigations on bovine neosporosis in Switzerland. *International Journal of Parasitology* 28, 679–691.

Hall, S., Ryan, M. and Buxton, D. (2000) The epidemiology of *Toxoplasma* infection. In: Joynson, D.H.M. and Wreghitt, T.G. (eds) *Toxoplasmosis, a Comprehensive Clinical Guide.* Cambridge University Press, Cambridge, UK, pp. 58–124.

Hartley, W.J. (1966) Some investigations into the epidemiology of ovine toxoplasmosis. *New Zealand Veterinary Journal* 14, 106–117.

Hartley, W.J. and Moyle, G.G. (1974) Further observations on the epidemiology of ovine toxoplasma infection. *Australian Journal of Experimental Biology and Medical Science* 52, 647–653.

Haessig, M., Sager, H., Reitt, K., Ziegler, D., Strabel, D. and Gottstein, B. (2003) *Neospora caninum* in sheep: a herd case report. *Veterinary Parasitology* 117, 213–220.

Ho-Yen, D.O. (2000) Infection in the immunocompetent. In: Joynson, D.H.M. and Wreghitt, T.G. (eds) *Toxoplasmosis, a Comprehensive Clinical Guide.* Cambridge University Press, Cambridge, UK, pp. 125–146.

Hurtado, A., Aduriz, G., Moreno, B., Barandika, J. and Garcia-Perez, A.L. (2001) Single tube nested PCR for the detection of *Toxoplasma gondii* in fetal tissues from naturally aborted ewes. *Veterinary Parasitology* 102, 17–27.

Innes, E.A. (1997) Toxoplasmosis: comparative species susceptibility and host immune response. *Comparative Immunology and Microbiology of Infectious Diseases* 20, 131–138.

Jackson, M.H. and Hutchison, W.M. (1989) The prevalence and source of *Toxoplasma* infection in the environment. In: Baker, J.R. and Muller, R. (eds) *Advances in Parasitology,* Vol. 28. Academic Press, London, pp. 55–105.

Jauregui, L.H., Higgins, J., Zarlenga, D., Dubey, J.P. and Lunney, J.K. (2001) Development of a real-time PCR assay for detection of *Toxoplasma gondii* in pig and mouse tissues. *Journal of Clinical Microbiology* 39, 2065–2071.

Jeffrey, M., O'Toole, D., Smith, T. and Bridges, A.W. (1988) Immunocytochemistry of ovine sporozoan encephalitis and encephalomyelitis. *Journal of Comparative Pathology* 98, 213–224.

Jenum, P.A., Stray-Pedersen, B. and Gundersen, A.G. (1997) Improved diagnosis of primary *Toxoplasma gondii* infection in early pregnancy by determination of antitoxoplasma immunoglobulin G avidity. *Journal of Clinical Microbiology* 35, 1972–1977.

Johnson, A.M. and Illana, S. (1991) Cloning of *Toxoplasma gondii* gene fragments encoding diagnostic antigens. *Gene* 99, 127–132.

Kijlstra, A., Eissen, O.A., Cornelissen, J., Munniksma, K., Eijck, I. and Kortbeek, T. (2004) *Toxoplasma gondii* infection in animal-friendly pig production systems. *Investigative Ophthalmology and Visual Science* 45, 3165–3169.

Lappin, M. (1999) Feline toxoplasmosis. *In Practice* 21, 578–589.

Lappin, M., Greene, C.E., Winston, S., Toll, S.L. and Epstein, M.E. (1989) Clinical feline toxoplasmosis. Serological diagnosis and therapeutic management of 15 cases. *Journal of Veterinary Internal Medicine* 3, 139–143.

Lappin, M., Burney, D.P., Dow, S.W. and Potter, T.A. (1996) Polymerase chain reaction for the detection of *Toxoplasma gondii* in aqueous humor of cats. *American Journal of Veterinary Research* 57, 1589–1593.

Lekutis, C., Ferguson, D.J., Grigg, M.E., Camps, M. and Boothroyd, J.C. (2001) Surface antigens of *Toxoplasma gondii*: variations on a theme. *International Journal for Parasitology* 112, 1–10.

Lin, M.H., Chen, T.C., Kuo, T., Tseng, C.C. and Tseng, C.P. (2000) Real-time PCR for quantitative detection of *Toxoplasma gondii. Journal of Clinical Microbiology* 38, 4121–4125.

Lind, P. and Buxton, D. (2000) Veterinary aspects of *Toxoplasma* infection. In: Ambroise-Thomas, P. and Petersen, E. (eds) *Congenital Toxoplasmosis; Scientific Background, Clinical Management and Control.* Springer, Berlin, pp. 261–269.

Lindquist, H.D.A., Bennett, J.W., Hester, J.D., Ware, M.W., Dubey, J.P. and Everson, W.V. (2003) Autofluorescence of *Toxoplasma gondii* and related Coccidian oocysts. *Journal of Parasitology* 89, 865–867.

Lundén, A., Carlsson, U. and Näslund, K. (1992) Toxoplasmosis and Border Disease in 54 Swedish sheep flocks. *Acta Veterinaria Scandinavica* 33, 175–184.

Lundén, A., Nasholm, A. and Uggla, A. (1994) Long-term study of *Toxoplasma gondii* infection in a Swedish sheep flock. *Acta Veterinaria Scandinavica* 35, 273–281.

Macdonald, D. (1980) The behaviour and ecology of farm cats. In: *The Ecology and Control of Feral Cats.* University Federation for Animal Welfare, Potters Bar, UK, pp. 23–29.

Mariuz, P. and Steigbigel, R.T. (2000) *Toxoplasma* infection in HIV-infected patients. In: Joynson, D.H.M. and Wreghitt, T.G. (eds) *Toxoplasmosis, a Comprehensive Clinical Guide.* Cambridge University Press, Cambridge, UK, pp. 147–177.

McColgan, C., Buxton, D. and Blewett, D.A. (1988) Titration of *Toxoplasma gondii* oocysts in non-pregnant sheep and the effects of subsequent challenge during pregnancy. *Veterinary Record* 123, 467–470.

Melvin, D. and Brooke, M.M. (1974) *Laboratory Procedures for the Diagnosis of Intestinal Parasites.* US Department of Health, Education and Welfare, Publication No. (CDC) 75 8282, Center for Disease Control, Atlanta, Georgia.

Miller, N.L., Frenkel, J.K. and Dubey, J.P. (1972) Oral infections with *Toxoplasma* cysts and oocysts in felines, and other mammals, and in birds. *Journal of Parasitology* 58, 928–937.

Munday, B.L. (1972) Transmission of *Toxoplasma* infection from chronically infected ewes to their lambs. *British Veterinary Journal* 128, lxxi–lxxii.

Munday, B.L. and Corbould, A. (1971) The application of the *Toxoplasma,* indirect, fluorescent-antibody test to sheep sera. *Australian Journal Medical Technology* 2, 3–6.

Nettleton, P.F. (2000) Border Disease. In: Martin, W.B. and Aitken, I.D. (eds) *Diseases of Sheep,* Blackwell Science. Oxford, UK, pp. 93–102.

Nicolle, C. and Manceaux, L. (1908) Sur une infection a corps de Leishman (ou organisms voisins) du gondii. *CR. Acad. Sci.* 147, 763 cited by Dubey, J.P. and Beattie, C.P. In: *Toxoplasmosis of Animals and Man* (1988), CRC Press, Boca Raton, Florida.

Obendorf, D.L., Statham, P. and Munday, B.L. (1990) Resistance to *Toxoplasma* abortion in female goats previously exposed to *Toxoplasma* infection. *Australian Veterinary Journal* 67, 233–234.

O'Donoghue, P.J., Riley, M.J. and Clarke, J.F. (1987) Serological survey for Toxoplasma infections in sheep. *Australian Veterinary Journal* 64, 40–45.

Oksanen, A., Asbakk, K., Nieminen, M., Norberg, H. and Nareaho, A. (1997) Antibodies against *Toxoplasma gondii* in Fennoscandian reindeer – association with the degree of domestication. *Parasitology International* 46, 255–261.

Owen, M.R. (1996) Molecular approaches to the epidemiology and diagnosis of ovine toxoplasmosis. PhD Thesis. University of Liverpool, UK.

Owen, M.R. and Trees, A.J. (1998) Vertical transmission of *Toxoplasma gondii* from chronically infected house (*Mus musculus*) and field (*Apodemus sylvaticus*) mice determined by polymerase chain reaction. *Parasitology* 116, 299–304.

Peach, W., Fowler, J. and Hay, J. (1989) Incidence of *Toxoplasma* infection in a population of European starlings *Sternus vulgaris* from central England. *Annals of Tropical Medicine and Parasitology* 83, 173–177.

Petersen, E. and Dubey, J.P. (2000) Biology of toxoplasmosis. In: Joynson, D.H.M. and Wreghitt, T.G. (eds) *Toxoplasmosis, a Comprehensive Clinical Guide.* Cambridge University Press, Cambridge, UK, pp. 1–42.

Plant, J.W., Richardson, N. and Moyle, G.G. (1974) *Toxoplasma* infection and abortion in sheep associated with feeding of grain contaminated with cat faeces. *Australian Veterinary Journal* 50, 19–21.

Reischl, U., Bretagne, S., Kruger, D., Ernault, P. and Costa, J.M. (2003) Comparison of two DNA targets for the diagnosis of Toxoplasmosis by real-time PCR using fluorescence resonance energy transfer hybridization probes. *BMC Infectious Diseases* 3, 7.

Rodger, S.M., Maley, S.W., Wright, S.E., Mackellar, A., Wesley, F., Sales, J. and Buxton, D. (2006) Role of endogenous transplacental transmission in toxoplasmosis in sheep. *Veterinary Record* 159, 768–772.

Rusbridge, C. (2001) The nervous system. In: Ramsey, I. and Tennant, B. (eds) *Manual of Canine and Feline Infectious Diseases*. British Small Animal Veterinary Association, Quedgeley, UK, pp. 231–249.

Sabin, A.B. and Feldman, H.A. (1948) Dyes as microchemical indicators of a new immunity phenomenon affecting a protozoon parasite (*Toxoplasma*). *Science* 108, 660–663.

Sager, H., Gloor, M., Tenter, A., Maley, S., Haessig, M. and Gottstein, B. (2003) Immunodiagnosis of primary *Toxoplasma gondii* infection in sheep by the use of a P30 IgG avidity ELISA. *Parasitology Research* 91, 171–174.

Sambrook, J., Fritsch, E.F. and Maniatis, T. (1989) *Molecular Cloning: a Laboratory Manual*. Cold Spring Harbor Laboratory Press, New York.

Sapero, J.J. and Lawless D.K. (1953) The 'MIF' stain-preservative technic for the identification of intenstinal protozoa. *American Journal of Tropical Medicine and Hygiene* 2, 613–619.

Savva, D., Morris, J.C., Johnson, J.D. and Holliman, R.E. (1990) Polymerase chain reaction for detection of *Toxoplasma gondii*. *Journal of Medical Microbiology* 32, 25–31.

Skjerve, E., Waldeland, H., Nesbakken, T. and Kapperud, G. (1998) Risk factors for the presence of antibodies to *Toxoplasma gondii* in Norwegian slaughter lambs. *Preventive Veterinary Medicine* 35, 219–227.

Tabor, R. (1980) General biology of feral cats. In: *The Ecology and Control of Feral Cats*. University Federation for Animal Welfare, Potters Bar, UK, pp. 5–11.

Tang, X., Bartlett, M.S., Smith, J.W. and Lee, C.H. (1997) A single-tube nested PCR for *Pneumocystis carinii* f. sp. *hominis*. *Journal of Clinical Microbiology* 35, 1597–1599.

Tenter, A.M., Vietmeyer, C. and Johnson, A.M. (1992) Development of ELISAs based on recombinant antigens for the detection of *Toxoplasma gondii*-specific antibodies in sheep and cats. *Veterinary Parasitology* 43, 189–201.

Tenter, A.M., Luton, K. and Johnson, A.M. (1994) Species-specific identification of *Sarcocystis* and *Toxoplasma* by PCR amplification of small subunit ribosomal RNA gene fragments. *Applied Parasitology* 35, 173–188.

Thulliez, P. (2000) Maternal and foetal infection. In: Joynson, D.H.M. and Wreghitt, T.G. (eds) *Toxoplasmosis, a Comprehensive Clinical Guide*. Cambridge University Press, Cambridge, UK, pp. 193–213.

Trees, A.J. and Williams, D.J.L. (2005) Endogenous and exogenous transplacental infection in *Neospora caninum* and *Toxoplasma gondii*. *Trends in Parasitology* 21, 558–561.

Trees, A.J., Crozier, S.J., Buxton, D. and Blewett, D.A. (1989) Serodiagnosis of ovine toxoplasmosis: an assessment of the latex agglutination test and the value of IgM specific titres after experimental oocyst-induced infections. *Research in Veterinary Science* 46, 67–72.

Truant, A.L., Elliott, S.H., Kelly, M.T. and Smith, J.H. (1981) Comparison of formalin–ethyl ether sedimentation, formalin–ethyl acetate sedimentation, and zinc sulfate flotation techniques for detection of intestinal parasites. *Journal of Clinical Microbiology* 13, 882–884.

Uggla, A. and Buxton, D. (1990) Immune responses against *Toxoplasma* and *Sarcocystis* infections in ruminants: diagnosis and prospects for vaccination. *Reviews of Science and Technology* 9, 441–462.

Uggla, A. and Hjort, M. (1984) A serological study on the prevalence of *Toxoplasma gondii* in meat-producing animals in Sweden. *Acta Veterinaria Scandinavica* 25, 567–576.

Uggla, A., Sjöland, L. and Dubey, J.P. (1987) Immunohistochemical demonstration of toxoplasmosis in fetuses and fetal membranes of sheep. *American Journal of Veterinary Research* 48, 348–351.

Van der Puije, W.N., Bosompem, K.M., Canacoo, E.A., Wastling, J.M. and Akanmori, B.D. (2000) The prevalence of anti-*Toxoplasma gondii* antibodies in Ghanaian sheep and goats. *Acta Tropica* 21, 21–26.

van Knapen, F., Franchimont, J.H. and van der Lugt, G. (1982) Prevalence of antibodies to *Toxoplasma* in farm animals in the Netherlands and its implication for meat inspection. *Veterinary Quarterly* 4, 101–105.

van Maanen, C., Wouda, W., Schares, G., von Blumröder, D., Conraths, F.J., Norton, R., Williams, D.J.L., Esteban-Redondo, I., Innes, E.A., Mattsson, J.G., Björkman, C., Fernández-García, A., Ortega-Mora, L.M., Müller, N., Sager, H. and Hemphill, A. (2004) An interlaboratory comparison of immunohistochemistry and PCR methods for detection of *Neospora caninum* in bovine foetal tissues. *Veterinary Parasitology* 126, 351–354.

Varea, M., Clavel, A., Doiz, O., Castillo, F.J., Rubio, M.C. and Gomez-Lus, R. (1998) Fuchin fluorescence and autofluorescence in *Cryptosporidium, Isospora* and *Cyclospora* oocysts. *International Journal for Parasitology* 28, 1881–1883.

Voller, A., Bidwell, D.E., Bartlett, A., Fleck, D.G., Perkins, M. and Oladehin, B. (1976) A microplate enzyme-immunoassay for *Toxoplasma* antibody. *Journal of Clinical Pathology* 29, 150–153.

Waldeland, H. (1977) Toxoplasmosis in sheep. Influence of various factors on the antibody contents. *Acta Veterinaria Scandinavica* 8, 237–247.

Wastling, J.M., Nicoll, S. and Buxton, D. (1993) Comparison of two gene amplification methods for the detection of *Toxoplasma gondii* in experimentally infected sheep. *Journal of Medical Microbiology* 38, 360–365.

Wastling, J.M., Harkins, D. and Buxton, D. (1994) Western blot analysis of the IgG response of sheep vaccinated with S48 *Toxoplasma gondii* (Toxovax). *Research in Veterinary Science* 57, 384–386.

Watson, W.A. and Beverley, J.K.A. (1971) Epizootics of toxoplasmosis causing ovine abortion. *Veterinary Record* 88, 120–124.

Werre, S.R., Jacobson, R.H., Bowman, D.D., Dubey, J.P. and Mohammed, H.O. (2002) Evaluation of kinetics and single-read enzyme-linked immunoassays for detection of *Toxoplasma gondii* antibodies in sheep. *Journal of Veterinary Diagnostic Investigation* 14, 225–230.

Williams, R.H., Morley, E.K., Hughes, J.M., Duncanson, P., Terry, R.S., Smith, J.E. and Hide, G. (2005) High levels of congenital transmission of *Toxoplasma gondii* in longitudinal and cross-sectional studies on sheep farms provides evidence of vertical transmission in ovine hosts. *Parasitology* 130, 301–307.

Wreghitt, T.G. and Joynson, D.H.M. (2000) *Toxoplasma* infection in immunosuppressed (HIV-negative) patients. In: Joynson, D.H.M. and Wreghitt, T.G. (eds) *Toxoplasmosis, a Comprehensive Clinical Guide.* Cambridge University Press, Cambridge, UK, pp. 178–192.

3.3 Sarcocystiosis

Anja R. Heckeroth and Astrid M. Tenter

General Considerations

The genus *Sarcocystis* (Lankester, 1882), is one of the larger genera of the protozoan phylum Apicomplexa (Levine, 1970), and is composed of about 200 species of heteroxenous tissue cyst-forming coccidia. *Sarcocystis* species are common parasites of a broad range of vertebrates, including mammals, birds and fish. The last taxonomic review of the genus *Sarcocystis* named 189 different species (Odening, 1998). Since then, several new species have been identified, and many more *Sarcocystis*-like parasites have been found in muscle or nervous tissue of various vertebrates, such as in water buffaloes.

Sarcocystis species differ in their life cycle, host specificity, morphology and pathogenicity. Virulent *Sarcocystis* species can cause disease, abortion or fetal abnormalities in their intermediate hosts, in particular in ruminants. Comprehensive reviews of the life cycles, biology, morphology, pathogenicity and immunology of *Sarcocystis* species have been written by Dubey *et al.* (1989a), Rommel (1989), Cawthorn and Speer (1990), Uggla and Buxton (1990) and Tenter (1995). Methods used for the isolation, *in vitro* cultivation and preservation of *Sarcocystis* species, as well as the preparation of *Sarcocystis* proteins, nucleic acids and chromosomes, have been reviewed by Rommel *et al.* (1995).

This chapter focuses on the diagnosis of *Sarcocystis* species of domestic ruminants (cattle, sheep, goats and water buffaloes). Unfortunately, there is still confusion about the nomenclature of the *Sarcocystis* species. Therefore, the names and frequently used synonyms of species reviewed here are listed in Table 3.3.1. This nomenclature is consistent with that used by Levine (1986) and Odening (1998), and conforms to the International Code of Zoological Nomenclature (Melville, 1984).

Table 3.3.1. Nomenclature of *Sarcocystis* species of domestic ruminants (Bovidae).

Intermediate host	*Sarcocystis* species	Synonym
Bovinae	*Sarcocystis buffalonis* (Huong *et al.*, 1997a)	? *Sarcocystis hirsuta* Moulé, 1888
	Sarcocystis cruzi (Hasselmann, 1923) Wenyon, 1926	*Sarcocystis bovicanis* Heydorn *et al.*, 1975
	Sarcocystis fusiformis (Railliet, 1897) Bernard and Bauche, 1912	–
	Sarcocystis hirsuta Moulé, 1888	*Sarcocystis bovifelis* Heydorn *et al.*, 1975
	Sarcocystis hominis (Railliet and Lucet, 1891) Dubey, 1976	*Sarcocystis bovihominis* Heydorn *et al.*, 1975
	Sarcocystis levinei Dissanaike and Kan, 1978	–
	? *Sarcocystis poephagi* Wei *et al.*, 1985	? *Sarcocystis hirsuta* Moulé, 1888; ? *Sarcocystis buffalonis* (Huong *et al.*, 1997)
	? *Sarcocystis poephagicanis* Wei *et al.*, 1985	? *Sarcocystis cruzi* (Hasselmann, 1923) Wenyon, 1926; ? *Sarcocystis levinei* Dissanaike and Kan, 1978
Caprinae	*Sarcocystis arieticanis* Heydorn, 1985	–
	Sarcocystis capracanis Fischer, 1979	–
	Sarcocystis gigantea (Railliet, 1886b) Ashford, 1977	*Sarcocystis ovifelis* Heydorn *et al.*, 1975
	Sarcocystis hircicanis Heydorn and Unterholzner, 1983	–
	Sarcocystis medusiformis Collins *et al.*, 1979	–
	Sarcocystis moulei Neveu-Lemaire, 1912	? *Sarcocystis caprifelis* El-Rafaii *et al.*, 1980; ? *Sarcocystis orientalis* Machul'skii and Miskaryan, 1958
	Sarcocystis tenella (Railliet, 1886a) Moulé, 1886	*Sarcocystis ovicanis* Heydorn *et al.*, 1975

?: Denotes *Sarcocystis* species of uncertain validity.

Biology, Transmission and Clinical Signs

Life cycle

Sarcocystis species are obligate intracellular protozoa with a typical coccidian life cycle consisting of alternating phases of merogony, gamogony and sporogony. Most *Sarcocystis* species are obligate heteroxenous, i.e. both an intermediate host and a definitive host need to be present for successful completion of the life cycle (Dubey *et al.*, 1989a; Tenter and Johnson, 1997). Asexual multiplication by merogony and tissue cyst formation take place in the intermediate host, sexual reproduction (gamogony) and sporogony in the definitive host. Sporogony results in the production of resistant sporocysts that are passed into the environment in the faeces of the definitive host. Horizontal transmission occurs from intermediate to definitive host and from definitive to intermediate host, but not between different intermediate hosts or between different definitive hosts (see Fig. 3.3.1).

The intermediate host, which is usually a herbivore or omnivore, becomes infected by ingesting sporocysts in contaminated food or water. Sporozoites are released from the sporocysts in the small intestine. In ruminants that serve as intermediate hosts for *Sarcocystis* species, the asexual development of the parasite consists of two to three generations of merogony, which takes place in the form of endopolygeny in vascular endothelial cells. This is followed by the formation of tissue cysts in muscles, and sometimes also in neural tissue (Heydorn and Gestrich, 1976; Leek *et al.*, 1977; Dubey *et al.*, 1982, 1998, 1989b; O'Donoghue and Ford, 1984; Orr *et al.*, 1984; Heydorn, 1985; Heydorn and Karaer, 1986; Dubey, 1988, 1993; Ghaffar *et al.*, 1989; Rommel, 1989; Cawthorn and Speer, 1990; Saito *et al.*, 1997).

First-generation meronts are often found in arterioles of the intestines and mesenteric lymph nodes, and in some species in vascular endothelial cells of the liver, kidney, lung, heart and skeletal muscles (see Table 3.3.2). Second-generation meronts have been seen in endothelial cells of capillaries of almost all internal organs. In addition, some species undergo endodyogeny in leucocytes.

Merozoites (syn. endozoites) of the terminal generation of merogony initiate the formation of tissue cysts in various striated muscles. Some species also form tissue cysts in cells of the central nervous system or in Purkinje fibres of the heart (see Tables 3.3.2, 3.3.3, 3.3.4, 3.3.5 and 3.3.6). In immature tissue cysts, further asexual reproduction takes place by repeated endodyogeny of metrocytes. Mature tissue cysts contain several thousands, in some species up to one million cystozoites (syn. bradyzoites, cyst merozoites), which do not divide further and are the terminal life cycle stage in the intermediate host. The cystozoites of *Sarcocystis* species are gametocytes and, thus, only develop further when ingested by a susceptible definitive host, which is always a carnivore.

In the definitive host, the cystozoites continue the sexual phase of the life cycle (gamogony), which takes place in cells of the small intestine and leads to the

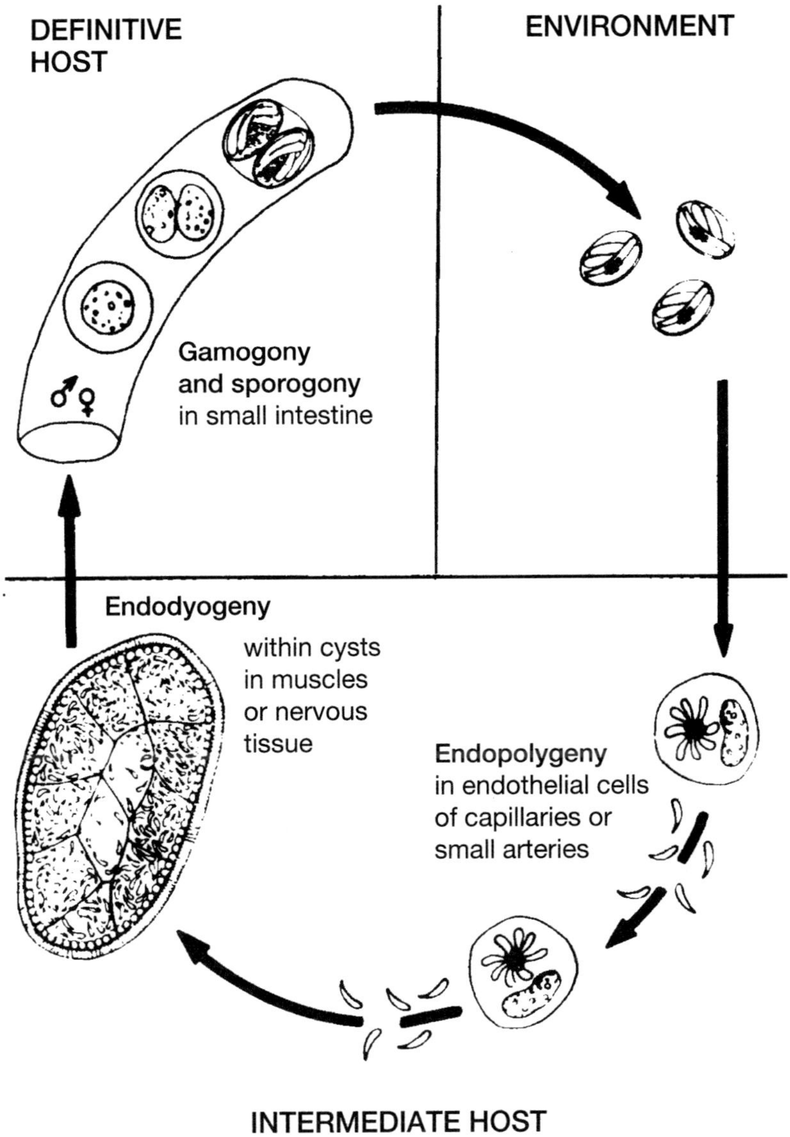

Fig. 3.3.1. Life cycle of the genus *Sarcocystis* (reproduced from Tenter, 1995). Copyright: Australian Society for Parasitology (*International Journal for Parasitology*).

formation of oocysts (Frenkel *et al.*, 1979; Levine, 1986; Dubey *et al.*, 1989; Cawthorn and Speer, 1990; Dubey, 1993; Odening, 1998). Oocysts sporulate in the *lamina propria* of the small intestine. Sporulated oocysts are the terminal life cycle stage in the definitive host (see Table 3.3.2). As the oocyst wall is thin, it often ruptures and free sporocysts are released into the intestinal lumen and passed into the environment in the faeces.

Host specificity and epidemiology of *Sarcocystis* species infecting ruminants

Sarcocystis species vary in their intermediate and definitive host spectrum. *Sarcocystis* species infecting Caprinae are either species-specific or genus-specific for their intermediate hosts (see Tables 3.3.2, 3.3.3 and 3.3.4), while *Sarcocystis* species infecting Bovinae usually use members of several closely related genera as intermediate hosts (see Tables 3.3.5 and 3.3.6). However, intermediate as well as definitive hosts can be infected by several different *Sarcocystis* species. Also, the occurrence of sympatric *Sarcocystis* species that use the same intermediate and the same definitive hosts is not unusual, and has been confirmed by experimental infections with defined sporocysts for goats and sheep (Heydorn and Unterholzner, 1983; Heydorn, 1985).

It is sometimes not clear how many different *Sarcocystis* species occur in the same host. For example, the number of *Sarcocystis* species that infect goats or water buffaloes is uncertain (see Table 3.3.1). Dubey *et al.* (1989a) synonymize *S. caprifelis* with *S. moulei*, while Levine (1986) regards these as being two different species. Some species, such as *S. mihoensis* in sheep, have been reported only once, and hence their validity lacks confirmation (Saito *et al.*, 1997). In addition, *Sarcocystis* species have been found in naturally infected yaks in China and India (Wei *et al.*, 1985; RangaRao *et al.*, 1994), but no cross-transmission studies to cattle have been carried out and the validity of *S. poephagi* and *S. poephagicanis* as independent species, or their identity with *Sarcocystis* species infecting cattle, needs to be confirmed.

Much of the confusion concerning the taxonomy and validity of *Sarcocystis* species is due to misinterpretations of original descriptions and inadequate new descriptions. Descriptions of new *Sarcocystis* species are often incomplete and based only on morphological characters, in particular the size of the tissue cyst and the structure of the tissue cyst wall. However, these characters are of little taxonomic value since they may vary during tissue cyst development, with the type, location and degenerative state of the host cell parasitized, as well as with the methods used for examination or fixation (Dubey *et al.*, 1989a). In addition, the tissue cysts of some *Sarcocystis* species, such as *S. gigantea* or *S. medusiformis* of sheep, continue to grow for several years even after they have reached maturity, i.e. after they contain infective cystozoites (Munday and Obendorf, 1984; Obendorf and Munday, 1987).

Table 3.3.2. Life-cycle, biological, morphological and clinical data of *Sarcocystis* species infecting cattle (*Bos taurus*).[a]

Character	*S. cruzi*	*S. hirsuta*	*S. hominis*
Geographical distribution:	Worldwide	Probably worldwide	Europe, Brazil and probably other parts of the world
Intermediate hosts:	*Bison bison, Bison bonasus, Bos gaurus, Bos javanicus, Bos taurus, Bubalus bubalis*	*Bison bison, Bison bonasus, Bos gaurus, Bos javanicus, Bos taurus*	*Bison bison, Bison bonasus, Bos gaurus, Bos javanicus, Bos taurus, Bubalus bubalis*
Degree of virulence for the intermediate host:	High	Intermediate	Low
First-generation merogony (dpi):[b]	7–26	7–23	–
Location of first-generation meronts:	Arterioles of many organs	Mesenteric and intestinal arterioles	–
Size of first-generation meront (μm):	41.0 × 17.5	37.2 × 22.3	–
Size of first-generation merozoite (μm):	6.3 × 1.5	5.1 × 1.2	–
Number of first-generation merozoites:	> 100	> 100	–
Second-generation merogony (dpi)	19–46	15–23	–
Location of second-generation meronts:	Capillaries of many organs	Capillaries of striated muscles	–
Size of second-generation meront (μm):	19.6 × 11.0	13.9 × 6.5	–

Continued

Table 3.3.2. – *Continued*

Character	*S. cruzi*	*S. hirsuta*	*S. hominis*
Size of second-generation merozoite (μm):	7.9 × 1.5	4.0 × 1.5	–
Number of second-generation merozoites:	4–37	3–35	–
Third-generation merogony	Yes	No	–
Location of third-generation meronts:	Leucocytes	–	–
Formation of tissue cysts (dpi):	45	30	–
Location of tissue cysts:	All striated muscles, CNS, Purkinje fibres	Probably all striated muscles	Probably all striated muscles
Time to maturity of tissue cyst:	70 dpi	75 dpi	–
Size of tissue cyst (μm):	≤ 500	≤8 × 1000	≤ 950
Morphology of tissue cyst wall:	Thin (< 1 μm), with hair-like villar protrusions (3.5 × 0.2–0.3 μm)	Thick (up to 7 μm), radially striated, with finger-like villar protrusions (6.7–8.0 × 1.5 μm)	Thick (up to 6 μm), radially striated, with finger-like villar protrusions (5.7–7.0 × 0.6–1.4 μm)
Size of bradyzoite (μm):	–	–	7–9
Definitive hosts:	*Canis familiaris* (dog), *Canis latrans (coyote)*, *Canis lupus* (wolf), *Procyon lotor* (raccoon), *Vulpes vulpes* (red fox)	*Felis catus* (domestic cat), *Felis silvestris* (wild cat)	*Pan troglodytes* (chimpanzee), *Homo sapiens* (human), *Macaca mulatta* (rhesus monkey), *Papio cynocephalus* (baboon)

Prepatent period (d):	7–33	7–10	9–10
Patency (months):	3	–	–
Size of sporocyst (μm):	14.5–17.0 × 9.0–11.0	11.0–14.0 × 7.0–9.0	14.7 × 9.3

[a] The data contained in this table have been taken from reviews of Dubey (1976); Frenkel *et al.* (1979); Fayer and Dubey (1986); Levine (1986); Dubey *et al.* (1989a); Fayer and Elsasser (1991); Lindsay *et al.* (1996); Taylor (2000); Markus *et al.* (2004); McEwen and Carman (2005); and original publications of Gomes and Lima (1982); Böttner *et al.* (1987); Dubey *et al.* (1988); Mitchell (1988); Vercruysse *et al.* (1989); Savini *et al.* (1992, 1994, 1996); Carvalho (1993); Fortier *et al.* (1993); Xiao *et al.* (1993); Mal'a and Baranova (1995); Claveria *et al.* (1997); Gunning *et al.* (2000).
[b] Days post-infection.
–: No data available.

Table 3.3.3. Biological and morphological data of *Sarcocystis* species infecting yak (*Bos grunniens*).[a]

Character	*S. poephagi*[b]	*S. poephagicanis*[c]
Geographical distribution:	China	China
Intermediate host:	*Bos grunniens*	*Bos grunniens*
Location of tissue cyst:	Skeletal muscles	Skeletal muscles
Size of tissue cyst (mm):	≤ 40.00	≤ 0.49
Definitive host:	Unknown	*Canis familiaris* (dog)

[a] The data contained in this table have been taken from a review of Dubey *et al.* (1989a) and original publications of Wei *et al.* (1985) and RangaRao *et al.* (1994).
[b] This species may be identical with *S. hirsuta* or *S. buffalonis* (see Tables 3.3.1, 3.3.2 and 3.3.4).
[c] This species may be identical with *S. cruzi* or *S. levinei* (see Tables 3.3.1, 3.3.2 and 3.3.4).

Table 3.3.4. Biological and morphological data of *Sarcocystis* species infecting water buffalo (*Bubalus bubalis*).

Character	*S. buffalonis*	*S. cruzi*	*S. fusiformis*	*S. hominis*	*S. levinei*
Geographical distribution:		Worldwide	Probably all countries with the distribution of the host	Europe, Brazil and probably other parts of the world	Asia, Europe, Egypt, Brazil
Intermediate hosts:	*Bubalus bubalis*	*Bison bison, Bison bonasus, Bos gaurus, Bos javanicus, Bos taurus, Bubalus bubalis*	*Bubalus bubalis*	*Bison bison, Bison bonasus, Bos gaurus, Bos javanicus, Bos taurus, Bubalus bubalis*	*Bubalus bubalis*
Degree of virulence for the intermediate host:			Non-virulent		Intermediate
First-generation merogony (dpi)[b]:	7–23	7–26	–	–	–
Location of first-generation meronts:	Mesenteric and intestinal arterioles	Arterioles of many organs	–	–	–
Size of first-generation meront (μm):	37.2 × 22.3	41.0 × 17.5	–	–	–
Size of first-generation merozoite (μm):	5.1 × 1.2	6.3 × 1.5	–	–	–
Number of first-generation merozoites:	> 100	> 100	–	–	–
Second-generation merogony (dpi):	15–23	19–46	–	–	–

Location of second-generation meronts:	Capillaries of striated muscles	Capillaries of many organs	–	–	–
Size of second-generation meront (μm):	13.9 × 11.0	19.6 × 11.0	–	–	–
Size of second-generation merozoite (μm):	4.0 × 1.5	7.9 × 1.5	–	–	–
Number of second-generation merozoites:	3–35	4–37	–	–	–
Third-generation merogony:	No	Yes	–	–	–
Location of third-generation meronts:	–	Leucocytes	–	–	–
Formation of tissue cysts (dpi):	30	45	–	–	–
Location of tissue cyst:	Probably all striated muscles	All striated muscles, CNS, Purkinje fibres	Oesophageal muscles	Probably all striated muscles	Probably all striated muscles
Time to maturity of tissue cyst (dpi):	75	70	–	–	–
Size of tissue cyst (μm):	up to 8 × 1	≤ 500	≤ 32 × 8	≤ 950	≤ 1200
Morphology of tissue cyst wall:	Thick (up to 7 μm), radially striated, with finger-like villar protrusions (6.7–8.0 × 1.5 μm)	Thin (< 1 μm), with hair-like villar protrusions (3.5 × 0.2–0.3 μm)	Thin, with cauliflower-like villar protrusions	Thick (up to 6 μm), radially striated, with finger-like villar protrusions (5.7–7.0 × 0.6–1.4 μm)	Thin (< 1 μm), with sloping villar protrusions

Continued

Table 3.3.4. – *Continued*

Character	*S. buffalonis*	*S. cruzi*	*S. fusiformis*	*S. hominis*	*S. levinei*
Size of bradyzoite (μm):	–	–	–	7–9	7–9
Definitive hosts	*Felis catus* (domestic cat), *Felis silvestris* (wild cat)	*Canis familiaris* (dog), *Canis latrans* (coyote), *Canis lupus* (wolf), *Procyon lotor* (raccoon), *Vulpes vulpes* (red fox)	*Felis catus* (domestic cat)	*Pan troglodytes* (chimpanzee), *Homo sapiens* (human), *Macaca mulatta* (rhesus monkey), *Papio cynocephalus* (baboon)	*Canis familiaris* (dog)
Prepatent period (d)	7–10	7–33	8–14	9–10	12–34
Patency (months)	–	3	–	–	–
Size of sporocyst (μm)	11.0–14.0 × 7.0–9.0	14.5–17.0 × 9.0–11.0	11.5–14.0 × 9.0–10.0	14.7 × 9.3	15.0–16.0 × 10.0

[a] The data contained in this table have been taken from reviews of Levine (1986); Dubey *et al.* (1989a); Markus *et al.* (2004) and original publications of Claveria *et al.* (1997); Dissanaike and Kan (1978); Xiao *et al.* (1993); Gupta *et al.* (1995); Huong *et al.* (1979a, b); Huong and Uggla (1999); Claveria and Cruz (2000); Yang *et al.* (2001) and Li *et al.* (2002).

[b] Days post-infection.

–: No data available.

Table 3.3.5. Biological and morphological data of *Sarcocystis* species infecting sheep (*Ovis aries*).[a]

Character	*S. arieticanis*	*S. gigantea*	*S. medusiformis*	*S. mihoensis*	*S. tenella*
Geographical distribution:	Probably worldwide	Worldwide	Australia, New Zealand, India, Iran	Japan	Worldwide
Intermediate hosts:	*Ovis ammon, Ovis aries*	*Ovis ammon, Ovis aries*	*Ovis ammon, Ovis aries*	*Ovis aries*	*Ovis ammon, Ovis aries*
Degree of virulence for the intermediate host:	Intermediate	Non-virulent	Non-virulent	–	High
First-generation merogony (dpi)[b]:	14–19	7–14	–	–	6–21
Location of first-generation meronts:	Mesenteric and mesenteric lymph node arteries	Capillaries and arterioles of lung, kidney and brain	–	–	Arterioles and arteries of many inner organs
Size of first-generation meront (µm):	45.0–80.0 × 35.0–50.0	21.0–44.0 × 5.0–12.0	–	–	13.0–45.0 × 12.0–32.0
Size of first-generation merozoite (µm):	6.8–7.5 × 2.3–3.0	–	–	–	7.0 × 1.5
Number of first-generation merozoites:	100–200	–	–	–	18–168
Second-generation merogony (dpi)	26–31	–	–	–	16–40

Continued

Table 3.3.5. – *Continued*

Character	*S. arieticanis*	*S. gigantea*	*S. medusiformis*	*S. mihoensis*	*S. tenella*
Location of second-generation meronts:	Capillaries of many organs	–	–	–	Capillaries of many organs
Size of second-generation meront (μm):	–	–	–	–	10.0–42.0 × 7.0–18.0
Size of second-generation merozoite (μm):	6.0–7.5 × 2.3–3.0	–	–	–	15.0 × 6.0
Number of second-generation merozoites:	50	–	–	–	18–54
Third-generation merogony (dpi):	Yes	–	–	–	36
Location of third-generation meronts:	Leukocytes	–	–	–	Leukocytes, macrophages of visceral lymph nodes, Kupffer cells
Size of third-generation meront (μm):	–	–	–	–	7.4 × 5.1

Number of third-generation merozoites:	–	–	–	–	6–9
Formation of tissue cysts (*dpi*):	31	40	188	–	35
Location of tissue cysts:	Probably all striated muscles	Predominantly oesophageal, laryngeal and lingual muscles	Diaphragmatic, abdominal and skeletal muscles	Striated muscles	All striated muscles, CNS, Purkinje fibres
Time to maturity of tissue cyst (*dpi*)	70	230 (continue to grow for up to 4 years)	1132 (continue to grow for several years)	–	70
Size of tissue cyst (μm):	≤ 900	≤15,000 × 5,000	≤8,000 × 200	1,300–2,100 × 200–300	≤ 700
Morphology of tissue cyst wall:	Thin (< 1 μm), with hair-like protrusions (5–11 μm)	Thin (< 2 μm), smooth, with cauliflower-like protrusions, with connective tissue secondary cyst wall	Thin (< 2 μm), with trapezoidal villar protrusions, no secondary wall	Thick (10–12 μm), radially striated	Thick (1–3 μm), cross-striated, with finger-like villar protrusions (3.5 × 0.5 μm)
Definitive hosts:	*Canis familiaris* (dog)	*Felis catus* (domestic cat)	*Felis catus* (domestic cat)	*Canis familiaris* (dog)	*Canis dingo* (dingo), *Canis familiaris* (dog), *Canis latrans* (coyote), *Vulpes vulpes* (red fox)

Continued

Table 3.3.5. – *Continued*

Character	*S. arieticanis*	*S. gigantea*	*S. medusiformis*	*S. mihoensis*	*S. tenella*
Prepatent period (d):	≥ 12	11–13	10–21	11	8–9
Patency:	Several months	Several months	Several months	–	Several months
Size of sporocyst (μm):	14.0–15.0 × 9.0–10.5	10.5–14.0 × 8.0–9.7	10.3–13.0 × 7.3–8.8	15.0–16.0 × 8.0–9.0	15.0–16.5 × 9.8–10.5

[a] The data contained in this table have been taken from reviews of Dubey (1976); Frenkel *et al.* (1979); Erber and Burgkart (1981); Levine (1986); Dubey *et al.* (1989a); Heckeroth and Tenter (1989b); Jeffrey (1993); Vaughan (1996); Buxton (1998); Buxton and Henderson (1998); Taylor (2000); Uggla (2000); Markus *et al.* (2004), and original publications of Gestrich *et al.* (1974); Collins *et al.* (1976); Heydorn and Gestrich (1976); Ashford (1977); Leek *et al.* (1977); Collins *et al.* (1979); Leek and Fayer (1979); Munday (1981, 1984, 1986); Dubey *et al.* (1982, 1989b); Munday and Obendorf (1984); O'Donoghue and Ford (1984, 1986); Heydorn (1985); Heydorn and Karaer (1986); Leek (1986); O'Toole *et al.* (1986, 1993); Obendorf and Munday (1987); Dubey (1988); McKenna and Charleston (1988); Schmidtova and Breza (1991); McNally (1993); Hinaidy and Egger (1994); Mal'a and Baranova (1995); Caldow *et al.* (2000); Sargison *et al.* (2000).

[b] Days post-infection.

–: No data available.

Table 3.3.6. Biological and morphological data of *Sarcocystis* species infecting goats (*Capra hircus*).[a]

Character	*S. capracanis*	*S. hircicanis*	*S. moulei*
Geographical distribution:	Probably worldwide	Europe, Asia	Europe, Asia, probably Egypt
Intermediate hosts:	*Capra hircus*	*Capra hircus*	*Capra hircus, Capra sibirica*
Degree of virulence for the intermediate host:	High	Intermediate	Non-virulent
First-generation merogony (dpi)[b]:	10–12	17–18	–
Location of first-generation meronts:	Mesenteric and mesenteric lymph node arteries	Arterioles of intestinal lymph nodes and liver	–
Size of first-generation meront (μm):	26.1 × 17.4	–	–
Size of first-generation merozoite (μm):	5.5–7.1 × 2.5	6.0–7.5 × 2.5–3.0	–
Number of first-generation merozoites:	≤ 80	≥ 100	–
Second-generation merogony (dpi):	14–24	28–35	–
Location of second-generation meronts:	Capillaries of many organs	Capillaries of many organs	–
Size of second-generation meront (μm):	16.0–18.8 × 10.1–11.0	–	–
Number of second-generation merozoites:	4–36	–	–
Size of second-generation merozoite (μm):	5.5 × 1.5	6.0–7.5 × 2.5–3.0	–

Continued

Table 3.3.6. – *Continued*

Character	*S. capracanis*	*S. hircicanis*	*S. moulei*
Intraleukocytic multiplication:	Yes	–	–
Formation of tissue cysts (*dpi*):	30	43	–
Location of tissue cyst:	All striated muscles, CNS, Purkinje fibres	Probably all striated muscles	Oesophageal muscles
Time to maturity of tissue cyst (dpi):	64	84	> 730 (2 years)
Size of tissue cyst (mm):	≤ 1 × 0.1	≤ 2.5	≤ 16
Morphology of tissue cyst wall:	Thick (2–3 μm), radially striated, with finger-like villar protrusions (≤ 3.5 μm)	Thin (< 1 μm), with hair-like villar protrusions	Thick (≤ 10 μm), with cauliflower-like protrusions, with connective tissue secondary cyst wall
Definitive hosts:	*Alopex corsac* (corsac fox), *Canis familiaris* (dog), *Canis latrans* (coyote), *Canis lupus* (wolf), *Cerdocyon thous* (crab-eating fox), *Vulpes vulpes* (red fox)	*Canis familiaris* (dog)	*Felis catus* (domestic cat)
Prepatent period (d):	9–12	12–15	9–10
Patency:	Several months	Several months	Several months
Size of sporocyst (μm):	12.0–16.0 × 9.0–11.0	15.0–17.3 × 10.5–11.3	11.6–13.1

[a] The data contained in this table have been taken from reviews of Levine (1986); Dubey *et al.* (1989a); Engeland *et al.* (1994); Buxton (1998); Markus *et al.* (2004), and original publications of Fischer (1979); El-Rafaii *et al.* (1980); Heydorn and Unterholzner (1983); Foreyt (1989); Ghaffar *et al.* (1989); Mackie *et al.* (1992); Orr *et al.* (1984); Sharma (1991); Mal'a and Baranova (1995); Mackie and Dubey (1996).

[b] Days post-infection.

–: No data available.

As there is no reliable way to determine the age of *Sarcocystis* tissue cysts in naturally infected animals, caution should be taken with respect to describing new species purely on the basis of their tissue cyst morphology. There is still a need for definitive transmission and cross-transmission experiments to validate species descriptions, in particular for those *Sarcocystis* species that infect Bovinae such as cattle, yaks, buffaloes and bison.

Clinical signs

Irrespective of the true number of valid *Sarcocystis* species that parasitize ruminants, there is no doubt that domestic ruminants are intermediate hosts for several *Sarcocystis* species of differing virulence (see Tables 3.3.2, 3.3.3, 3.3.4, 3.3.5 and 3.3.6). In intermediate hosts, virulent *Sarcocystis* species can cause acute disease during the early phase of infection and chronic disease during the late phase of infection. Little is known about the mechanisms that cause virulence of *Sarcocystis* species. Most virulent *Sarcocystis* species cause disease only in their intermediate, but not in their definitive, hosts. In general, *Sarcocystis* species transmitted by canids are more virulent than those transmitted by felids (see Tables 3.3.2, 3.3.5 and 3.3.6).

The severity of clinical symptoms caused by virulent *Sarcocystis* species depends on the dose of ingested sporocysts and the immune status of the host. Clinical symptoms during the early merogonic phases, which take place in vascular endothelial cells of almost all inner organs, are usually more severe than those observed during tissue cyst formation and development in muscle or nervous tissue (Dubey *et al.*, 1989b; Cawthorn and Speer, 1990; Uggla and Buxton, 1990; Dubey and Rommel, 1992; O'Donoghue and Rommel, 1992). However, there are no clinical symptoms that are specific for sarcocystiosis.

Acute sarcocystiosis

Sarcocystis species infecting sheep, goats and cattle appear to be the most virulent species. The uptake of a high dose of sporocysts of a virulent *Sarcocystis* species, in particular of *S. tenella*, *S. capracanis* or *S. cruzi*, can lead to acute sarcocystiosis in previously uninfected animals that have not yet developed immunity to the respective species.

In addition to several non-specific symptoms, signs of acute sarcocystiosis include haemorrhagic diathesis, encephalitis and encephalomyelitis, which can cause the death of the animal. In pregnant animals, acute sarcocystiosis frequently results in fetal death, abortion or premature birth of the offspring (Dubey *et al.*, 1989b; Dubey and Rommel, 1992; Jeffrey, 1993).

Animals surviving a primary *Sarcocystis* infection usually acquire immunity that protects them against acute disease after challenge with the homologous species, but not after infection with a virulent heterologous species (Uggla and Buxton, 1990; O'Donoghue and Rommel, 1992).

Chronic sarcocystiosis

Chronic sarcocystiosis can result from the ingestion of a low dose of sporocysts of a virulent *Sarcocystis* species and can cause economic losses in the livestock industry due to reduced quality and quantity of meat or wool in cattle and sheep (Fayer and Elsasser, 1991; O'Donoghue and Rommel, 1992). Additional economic losses are caused by infections with *Sarcocystis* species that form macroscopic cysts in cattle or sheep, resulting in condemnation of whole carcases or affected parts after slaughter. Infections with *Sarcocystis* species have been estimated to account for annual losses of several million dollars due to downgrading of beef in the USA (Fayer and Dubey, 1986).

Diagnostic Techniques

Diagnosis of *Sarcocystis* infections in ruminants can be impeded by several properties of the parasites. Ruminants can be parasitized by several different species with similarities in biology and morphology (see Tables 3.3.2, 3.3.3, 3.3.4, 3.3.5 and 3.3.6). Traditional antigen preparations derived from virulent *Sarcocystis* species are highly cross-reactive with antibodies directed against non-virulent species (see Fig. 3.3.2).

As a consequence, most of the currently available immunological tests are only genus-specific and can not differentiate between virulent and non-virulent *Sarcocystis* species. Diagnosis of acute sarcocystiosis is further limited by the late onset of seroconversion and, thus, serological tests usually only detect chronic sarcocystiosis.

Therefore, diagnosis of acute sarcocystiosis and *Sarcocystis*-induced abortion has been based mainly on post-mortem examination, i.e. after the animal had succumbed to the disease. However, in recent years, molecular biological methods have facilitated the development of species-specific tests for the diagnosis of *S. cruzi* in cattle and the diagnosis and differentiation of *S. tenella* and *S. arieticanis* in sheep *intra vitam* during the acute phase of the disease. These tests allow, for the first time, comprehensive studies on the epidemiology of infections with virulent *Sarcocystis* species in sheep and cattle, as well as the importance of abortions caused by *S. cruzi*, *S. tenella* or *S. arieticanis*.

Here, we have compiled a collection of methods that have been used in the diagnosis of sarcocystiosis and *Sarcocystis*-induced abortion in ruminants. We also describe new strategies that may lead to the development of new diagnostic methods. Methods that are used regularly and with success in our laboratories are outlined in detail. In addition, we have given references for those methods that have been used successfully by other researchers, but with which we have no experience ourselves.

Because of the diversity among species in the genus *Sarcocystis*, it should be noted that it may be necessary to optimize conditions if a method described here

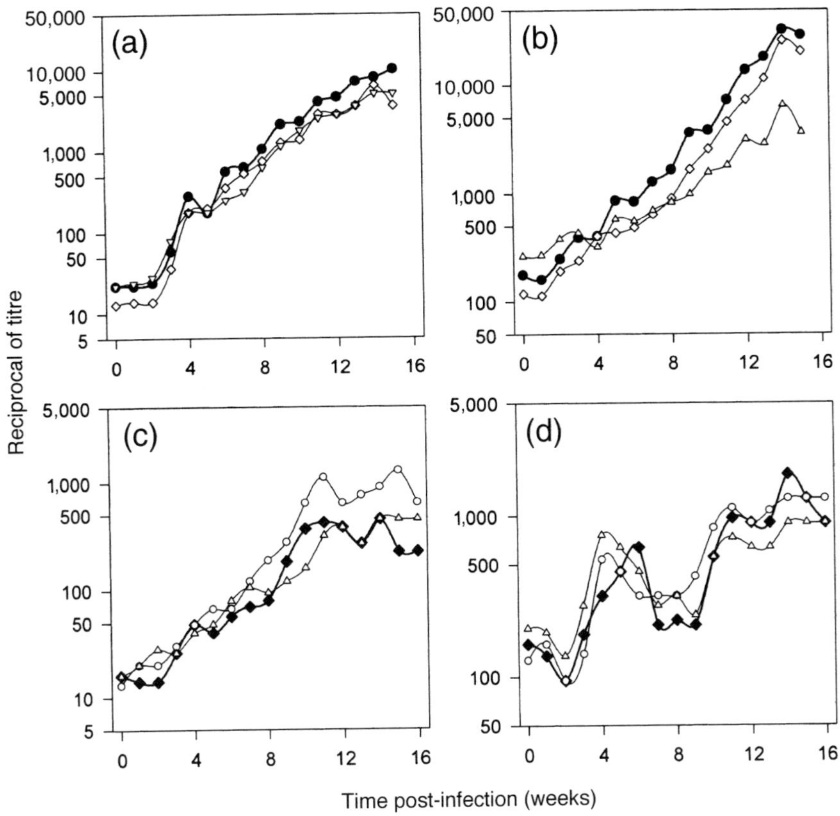

Fig. 3.3.2. Detection of *Sarcocystis*-specific antibodies in sera of sheep infected with *S. tenella* (a, b) and *S. arieticanis* (c, d) in an IFAT (a, c) and an ELISA (b, d). ●, antigen derived from cystozoites of *S. tenella*; ♦, antigen derived from cystozoites of *S. arieticanis*; Δ, antigen derived from cystozoites of *S. gigantea*. Filled symbols show reactions with antigen of the homologous species; open symbols show reactions with antigens of the heterologous species. Data shown are geometric mean titres of reactions measured for eight sheep infected experimentally with sporocysts of *S. tenella* and five sheep infected experimentally with sporocysts of *S. arieticanis* (reproduced from Tenter, 1995). Copyright: Australian Society for Parasitology (*International Journal for Parasitology*).

is used for a *Sarcocystis* species different from that it was originally developed for, and not all methods described in this chapter may be useful for each *Sarcocystis* species.

Detection of *Sarcocystis* species in intermediate hosts (ruminants)

Detection of tissue cysts

The isolation of intact tissue cysts from host muscle is difficult, and although several methods have been described (Markus, 1979; Agarwal *et al.*, 1987; Toparlak, 1987), their success depends greatly on the size of the tissue cyst formed by the *Sarcocystis* species under study and on the degree of infection of the host.

If infection levels are high, tissue cysts ≥ 500 μm may be isolated by dissecting muscle samples using a stereo-microscope and recovering the tissue cysts with dissecting needles. To isolate smaller tissue cysts, the following method may be used:

1. About 10 g of muscle samples are minced for 30 s in 80 ml saline using a blender.
2. The suspension is passed consecutively through 1.5 mm and 1.0 mm aperture sieves, and then centrifuged at 3000 *g* for 2 min.
3. The supernatant fluid is carefully removed from the top of the tube as far as the bottom 3 ml of fluid.
4. This is then mixed thoroughly with the sediment, and drops of this suspension are examined microscopically (magnification 15–50 ×) for intact tissue cysts (Rommel *et al.*, 1995).

This method is not very sensitive if infection levels are low, but if tissue cysts are detected it may allow the differentiation of *Sarcocystis* species based on tissue cyst wall morphology (see Tables 3.3.2, 3.3.3, 3.3.4, 3.3.5 and 3.3.6).

Detection and isolation of cystozoites

Several epidemiological studies on *Sarcocystis* infections in ruminants have been carried out by enzymatic digestion of muscle samples derived from animals at the time of slaughter and subsequent microscopic examination of the muscle digest for cystozoites. Enzymatic digestion is a sensitive method for detection of infection with any *Sarcocystis* species, even if infection levels are low.

However, it does not allow species differentiation, because the tissue cyst walls are also digested and the cystozoites of different *Sarcocystis* species are morphologically too similar to be differentiated from each other (Dubey *et al.*, 1989a; Rommel *et al.*, 1995). Hence, this method is not suitable for collection of accurate epidemiological data on rates of infection with virulent *Sarcocystis* species in ruminants.

ISOLATION OF CYSTOZOITES FOR ANTIGEN AND NUCLEIC ACID PREPARATIONS

Enzymatic digestion is also used to obtain *Sarcocystis* cystozoites for antigen preparations used in serological tests or for extraction of nucleic acids used in molecular biological assays. To obtain optimal yields of cystozoites, it is important to know the maturation time for the tissue cyst of the *Sarcocystis* species under study (see Table 3.3.2).

Tissue cyst maturation times vary between the different *Sarcocystis* species, and the tissue cysts of some *Sarcocystis* species degenerate over time. Immature cysts and degenerate tissue cysts do not contain cystozoites. Therefore, it would not be possible to isolate cystozoites from animals that are slaughtered too early or too late after initiation of infection. For most *Sarcocystis* species forming microscopic tissue cysts in ruminants, e.g. *S. tenella*, *S. arieticanis*, *S. capracanis* or *S. hircicanis*, 85–100 dpi (days post-infection) is an optimal time for the isolation of cystozoites. By contrast, the tissue cysts of *Sarcocystis* species that form macroscopic tissue cysts, e.g. *S. gigantea* or *S. moulei*, need 1–4 years to mature.

The enzymatic digestion method used for the isolation of cystozoites depends greatly on the size of the tissue cysts from which the cystozoites have to be liberated, on the infection level of the host and on the fat content of the host muscles. Because some of the methods that involve enzymatic digestion of large amounts of muscle samples from ruminants are very time-consuming, it is advisable to keep the muscle samples from which cystozoites are going to be isolated in phosphate-buffered saline (PBS, pH 7.2–7.4) at 4°C until they can be processed. This will keep the tissue cysts intact for at least 48 h.

Tryptic digestion of host tissue is our method of choice for detection of *Sarcocystis* species forming microscopic tissue cysts in the muscles of cattle, sheep and goats, e.g. for cystozoites of *S. cruzi*, *S. tenella*, *S. arieticanis*, *S. capracanis* or *S. hircicanis* (Rommel *et al.*, 1995). This method has been optimized in such a way that most of the host tissue is digested, while the majority of the isolated cystozoites stay intact (see Fig. 3.3.3).

Cystozoites isolated by this method can be used for extraction of proteins or nucleic acids for further studies. A comparison of proteins and RNA of *Sarcocystis* cystozoites isolated by three different methods has shown that short-time digestion of host tissue with trypsin does not adversely affect the biological activity of proteins or nucleic acids derived from the isolated cystozoites (Tenter *et al.*, 1991a). By using this method, 10^{10}–10^{11} cystozoites of *S. tenella* and 10^{8}–10^{9} cystozoites of *S. arieticanis* can be isolated from infected sheep.

DETECTION OF *SARCOCYSTIS* SPECIES BY TRYPTIC DIGESTION OF HOST TISSUE

1. Excise skeletal muscles from the carcasses of slaughtered animals. Remove all fat, ligaments and connective tissue, to prevent these interfering with the digestion.

2. Finely mince 30–50 g of the muscle samples with 150 ml of digestion fluid (see below) for 20 s at high speed in a 1 l laboratory blender (Waring®, Dynamics Corporation of America, New Hartford, Connecticut, USA). The ratio of muscle to digestion fluid depends on the type of muscles and the amount of fat and connective tissue present. Normally, it is best not to exceed a ratio of 1:4 (w/v). For some types of muscles it may be necessary to decrease this ratio to 1:5 (w/v) or 1:6 (w/v). The total volume of muscle samples and digestion fluid in the

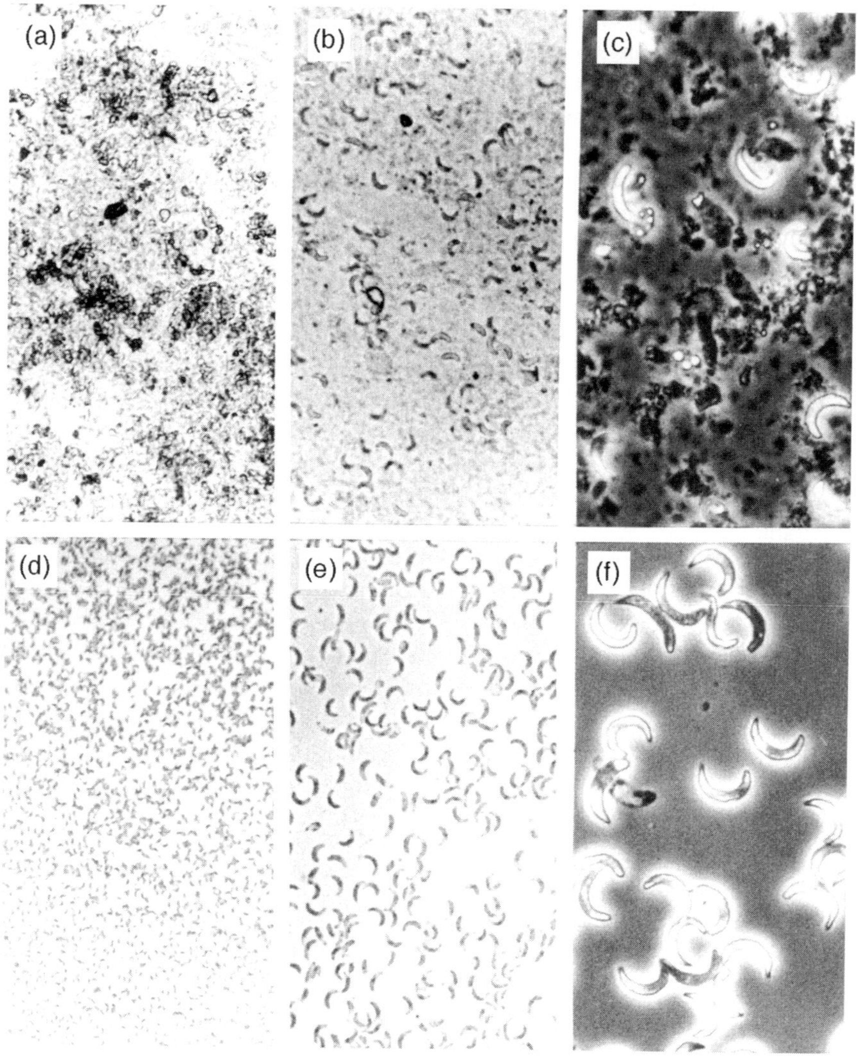

Fig. 3.3.3. Isolation and purification of *Sarcocystis* cystozoites. Preparation of *Sarcocystis* cystozoites after digestion of infected host muscle with trypsin (a–c). Preparation of *Sarcocystis* cystozoites after purification of the digest by density gradient centrifugation using Persoll (d–f).

blender should not exceed 20% of the blender's capacity, as otherwise the muscle samples will not be mixed properly with the digestion fluid.

- Digestion fluid: 2.5 g trypsin (Trypsin 1:250; Difco Laboratories, Detroit, Michigan, USA) per litre 0.01 M phosphate-buffered 0.15 M saline (PBS) with a pH value of 7.2–7.4.

3. Stir the tissue suspension at 37°C for 10–15 min in a magnetic water bath. This step should not exceed this length of time, otherwise any cystozoites present will be released from the tissue cysts and may be affected by the digestion process.
4. Filter the digest through three layers of gauze. Add 0.5–1.0 volume of cold PBS and centrifuge (2500 *g*, 15 min, 10°C) the filtered suspension immediately to stop the digestion.
5. Discard the top 75% of the supernatant fluid. Resuspend the bottom 25% contained in the centrifuge tube in fresh PBS and centrifuge as above.
6. Discard the supernatant fluid and wash the sediment three to five times (i.e. until the supernatant fluid is clear) by resuspension in PBS and centrifugation as above to remove all of the digestion fluid.

The isolate containing *Sarcocystis* tissue cysts and cystozoites can now be stored at 4°C in PBS for up to 1 d. It can then be combined with isolates obtained from more muscle samples for further purification of cystozoites by density gradient centrifugation (see Chapter 2, this volume). If the infection level of the animal host is low, i.e. if a large amount of undigested host tissue is still present, the isolate may be digested a second time to reduce the amount of material that has to be purified by density gradient centrifugation. For this, proceed as follows:

1. Suspend the isolate in PBS and mix with an equal volume of 2 × digestion fluid (5 g trypsin/l PBS). Stir the suspension at 37°C for 10–20 min in a magnetic water bath. Check the result of the digestion every 5 min by microscopical examination.
2. Filter the digest through three layers of gauze, add 0.5–1.0 volume of fresh PBS and centrifuge (2500 *g*, 15 min, 10°C) immediately to stop the digestion.
3. Discard the supernatant fluid. Wash the sediment three times by resuspension in PBS and centrifugation as above to remove all of the digestion fluid.
4. Purify cystozoites from tissue debris by density gradient centrifugation (see Chapter 2, this volume).

Mechanical isolation of cystozoites can be achieved for tissue cysts that are macroscopically visible, such as those of *S. gigantea* or *S. moulei* (Tenter *et al.*, 1991). The following method is quick and convenient for isolation of cystozoites from macroscopically visible tissue cysts. To avoid contamination of the isolated cystozoites with bacteria, which are always present when oesophageal muscles are being dissected, all steps should be carried out using aseptic or sterile conditions. Equipment and buffers should be autoclaved.

1. Excise intact *Sarcocystis* tissue cysts from host tissue using scissors and transfer the tissue cysts into a dish containing PBS, pH 7.2–7.4.
2. Wash the tissue cysts repeatedly in the dish with fresh PBS until the buffer stays clear. The following steps should ideally be carried out in a tissue-culture cabinet.
3. Finely dice the tissue cysts into pieces with scissors in PBS.

4. Transfer the PBS containing the diced tissue cysts to a glass homogenizer. Use ten to 20 gentle strokes with the homogenizer to liberate the cystozoites from the tissue cyst walls.
5. Transfer the homogenate to a centrifuge tube, and wash the cystozoites three to five times by centrifugation and resuspension in PBS until the supernatant buffer is clear.
6. Purify cystozoites from tissue cyst wall and host tissue debris by density gradient centrifugation (see Chapter 2, this volume).

PURIFICATION OF CYSTOZOITES FOR ANTIGEN AND NUCLEIC ACID PREPARATIONS Our method of choice for the purification of *Sarcocystis* cystozoites from tissue cyst wall and host tissue debris is a discontinuous density gradient centrifugation. This method yields preparations containing > 95% cystozoites (Tenter, 1987). If the cystozoites have been isolated by digestion of host tissue, it is possible to obtain preparations containing > 99.9% cystozoites (see Fig. 3.3.3). No host cell nuclei were detected in Giemsa-stained smears of *S. tenella* and *S. arieticanis* cystozoites that had been obtained by digestion of sheep muscles with trypsin and subsequent purification of the isolated cystozoites by density gradient centrifugation.

It is important to know that the density of *Sarcocystis* cystozoites varies with the age of the tissue cyst and with the *Sarcocystis* species. Therefore, the composition of discontinuous density gradients for purification of cystozoites has to be optimized accordingly. However, the following method can be regarded as a general guideline. All steps should be carried out using sterile conditions:

1. Prepare an isotonic stock solution of Percoll® (density 1.13 g/ml, osmolality 13–16 mosmol/kg H_2O) by adding 1 volume of 10 × PBS (pH 7.2–7.4) to 9 volumes of Percoll. For the bottom layer of the gradient, dilute the isotonic Percoll stock solution with 1 × PBS to give a solution of 80–90% Percoll. For the top layer of the gradient, dilute the isotonic Percoll stock solution with 1 × PBS to give a solution of 40–50% Percoll.
2. Pipette 4 ml of high-density Percoll solution (80–90%) into 15 ml conical centrifuge tubes and overlay them with 4 ml of low-density Percoll solution (40–50%) to prepare discontinuous density gradients. The number of tubes needed to purify about 10^9 cystozoites depends greatly on the *Sarcocystis* species, on the method used for isolation of the cystozoites and on the degree of contamination with host tissue. For example, two tubes may be sufficient for the purification of *S. gigantea* cystozoites, whereas 200 tubes or more may be necessary for the purification of *S. arieticanis* cystozoites.
3. Homogenize preparations of *Sarcocystis* cystozoites isolated by one of the methods described above by ten to twenty gentle strokes in a glass homogenizer. This should remove all lumps in the preparation to prevent clotting at the interphases of the gradient. Dilute the homogenate with PBS to give a transparent suspension. If the suspension from which cystozoites are to be purified is too concentrated, clotting may occur during centrifugation.

4. Overlay the Percoll gradients with 7 ml of the homogenized suspension and centrifuge the tubes for 30–40 min at 2500 *g* and 18°C. After centrifugation, a white layer containing the purified cystozoites should be found in the conical part of the centrifuge tube. Host cell debris is usually found at the interphase between the top, low-density Percoll layer and the supernatant PBS, while some tissue cyst wall debris forms a pellet underneath the bottom, high-density Percoll layer. If cystozoites and debris have not separated properly, the density of the two Percoll layers has to be optimized.
5. Remove and discard the top 12 ml of fluid from the centrifuge tubes.
6. Collect the layer containing the cystozoites and combine it with those collected from other density gradient tubes. Add at least 1/3 volume of PBS and centrifuge (2500 *g*, 10 min, 15°C) to remove the Percoll.
7. After centrifugation, the purified cystozoites should be pelleted at the bottom of the tube. Resuspend the pellet in PBS and wash the cystozoites three times by centrifugation (2500 *g*, 10 min, 4°C) and resuspend in PBS.

Isolated cystozoites do not survive freezing. To preserve the molecular composition of purified cystozoites, proceed as follows:

1. Pellet the cystozoites by centrifugation in a cryopreservation tube.
2. Completely remove and discard the supernatant buffer.
3. Snap-freeze and store the pellet in liquid nitrogen.

This method of storage allows the use of the cystozoites for extraction of proteins and antigens, as well as nucleic acids. The respective extraction buffer is added to the pellet immediately after removal from storage. Cystozoites stored by this method can also be used for extraction and purification of messenger RNA; however, in these cases it may be preferable to add 1 ml of 6 M guanidine hydrochloride/0.2 M sodium acetate (pH 5.2)/1 mM 2-mercaptoethanol to the cystozoite pellet before storage in liquid nitrogen.

Pathology

HISTOPATHOLOGY Methods for the detection and identification of life cycle stages of the vascular phase of *Sarcocystis* species (meronts and merozoites) in histological sections have been described by Jeffrey *et al.* (1988) and Granstrom *et al.* (1991). Histopathological evidence of an active infection may present with inflammatory changes in the central nervous system (CNS), sometimes with prominent lesions in the spinal cord. Typically, acute infection induces a non-suppurative meningoencephalitis, associated with sarcocyst-like merozoites and meronts, focal microgliosis and mild oedematous changes (Jeffrey *et al.*, 1988; Dubey *et al.*, 1989b; O'Toole *et al.*, 1993).

Mackie *et al.* (1992) reported a stillborn goat kid with a multifocal, necrotizing, non-suppurative encephalitis with associated vascular endothelial cells featuring sarcocyst-like meronts containing merozoites. In other tissues, the heart, tongue and liver may harbour mononuclear cell inflammation, while tissues such as skeletal muscle, lung and kidney are usually less affected (Uggla and Buxton, 1990), although in a spontaneous outbreak of acute myopathy in sheep linked with a sarcocystis infection, histopathological lesions were confined largely to skeletal muscles and consisted of a non-suppurative myositis associated with mature, immature and degenerate sarcocysts (Jeffrey *et al.*, 1989).

As with toxoplasmosis and neosporosis, immunohistochemistry is a valuable aid for confirming a diagnosis (Jeffrey *et al.*, 1988; O'Toole *et al.*, 1993).

IMMUNOHISTOCHEMISTRY Several immunohistochemical methods have been developed for the improvement of the sensitivity and specificity of histological detection of life cycle stages of *Sarcocystis* species (Smith and Herbert, 1986; Jeffrey *et al.*, 1988; Uggla and Buxton, 1990; Hamir *et al.*, 1993). However, these methods are complicated by the fact that polyclonal antisera derived from animals infected experimentally with any species of *Sarcocystis* show a high cross-reactivity with antigens of heterologous *Sarcocystis* species (see Fig. 3.3.2). In addition, immune sera of rabbits immunized with traditionally derived *Sarcocystis* antigens may also cross-react with antigens of other cyst-forming coccidia infecting sheep, such as *Toxoplasma gondii* (Jeffrey *et al.*, 1988; Uggla and Buxton, 1990).

Therefore, a species-specific diagnosis of *Sarcocystis* stages can be made only when specific monoclonal antibodies are used for immunohistochemistry (Uggla and Buxton, 1990). However, the repertoire of monoclonal antibodies that are directed against *Sarcocystis* species of ruminants is limited. Thus far, monoclonal antibodies have been produced only for *S. cruzi*, *S. tenella*, *S. arieticanis* and *S. gigantea*, but most of these are cross-reactive with antigens of heterologous *Sarcocystis* species (Burgess *et al.*, 1988; O'Donoghue *et al.*, 1990). Species-specific monoclonal antibodies have been developed only for *S. tenella* and *S. gigantea* (O'Donoghue *et al.*, 1990; O'Donoghue and Rommel, 1992; Tenter, 1995).

Serological methods

Several immunological tests have been developed for the serological diagnosis of *Sarcocystis* infections in ruminants (Dubey *et al.*, 1989a; Uggla and Buxton, 1990). The most reliable and commonly used tests are the enzyme-linked immunosorbent assay (ELISA) and the indirect fluorescent antibody test (IFAT) (Uggla and Buxton, 1990; O'Donoghue and Rommel, 1992; Rommel *et al.*, 1995). These tests show a high sensitivity for the late phase of infection. However, they are not species-specific because traditional antigen preparations derived from cystozoites or merozoites of any *Sarcocystis* species are highly cross-reactive with antibodies directed against heterologous *Sarcocystis* species (Reiter *et al.*, 1981; Weiland *et al.*, 1982; Gasbarre *et al.*, 1984; O'Donoghue and Weyreter, 1984; Smith and Herbert, 1986; Savini *et al.*, 1994a; Tenter, 1995).

ELISA Antigens have been prepared by destruction of purified cystozoites or merozoites by several freeze/thaw cycles only (Heckeroth *et al.*, 1999b), by freezing/thawing followed by ultrasonication (Savini *et al.*, 1994a) or by disruption in a homogenizer and subsequent dialysation (Gasbarre *et al.*, 1984). Intact cystozoites have also been shown to be a suitable antigen (Morsy *et al.*, 1994). In the following, an ELISA based on *S. tenella* and *S. arieticanis* antigens is described as an example (see Fig. 3.3.4).

1. Suspend cystozoites from *S. tenella* and *S. arieticanis*, derived from experimentally infected sheep and isolated as described earlier, in PBS (pH 7.4).
2. Disrupt them by three freeze/thaw (–80°C/+37°C) cycles, and store at –20°C.
3. Before use in the ELISA, the cystozoite antigen preparation should be solubilized by adding an equal volume of 8 M urea and incubating the mixture at 22°C for 15 min.
4. Coat ELISA plates (NUNC Maxisorb™) with either the *S. tenella* antigen preparation (1.5 μg protein/well) or the *S. arieticanis* antigen preparation (1.0 μg protein/well) in a carbonate/bicarbonate buffer (pH 9.6) and incubate at 4–8°C overnight.
5. After washing with PBS (pH 7.4), block the plates with PBS/5% BSA for 1 h at 37°C and wash again.
6. Dilute control and test sera with PBS/Tween 20/2% BSA to a starting concentration of 1:40 and dilute further in 1:2 steps in the respective buffer.
7. After an incubation at 37°C for 1 h, wash the plates and incubate again (37°C, 1 h) with horseradish peroxidase (HRPO)-conjugated rabbit anti-sheep IgG (H+L) in PBS/Tween 20/2% BSA.
8. After another washing step, incubate the plates with orthophenylene diamine (OPD) for 10 min at ambient temperature in the dark, and then stop the reaction with 2.5 M H_2SO_4.
9. Determine the optical density (OD) at 492 nm.
10. Ensure comparability of different tests by calculation of index values (Index = OD test serum – OD blank/OD positive control – OD blank).
11. Estimate sensitivity and specificity by a two-graph receiver operating characteristics program (TG-ROC; Greiner, 1995, 1996; Greiner *et al.*, 1995) using sera known to be negative (*n*, 31) and positive (*n*, 14) for *S. tenella* or *S. arieticanis*.

In previous trials at a dilution of 1:320, sensitivity and specificity were 100% and 100% for the *S. tenella*-ELISA and 92.9% and 96.8% for the *S. arieticanis*-ELISA, respectively.

IFAT For preparation of IFAT antigen, intact cystozoites are used after purification with one of the above-mentioned methods (Reiter *et al.*, 1981; Tenter, 1987). Below, the antigen preparation and the test procedure for a *S. muris* IFAT are described as an example. The method can also be used for other *Sarcocystis* species.

1. Purify cystozoites derived from experimentally infected mice by Percoll gradient centrifugation.

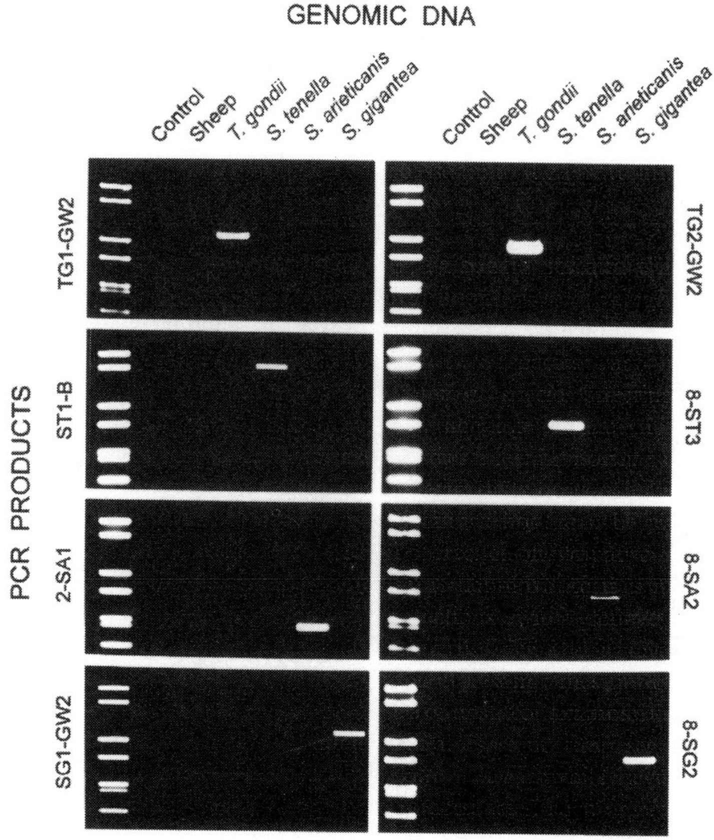

Fig. 3.3.4. Species-specific PCR products derived from the 18s rDNA of *S. tenella* (ST1–B, 8–ST3), *S. arieticanis* (2–SA1, 8–SA2) and *S. gigantea* (SG1–GW2, 8–SG2). PCR products derived from the 18s rDNA of *T. gondii* (TG1–GW2, TG2–GW2) were used as a control. Control, negative control without endogenous DNA. Marker, pBR322 fragments restricted with either *Bst* NI or *Msp* I: 1058, 929, 622, 527, 404, 383, 307 bp (reproduced from Tenter, 1995). Copyright: Australian Society for Parasitology (*International Journal for Parasitology*).

2. Isolated cystozoites are then diluted in PBS (pH 7.2) to a concentration of appr. 1.5–2.0 × 10^3 cystozoites per ml and dropped (one drop/approx. 4 ml per circle) on IFAT microscope slides (15-well multitest slides; Flow Laboratories, McLean, Virginia, USA).

3. Dry the slides overnight at 27°C, wrap in aluminium foil with tissue paper separating each slide and store at –80°C.

4. When required, batches of these prepared slides are thawed at ambient temperature in a desiccator for 2 h, before unwrapping and placing in a moist chamber.

5. Dilute control and test sera in cold PBS to a starting concentration of 1:40 and dilute further in 1:2 steps in PBS. Pipette one drop (~ 4 μl) of each serum dilution into each respective circle (well) and incubate the slides at 37°C for 30 min and then wash the sera off with PBS.
6. Rinse the slides twice in fresh PBS on a magnetic stirrer (medium speed) and then rinse once in distilled water.
7. Air-dry the slides and pipette one drop (~2 μl) of FITC-conjugated rabbit anti-mouse IgG (H+L) into each circle.
8. Incubate, wash and dry the slides as before and then overlay with glycerine-buffer (9 volumes glycerine + 1 volume PBS) and cover with a coverslip.

The degree of fluorescence is scored as 3+, 2+, 1+, trace and negative. The highest serum dilution showing a 1+ fluorescence is taken as the IFAT titre.

The optimal working dilution for the conjugate is determined by block titration against positive and negative reference sera. For this titration, the conjugate must be diluted in PBS containing 25% Evans blue stock solution (0.25 g Evan's blue in 999.9 ml PBS) as counterstain. The conjugate is then used at a dilution which gives no loss in titre of the positive reference serum and which at the same time gives no reaction with the negative reference serum (1:32–1:64). The conjugate is diluted in cold PBS and stored frozen (–80°C for long storage, –20°C when in use) in aliquots. Evans blue stock solution is added to this dilution after thawing at 4°C.

Molecular biological methods

In recent years, the advent of new molecular biological techniques has provided new diagnostic means for parasitic infections. Direct diagnostic tests based on the detection of parasite molecules by specific DNA probes or PCR techniques provide a valuable alternative to indirect tests based on the detection of parasite-specific antibodies, and may be able to overcome the diagnostic drawbacks caused by delayed detection of antibodies at the time of acute sarcocystiosis or by the lack of antibody responses in definitive hosts of *Sarcocystis* spp.

EXTRACTION OF NUCLEIC ACIDS Nucleic acids can be extracted from *Sarcocystis* cystozoites isolated mechanically or by digestion of host tissue with trypsin or pepsin (see above). Enzymatic digestion of host tissue does not adversely affect the biological activity of nucleic acids extracted from the isolated cystozoites (Tenter *et al.*, 1991a). Commercially available and widely used extraction kits are not only suitable for extraction of nucleic acids from purified material, but also from clinical samples such as tissue and blood.

However, in order to obtain large amounts of purified DNA for assay development and validation, digestion with proteinase K, extraction with phenol/chloroform followed by an ethanol precipitation can be used (Tenter *et al.*, 1994; Joachim *et al.*, 1996). As an example, a method for DNA extraction from purified cystozoites is described:

1. Resuspend cystozoite pellets in 10 ml cold buffer (10 mM Tris HCl, 10 mM EDTA, pH 8.0) and lyse with sodium dodecyl sulphate at a final concentration of 1%.
2. After the addition of 200 µg/ml proteinase K, incubate the suspension at 56°C for 2 h with periodic mixing.
3. Add fresh proteinase K to a final concentration of 400 µg/ml and incubate at 56°C for a further 2 h with periodic mixing.
4. After adding 0.2 volumes of 6 M sodium perchlorate, add an equal volume of phenol/chloroform/isoamylalcohol (25:24:1/v:v:v) and separate the two phases by centrifugation at 10,000 *g* for 20 min at 10°C in a fixed-angle rotor centrifuge.
5. Repeat the extraction step at least twice, until the interphase is clear, and follow by a single chloroform extraction.
6. The aqueous phase is transferred to a new tube before adding RNAse to a final concentration of 0.2 mg/ml and incubating for 15 min on ice.
7. Precipitate the DNA with either 0.5 volumes of 7.5 M ammonium acetate (pH 7.5) and 3 volumes of ice-cold 100% ethanol or 0.1 volumes of 3 M sodium acetate (pH 5.5) and 2.5 volumes of ice-cold 100% ethanol.
8. Allow the DNA to precipitate at –20°C overnight or at –70°C for several hours.
9. Pellet the DNA by centrifugation at 10,000 *g* for 10 min at 0°C and then wash the pellet twice with ice-cold 70% ethanol and once with ice-cold 100% ethanol and allow it to dry at ambient temperature.
10. Dissolve the DNA in TE buffer (10 mM Tris HCl, 1 mM EDTA, pH 8.0) and store at 4°C until use. Determine the concentration of DNA by measuring the UV adsorbance at 260 nm.

In principle, parasitic DNA can be extracted from blood samples by the above-mentioned method as well as with commercially available kits. However, since these methods are time-consuming and/or expensive, and thus not applicable to diagnostic samples, a quick and easy method for extraction of *Sarcocystis* DNA from blood 'buffy coat' cells derived from experimentally infected sheep was established (Heckeroth *et al.*, 1999a):

1. Centrifuge blood samples in haematocrit capillaries at 5500 *g* for 4 min. This centrifugation should increase the numbers of parasites in the buffy coat and/or the erythrocyte fraction beneath the buffy coat. Therefore, the buffy coats and about 1 mm of the adjoining erythrocyte fraction are isolated from six haematocrit capillaries. Combine these samples (equivalent to about 400 µl of the initial blood sample) and store at –20°C.
2. For DNA extraction, thaw the samples at ambient temperature and suspended in 1 ml erythrocyte lysis buffer containing 0.32 M sucrose, 10 mM Tris-HCl (pH 7.5), 5 mM magnesium chloride and 1% Triton X-100. After centrifugation at 1400 *g* for 5 minutes discard the supernatant.
3. Resuspend the pellet in 1 ml erythrocyte lysis buffer and wash by centrifugation as before.

4. Suspend the pellet in 100 μl TNN buffer containing 0.5% Tween 20, 0.5% Nonidet P-40 and 10 mM NaOH.
5. Boil this preparation for 5 min and centrifuge at high speed for 2 min to remove any remaining cell fragments.
6. Add 10 μl of the supernatant directly to an initial reaction mixture of the nested PCR (see Chapter 2, this volume).

This method may also be applied to whole blood samples. Lysis of erythrocytes and removal of haemoglobin – known to inhibit the PCR reaction – is critical and, therefore, the preparation should be washed with erythrocyte lysis buffer until it appears clear.

HYBRIDIZATION ASSAYS Species-specific oligonucleotides for *S. cruzi* and *S. tenella* should be synthesized after sequence comparison of the 18S rRNA of various *Sarcocystis* species and *T. gondii* (Holmdahl *et al.*, 1993) and used for dot-blots targeting RNA from the respective species. The method is as follows:

1. Dissolve the RNA in ice-cold 10 mM NaOH with 1 mM EDTA and immediately apply to blotting membranes.
2. After rinsing in 2 × SSC (1 × SSC is 0.15 mM NaCl and 15 mM sodium citrate) with 0.1% SDS, air-dry the membranes and store at ambient temperature until required.
3. After 30 min of prehybridization in 2 × SSC/2% SDS, 0.15 mg/ml denatured herring sperm DNA and 10 × Denhardt's solution, add the radiolabelled probe and incubate for 2 h.
4. After hybridization, rinse the membranes with 2 × SSC at different temperatures and visualize the hybridization signals by autoradiography for 6–24 h at –70°C.
5. The assay should be able to detect 25 ng RNA. Use clones selected from a *S. cruzi* sporozoite genomic library to detect merozoite DNA in blood samples (Ndiritu *et al.*, 1996).
6. Resuspend buffy coat pellets in 50 volumes of cold cell lysis buffer (10 mM Tris HCl (pH 7.4), 10 mM NaCl, 10 mM EDTA).
7. Add 10% SDS to the buffer to a final concentration of 1% and proteinase K to a final concentration of 200 μg/ml and incubate the suspension at 37°C overnight.
8. Extract the DNA with the phenol/chloroform/isoamylalcohol method as described above.
9. Load the DNA in TE buffer on to nylon membrane filters and bake in a vacuum oven at 80°C for 2 h.
10. Treat the filters with prehybridization buffer (5 × SSC, 5 × Denhardt's, 50 mM sodium phosphate (pH 6.5), 0.1% w/v SDS, 250 μg/ml salmon sperm DNA, 50% v/v formamide, 100 μg/ml tRNA) overnight at 42°C.
11. After draining the buffer, hybridize the filters with the radiolabelled DNA probe.

12. Carry out high-stringency washing to remove excess and non-specifically bound probe (2 × SSC/0.1%SDS twice for 15 min at room temperature, 0.1 × SSC/0.1% SDS twice for 15 min at room temperature and 0.1 × SSC/0.1% SDS twice for 30 min at 55°C).

Hybridization signals may be detected by autoradiography at –70°C for 48 h. Although the probe will hybridize to *S. cruzi* and *S. hirsuta* DNA and thus is only genus-specific, detection of *S. cruzi* merozoite DNA in blood samples is possible. The detection limit of this assay is around 27 pg DNA.

RAPD-PCR Different assays using randomly designed primers have been described for the detection of *Sarcocystis* DNA in sheep (Joachim *et al.*, 1996; Aledmir and Dik, 2003) and cattle (MacPherson and Gajadhar, 1994; Gueclue *et al.*, 2004). Primer sequences and PCR conditions are summarized in Table 3.3.7.

As an example, a RAPD-PCR for detection of *S. tenella*, *S. arieticanis* and *S. gigantea* is described (Joachim *et al.*, 1996):

1. Thirty nanograms of genomic DNA derived from cystozoites from experimentally infected sheep is used as a template.
2. The reaction is carried out in a final volume of 30 μl, containing 3.33 mM magnesium chloride, 67 mM Tris HCl (pH 8.8), 16.6 mM ammonium sulphate, 0.2 mg/ml gelatine and 0.45% Triton X-100.
3. Each reaction tube should contain 1.4 U of Taq polymerase, 1 pM primer and 0.2 mM each of dATP, dCTP, dGTP and dTTP.

The OPF-10 primer has been shown to reveal distinct fragment patterns for the three parasites. One DNA fragment amplified with DNA from *S. tenella* and *S. arieticanis* (but not with *S. gigantea* or the closely related *T. gondii*) was isolated, radiolabelled and used as a probe in a Southern blot. The probe was shown to be specific for the virulent *Sarcocystis* species (*S. tenella* and *S. arieticanis*) in sheep.

PCR-RFLP Based on the 18S rDNA from *S. fusiformis* (Holmdahl *et al.*, 1994), Li *et al.* (2002) have developed a PCR followed by restriction of PCR products by 12 different endonucleases (see Table 3.3.7):

1. The PCR is performed in 50 μ, using 10 pM of each primer and 1.25 U *Taq*polymerase.
2. In a 20 μl reaction, 2–5 μl PCR product should be mixed with 5–10 U of restriction enzyme and the appropriate restriction enzyme buffer.
3. Incubate the reactions for 4–12 h at 37°C or 65°C and terminate by the addition of loading buffer (50% glycerol, 0.1 M EDTA, 0.01% bromphenolblue, pH 8.0).
4. Electrophorese the samples on 2% agarose gels at 60 V for 120 min.

This PCR-RFLP system should give evidence that *Sarcocystis* species derived from different intermediate hosts such as water buffaloes and cattle belong to the same species (Yang *et al.*, 2002).

Table 3.3.7. Set-up of different PCR assays used for the detection or differentiation of *Sarcocystis* species in domestic ruminants.

PCR type	Target	Designation of primers	Sequence of primers	Template	PCR conditions	Reference
RAPD-PCR	*S. arieticanis, S. gigantea, S. tenella*	10-mer primer, e.g. OPF-10	5'–GGAAGCTTGG–3'	Extracted genomic DNA	45 cycles at 94°C for 1 min, 36°C for 1 min, 72°C for 2 min	Joachim *et al.* (1996)
RAPD-PCR	*S. cruzi*	TGA; or TGB; or TGC; or TGD; or TGE	5'-CCAGGGGAAGAGGCAT-3'; 5'-AGACCGAAAGTCAACGCGAC-3'; 5'-GTGGAGAAATCCAGAA-3'; 5'-CTGGAAAAAACTCCAC-3'; 5'-GCACGAACGCGCCACAAA-3'	Extracted genomic DNA	94°C for 10 min, then 45 cycles at 94°C for 1.5 min, 42°C for 1.5 min, 72°C for 1.5 min; final extension step at 72°C for 10 min	MacPerson and Gajadhar (1994)
RAPD-PCR	*S. arieticanis, S. gigantea, S. tenella*	OSA–01; or OSA–02; or OSA–03	5'-CCAAGCTTCC-3'; 5'-GCAAGCTTGG-3'; 5'-CCAGTACTCC-3'	DNA extract derived from host muscles	-	Aldemir and Dik (2003)
RAPD-PCR	*S. cruzi, S. hirsuta, S. hominis*	OSA–04; or OSA–05; or OSA–06; or OSA–07; or OSA–08	5'-CCAGGGGAAGAGGCAT-3'; 5'-AGACCGAAAGTCAACGCGC-3'; 5'-GTGGAGAAATCCAGAA-3'; 5'-CTGGAAAAAACTCCAC-3'; 5'-GCACGAACGCGCCACAAA-3'	Extracted genomic DNA	45 cycles at 94°C for 1 min, 42°C for 1.5 min, 72°C for 2 min	Güçlü *et al.* (2004)

Continued

Table 3.3.7. – *Continued*

PCR type	Target	Designation of primers	Sequence of primers	Template	PCR conditions	Reference
PCR-RFLP	*S. fusiformis, S. hirsuta, S. hominis, S. sinensis*	18S9L (forward), 18S1H (reverse)	5'-GGATAACCTGGTAATTCTATG-3'; 5'GGCAAATGCTTTCGCAGTAG-3'	–	97°C for 3 min, then 40 cycles at 94°C for 40 s, 56°C for 60 s, 72°C for 80 s; final extension step at 72°C for 5 min	Li *et al.* (2002); Yang *et al.* (2002)
PCR	*S. arieticanis*	2 (forward), SA1 (reverse); or 8 (forward), SA2 (reverse)	5'-AGGGTTCGATTCCGGAG-3'; 5'-GCGGGAAGAGGAGAAT-3'; 5'-TTTGACTCAACACGGG-3'; 5'-TGAAACGGCGCGTAGA-3'	Extracted genomic DNA	93°C for 5 min, then 26 or 36 cycles at 93°C for 2 min, 57°C for 2 min, 72°C for 2 min,; inal extension step at 72°C for 5 min	Tenter *et al.* (1994)

PCR[a]	*S. gigantea*	SG1 (forward), GW2 (reverse); or 8 (forward), SG2 (reverse)	5'-ACCAACAGTGCGCCAT-3'; 5'-TCAGTCCTAGAAACCAACAAA-3'; 5'-TTTGACTCAACACGGG-3'; 5'-AAAAGGAACCGCGACA-3'	Extracted genomic DNA	93°C for 5 min, then 26 or 36 cycles at 93°C for 2 min, 57°C for 2 min, 72°C for 2 min; final extension step at 72°C for 5 min	Tenter *et al.* (1994)
PCR[b]	*S. tenella*	ST1 (forward), B (reverse); or 8 (forward), ST3 (reverse)	5'-GGATCGCATTATGGTCAT-3'; 5'-CCGTCAATTCMTTTRAGTTT-3'; 5'-TTTGACTCAACACGGG-3'; 5'-CGTTGCCGCGCGTTAA-3'	Extracted genomic DNA	93°C for 5 min, then 26 or 36 cycles at 93°C for 2 min, 57°C for 2 min, 72°C for 2 min; final extension step at 72°C for 5 min	Tenter *et al.* (1994)

Continued

Table 3.3.7. – *Continued*

PCR type	Target	Designation of primers	Sequence of primers	Template	PCR conditions	Reference
Nested PCR	*S. arieticanis*	External primers: STA (forward) SA2 (reverse); nested primers: 2 (forward) SA1 (reverse)	5'-TTTCGCAAGGAAGAGGA-3'; 5'-TGAAACGGCGCGTAGA-3'; 5'-AGGGTTCGATTCCGGAG-3'; 5'-GCGGGAAGAGGAGAAT-3'	DNA extract derived from peripheral blood samples	94°C for 4 min, then 26 cycles at 93°C for 2 min, 57°C for 2 min, 72°C for 2 min; final extension step at 72°C for 5 min	Heckeroth and Tenter (1999)
Nested PCR	*S. tenella*	External primers: ST1 (forward) AP2 (reverse); nested primers: 8 (forward) ST3 (reverse)	5'-GGATCGCATTATGGTCAT-3'; 5'-CCCGGGATCCAAGCTTGATCCT TCTGCAGGTTCACCTAC-3' 5'-TTTGACTCAACACGGG-3'; 5'-CGTTGCCGCGCGTTAA-3'	DNA extract derived from peripheral blood samples	94°C for 4 min, then 26 cycles at 93°C for 2 min, 57°C for 2 min, 72°C for 2 min; final extension step at 72°C for 5 min	Heckeroth and Tenter (1999)

–: No data available.

[a] This PCR can also be used to detect *S. moulei* in goats.

[b] This PCR can also be used to detect *S. capracanis* in goats.

PCR Sequences of the 18S rDNA of three *Sarcocystis* species that infect sheep in Europe, i.e. *S. tenella*, *S. arieticanis* and *S. gigantea*, as well as closely related cyst-forming coccidia, i.e. *T. gondii* and *N. caninum*, have been used for the generation of species-specific oligonucleotides that can be used as primers for specific amplification of 18S rDNA fragments from genomic DNA templates of the homologous species by PCR (see Fig. 3.3.4; Tenter *et al.*, 1994).

In order to obtain optimal sensitivity for diagnostic applications, nested PCRs combining the above-mentioned primers and which consist of two consecutive amplifications have been developed for *S. tenella* (ST-Nested-PCR) and *S. arieticanis* (SA-Nested-PCR), validated and adapted for clinical samples (Heckeroth and Tenter, 1999a):

1. After optimization of individual PCR parameters, single-step amplifications are carried out in 100 μl reaction volumes containing 10 mM Tris-HCl, 50 mM potassium chloride, 0.1% Triton X-100, 1.75 mM magnesium chloride, 0.1 mM each of dGTP, dATP, dTTP and dCTP, 100 pmol of each primer and the template DNA.
2. After initial denaturation of the template DNA at 94°C for 4 min, 1.5 U of *Taq* polymerase should be added to the reaction mixture, which is then overlaid with mineral oil to prevent evaporation.

Under laboratory conditions, the ST-Nested-PCR and SA-Nested-PCR can detect 100 fg and 10 fg of genomic DNA, respectively, which corresponds to less than one parasite (see Fig. 3.3.5). These assays enable the diagnosis and differentiation of infections with *S. tenella* and *S. arieticanis* in sheep *intra vitam* during the acute phase of the disease, as has been proved by examination of blood samples from experimentally infected sheep (see Figs 3.3.5, 3.3.6 and 3.3.7). PCR results directly correlate with clinical symptoms after experimental infection, which indicates that circulating merozoites are being detected.

Detection of ruminant *Sarcocystis* species in definitive hosts

Sporocysts can be detected in, and collected from, faeces of definitive hosts by coproscopic methods. Faecal material is subjected to a flotation procedure with a solution of either NaCl/$ZnCl_2$ or $ZnSO_4$ (specific gravity 1.3).

The sporocysts are aspirated from the surface with a Pasteur pipette, the tip of which is broken off to give an oval mouth. When working with larger volumes, the sporocysts may be aspirated using a 10–20 ml glass pipette with pipette bulb. The opening of the pipette is immersed in the flotation fluid in such a way that only half of it is submerged, to ensure that only the surface is aspirated. If the mouth of the pipette is totally immersed, only flotation fluid may be aspirated, leaving the sporocysts on the surface of the fluid. Aspirated sporocysts are washed at least three times by centrifugation and resuspension in distilled water.

It is important to be aware that only low numbers of sporocysts are excreted in the faeces of definitive hosts (see Fig. 3.3.8). Therefore, large amounts of

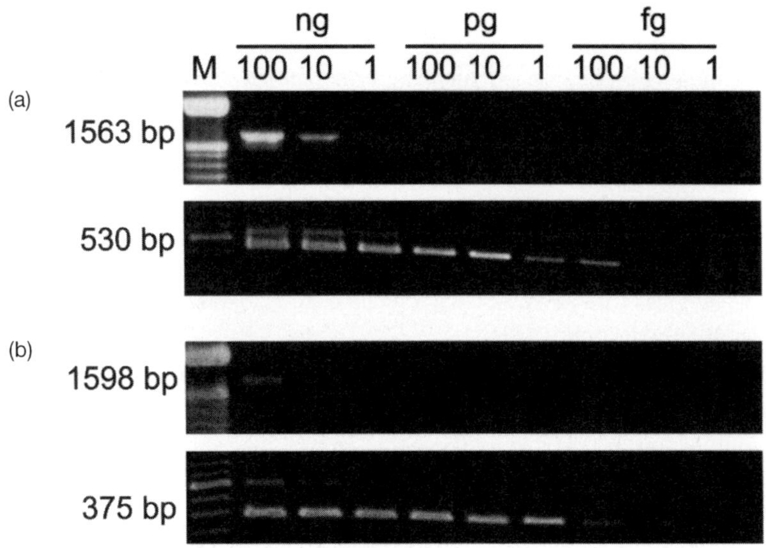

Fig. 3.3.5. Sensitivity of the ST-nested-PCR (a) and SA-nested-PCR (b) as assessed with genomic DNA templates of *S. tenella* (a) and *S. arieticanis* (b). Upper gel image: sensitivity obtained by the initial amplification; lower gel image: sensitivity obtained by the nested amplification. M, 100 bp marker showing a double-intensity band at 600 bp (reproduced from Heckeroth and Tenter, 1999a). Copyright: Australian Society for Parasitology (*International Journal for Parasitology*).

faeces have to be processed to obtain a reasonable number of sporocysts for further characterization. Other authors have described methods that employ NaCl solution (specific gravity 1.2) for the isolation of sporocysts from faeces (McKenna and Charleston, 1988b).

If transmission studies are planned, sporocysts can be stored at 4°C in water containing penicillin (200 IU/ml), streptomycin (200 μg/ml) and amphotericin (1 μg/ml). They remain viable for at least 6 months. The addition of potassium dichromate, which is recommended as a preservative for oocysts of other genera of coccidia, may kill *Sarcocystis* sporocysts (Leek and Fayer, 1979). Sporocysts of *S. cruzi* were still infective for calves after storage in water at −22°C for 10 d, and *S. tenella* sporocysts were still infective for lambs after storage in HBSS-PMSF medium at −15–20°C for 19 months (Leek, 1986).

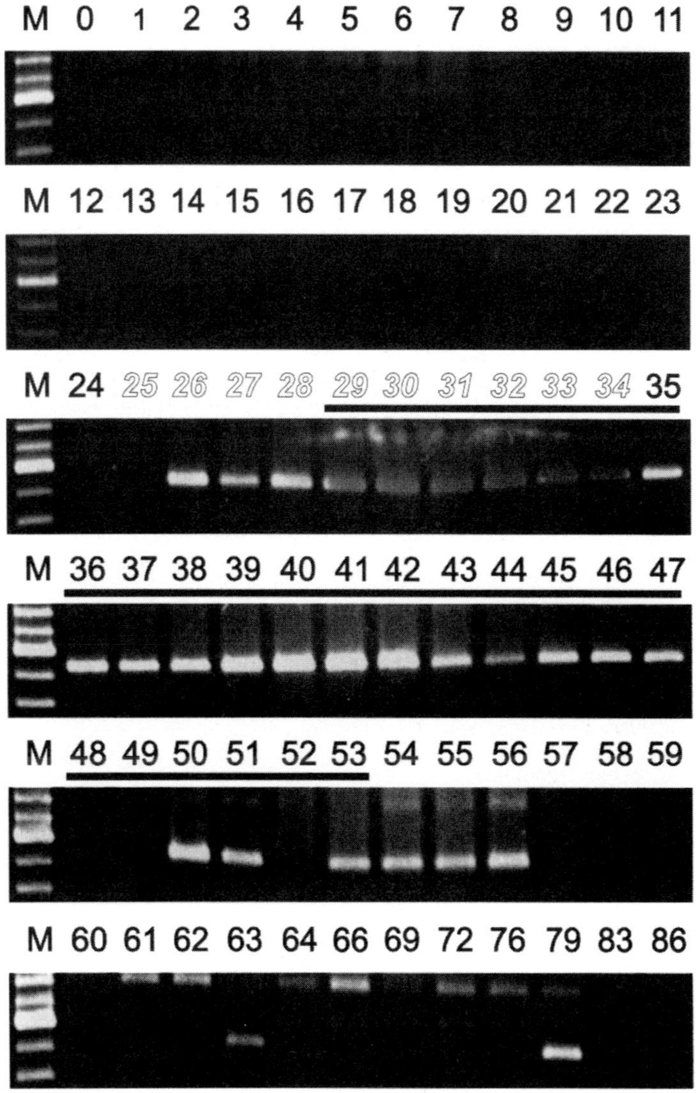

Fig. 3.3.6. Results obtained by the ST-nested-PCR in a sheep infected experimentally with 50,000 sporocysts of *S. tenella*. Numbers above the gel images indicate days post-infection. The line below these numbers indicates days with PCV < 25%. Days with fever are shown in open bold and italics. Seroconversion in an ELISA using cystozoite-derived antigen of *S. tenella* was observed as late as day 90 post-infection. M, 100 bp marker showing a double-intensity band at 600 bp (reproduced from Heckeroth and Tenter, 1999b). Copyright: Australian Society for Parasitology (*International Journal for Parasitology*).

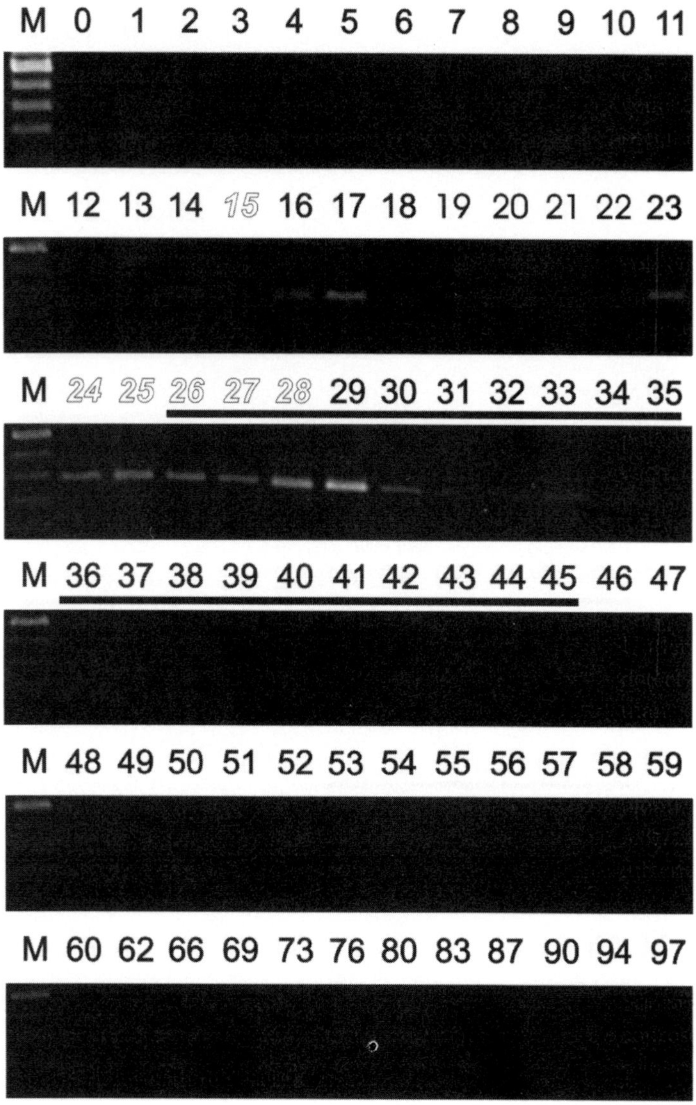

Fig. 3.3.7. Results obtained by the SA-nested-PCR in a sheep infected experimentally with 5000 sporocysts of *S. arieticanis*. Numbers above the gel images indicate days post infection. The line below these numbers indicates days with PCV < 25%. Days with fever are shown in open bold and italics. Seroconversion in an ELISA using cystozoite-derived antigen of *S. arieticanis* was observed at day 29 post-infection. M, 100 bp marker showing a double-intensity band at 600 bp (reproduced from Heckeroth and Tenter, 1999b). Copyright: Australian Society for Parasitology (*International Journal for Parasitology*).

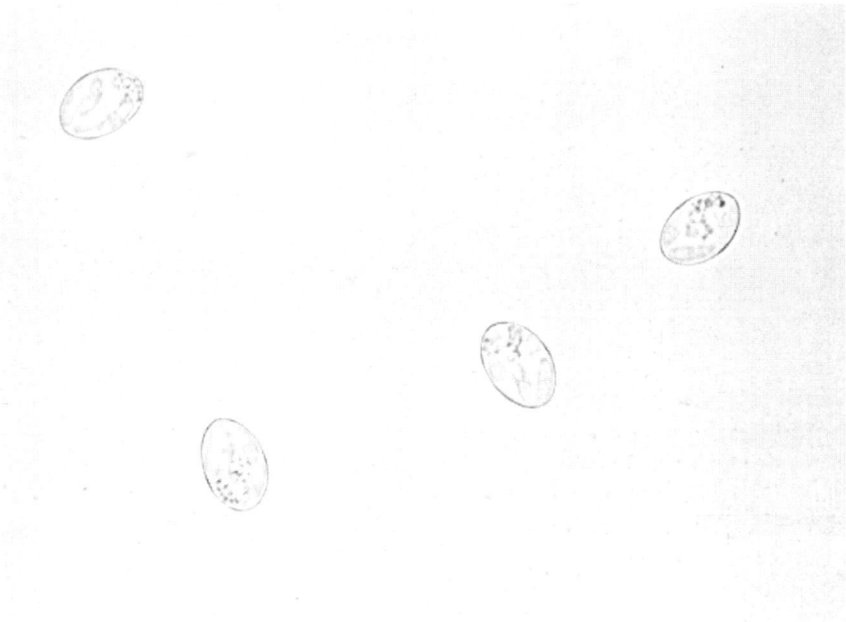

Fig. 3.3.8. Sporocysts of *Sarcocystis* species collected from the faeces of a definitive host by flotation procedure (reproduced from Schnieder *et al.*, 2004). Copyright: Institut für Parasitologie, TiHo Hannover, Germany.

Diagnostic Applications and Recommended Diagnostic Schemes in Ruminants

Fetal diagnosis

In *Sarcocystis*-induced abortions, meronts associated with haemorrhages and focal necrosis can be found in maternal tissues of the placentome, but are not consistently found in fetal membranes or tissues (Uggla and Buxton, 1990; Dubey and Rommel, 1992).

Post-natal detection of infection

Clinical diagnosis

The severity of clinical symptoms caused by virulent *Sarcocystis* species depends on both the dose of ingested sporocysts and the immune status of the host. Clinical symptoms during the early merogonic phases in vascular endothelial cells of internal organs are usually more severe than those observed during the formation and development of tissue cysts in muscle or nervous tissue (Dubey *et al.*, 1989; Cawthorn and Speer, 1990; Uggla and Buxton, 1990; Dubey and Rommel, 1992; O'Donoghue and Rommel, 1992).

The uptake of a high dose of sporocysts of a virulent *Sarcocystis* species can lead to acute sarcocystiosis in previously uninfected animals that have not yet developed immunity to the respective species. In addition to several non-specific symptoms such as fever, anorexia, tachypnoea, tachycardia and anaemia, signs of acute sarcocystiosis include central nervous symptoms caused by encephalitis and encephalomyelitis, and haemorrhagic diathesis – which can cause the death of the animal (Gestrich *et al.*, 1974; Heydorn *et al.*, 1976; Leek *et al.*, 1977; Heydorn, 1985; Heydorn and Karaer, 1986; Dubey, 1988; Jeffrey, 1993).

In pregnant sheep, goats and cattle, acute sarcocystiosis frequently results in fetal death, abortion or premature birth of the offspring (Leek and Fayer, 1978; Munday, 1981; Fayer and Dubey, 1988; Jeffrey, 1993). Animals surviving a primary *Sarcocystis* infection usually acquire immunity that protects them against acute disease after challenge with the homologous species, but not after infection with a heterologous virulent species (Uggla and Buxton, 1990; O'Donoghue and Rommel, 1992).

Chronic sarcocystiosis can result from the ingestion of a low dose of sporocysts of a virulent *Sarcocystis* species and may cause economic loss due to reduced quality and quantity of meat, milk or wool (Leek *et al.*, 1977; Erber and Burgkart, 1981; Munday, 1984, 1986; O'Donoghue and Rommel, 1992). Additional economic losses in the livestock industry are caused by macroscopic *Sarcocystis* cysts, resulting in condemnation of whole carcasses or affected parts after slaughter (Collins *et al.*, 1976; O'Donoghue and Ford, 1986; O'Donoghue and Rommel, 1992).

Pathology and histopathology

The most dominant signs of acute sarcocystiosis observed at post-mortem examination are capillary haemorrhages in all organs, particularly in the tongue, heart and skeletal muscles (see Fig. 3.3.9). Encephalitis, with haemorrhages and focal gliosis, can be found in some cases. By histological methods, meronts are predominantly found in the brain, tongue, heart and kidney (see Figs 3.3.10 and 3.3.11).

However, the meronts of different *Sarcocystis* species can not be differentiated morphologically. In addition, meronts may already have disappeared by the time the animal succumbs to the disease (O'Toole *et al.*, 1986; Dubey, 1988; Dubey *et al.*, 1989b; Uggla and Buxton, 1990). In *Sarcocystis*-induced abortions, meronts associated with haemorrhages and focal necrosis can be found in maternal tissues of the placentome, but are not consistently found in fetal membranes or tissues (Uggla and Buxton, 1990).

In chronic sarcocystiosis, tissue cysts can be found microscopically in skeletal or heart muscle (see Figs 3.3.12, 3.3.13, 3.3.14 and 3.3.15). In some cases, the *Sarcocystis* species can be identified in stained histological sections or squash

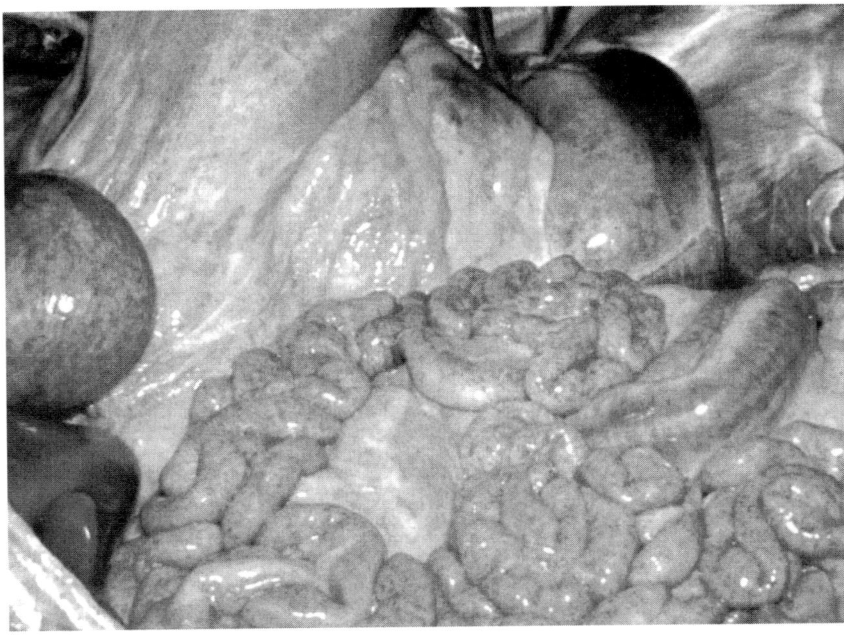

Fig. 3.3.9. Acute sarcocystiosis in a calf infected with *S. cruzi* (reproduced from Schnieder *et al.*, 2004). Copyright: Institut für Parasitologie, TiHo Hannover, Germany.

preparations based on the morphology of the wall of mature tissue cysts, for example *S. tenella* versus *S. arieticanis* (see Table 3.3.4; Dubey *et al.*, 1989a).

However, in many cases, a species-specific diagnosis of *Sarcocystis* tissue cysts requires immunohistochemical or electron microscopic methods. As only a small section of muscle tissue can be examined by these methods, they lack sensitivity and also may not allow detection of concurrent infections with different *Sarcocystis* species (Dubey *et al.*, 1989a; Uggla and Buxton, 1990).

Serology

The high cross-reactivity among different *Sarcocystis* species hampers the development of species-specific diagnostic methods for *Sarcocystis* infections in ruminants. Therefore, it is currently not possible to differentiate between infections with virulent and non-virulent *Sarcocystis* species in ruminants *intra vitam*. In addition, the immunological diagnosis of acute sarcocystiosis or *Sarcocystis*-induced abortion is complicated by the fact that humoral antibody levels measured with traditional antigen preparations are usually very low during the early phase of infection.

Therefore, it is often not possible to detect *Sarcocystis*-specific antibodies at the time when clinical disease becomes obvious (Uggla and Buxton, 1990;

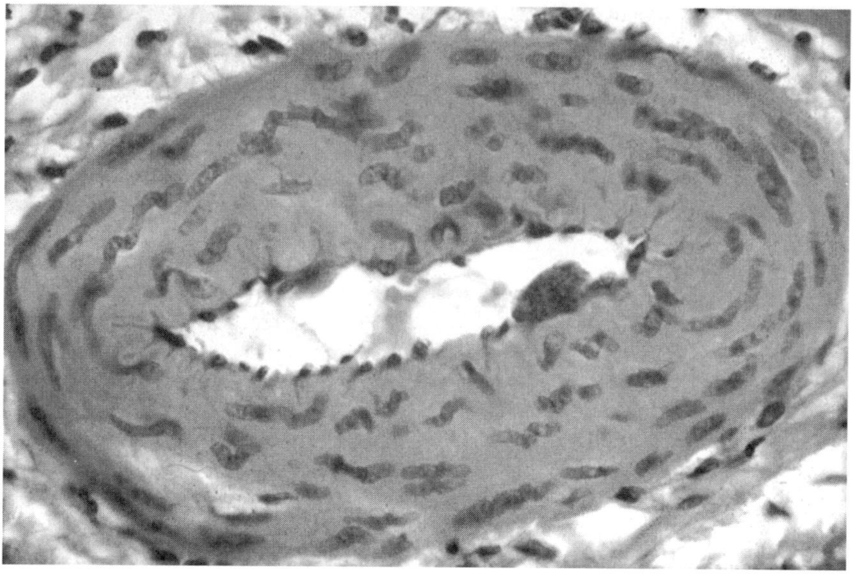

Fig. 3.3.10. First-generation meront in the kidney of a goat infected with *S. capracanis* (12 d post-infection). Copyright: Institut für Parasitologie, TiHo Hannover, Germany.

O'Donoghue and Rommel, 1992; Tenter, 1995). However, for epidemiological studies with regard to risk factor assessment at the herd level, serological tests might be appropriate.

Molecular biology

Ribosomal RNA molecules are the most abundant macromolecules in cellular organisms. Growing cells contain 10–50 times more RNA than DNA. rRNA constitutes about 90–95% of the total cellular RNA, and about 30% of this is 18S rRNA. Therefore, some authors have suggested that diagnostic assays targeting naturally abundant rRNA should be much more sensitive and accurate than assays targeting rRNA genes (Waters and McCutchan, 1990; Gajadhar *et al.*, 1992).

However, the differentiation of *Sarcocystis* spp. under natural conditions requires a test system that is applicable to the examination of a broad range of different specimens. For example, diagnosis of abortion or of fatal acute disease caused by these parasites requires their detection in tissue samples taken from aborted fetuses, placentae or internal organs at post-mortem examination.

By contrast, epidemiological studies on infections with *Sarcocystis* spp. require a diagnostic method that is suited for the examination of a large number of samples, such as blood or buffy coat, derived from animals *intra vitam*. All of these samples are usually several hours to days old before they can be examined.

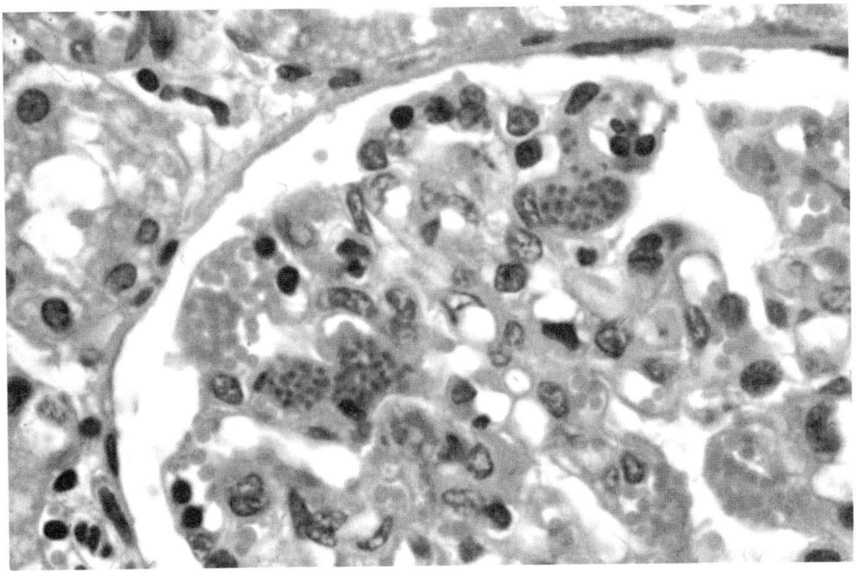

Fig. 3.3.11. Second-generation meronts in the kidney of a goat infected with *S. capracanis* (20 d post-infection). Copyright: Institut für Parasitologie, TiHo Hannover, Germany.

Therefore, it may not be feasible to base a diagnostic test designed for the examination of field samples, or for the examination of large numbers of samples for epidemiological studies, on RNA that is not a stable target and is easily degraded by enzymatic digestion.

In addition, standardization of RNA hybridization assays is difficult (Gajadhar *et al.*, 1992; MacPherson and Gajadhar, 1993), but would be a pre-requisite for the comparison of epidemiological data collected at different times and places. Another disadvantage of hybridization assays is the necessity of using radioactively labelled oligonucleotide probes to obtain reasonable sensitivities (Gajadhar *et al.*, 1992; Holmdahl *et al.*, 1993; MacPherson and Gajadhar, 1993), which restricts the use of these tests to a limited number of specialized laboratories and is impractical for many diagnostic applications and for prolonged epidemiological studies.

RAPD-PCR fingerprinting has been used to distinguish different *Sarcocystis* species which might be prevalent in one intermediate host, i.e. sheep and cattle (Aldemir and Dik, 2003; Gueclue *et al.*, 2004). Both RAPD-PCRs have given distinct fragment patterns for *S. tenella*, *S. arieticanis* and *S. gigantea* and *S. cruzi*, *S. hirsuta* and *S. hominis*, respectively. Amplification of short regions of genomic DNA, without prior knowledge of sequences, is a great advantage of this approach.

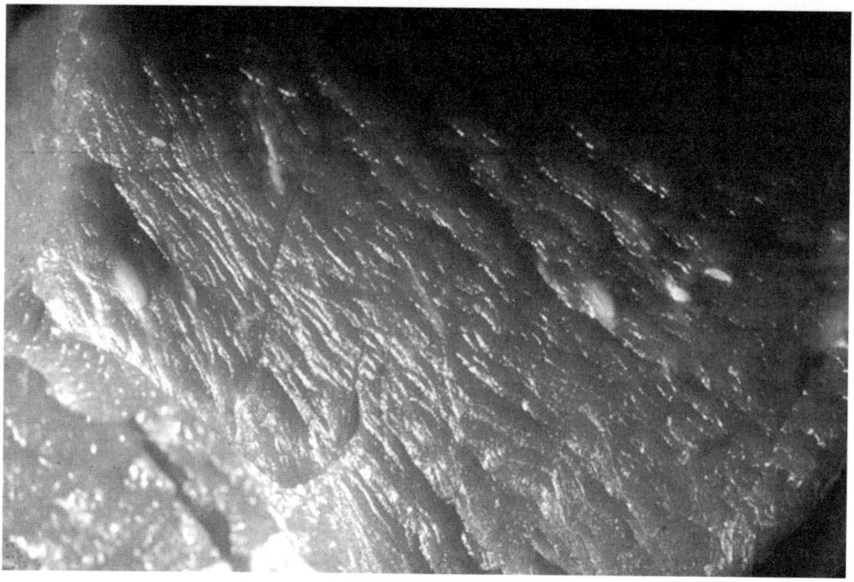

Fig. 3.3.12. Macroscopically visible tissue cysts of *Sarcocystis* (*S. hirsuta* and/or *S. hominis*) in skeletal muscle of cattle (reproduced from Schnieder *et al.*, 2004). Copyright: Institut für Parasitologie, TiHo Hannover, Germany.

However, RAPD-PCR techniques may amplify more than one fragment and distinction between *Sarcocystis* species by pattern analysis might be difficult. Furthermore, simultaneous infection with more than one species may reveal as yet unknown fragment patterns that cannot be ascribed to the respective species. To overcome this disadvantage, probes derived from RAPD-PCR fragment patterns have been used in Southern blots (Joachim *et al.*, 1996). However, this method, being time-consuming and expensive, has not been developed as a diagnostic test.

Probably the most successful strategy in the development of new diagnostic methods for *Sarcocystis* spp. has been the recent analysis of variable regions of the 18S rDNA of *Sarcocystis* spp. to identify species-specific sequences that can be targeted by corresponding synthetic oligonucleotides in RNA hybridization assays or the PCR (Gajadhar *et al.*, 1992; Holmdahl *et al.*, 1993; Tenter *et al.*, 1994; Heckeroth and Tenter, 1999a; Yang *et al.*, 2002).

The sequence of 18S rDNA is evolutionarily conserved among related organisms, particularly in the regions that determine the core of the secondary structure of the molecule. However, in other regions, the 18S rDNA sequences vary even between closely related taxa. This double feature of conservation and variation in different regions of the molecule, together with the universal and abundant nature of 18S rDNA, has made this molecule a useful basis for the development of effective diagnostic tools.

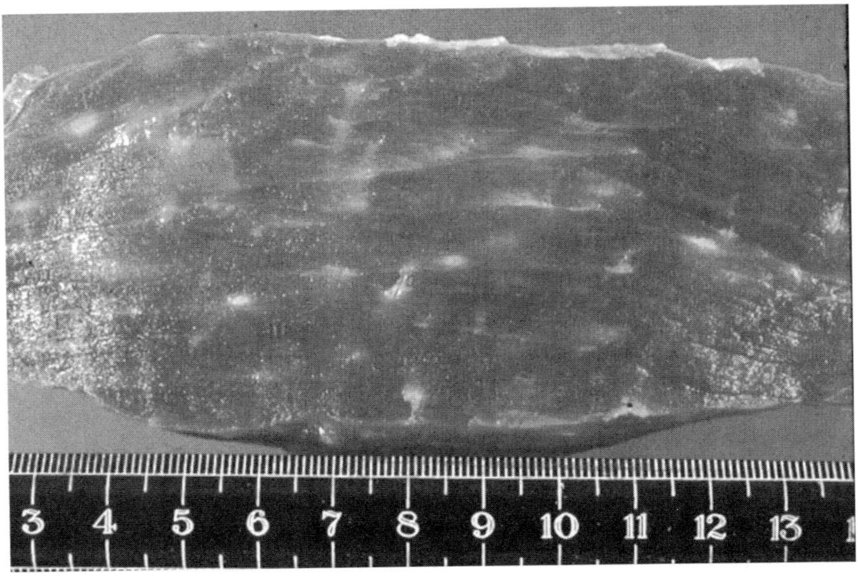

Fig. 3.3.13. Macroscopically visible tissue cysts of *Sarcocystis* (*S. hirsuta* and/or *S. hominis*) in skeletal muscle of cattle. Copyright: Institut für Parasitologie, TiHo Hannover, Germany.

A nested-PCR assay, targeting the 18S rDNA, has been established as a species-specific and sensitive diagnostic test for the detection of *S. tenella* or *S. arieticanis* in sheep. DNA templates can be expected to be stable targets for PCR amplification, even when diagnostic samples have been stored at ambient temperature for some time. In the genome of *T. gondii*, the ribosomal transcription unit is present in 110 copies (Guay *et al.*, 1992). Therefore, it can be assumed that a similar high number of copies exists in the genome of closely related cyst-forming coccidia (Tenter and Johnson, 1997) and that a PCR assay based on rRNA genes of *Sarcocystis* species has a sufficient sensitivity for diagnostic applications.

Under natural conditions, sheep may be infected with different virulent and non-virulent *Sarcocystis* species at the same time. Diagnosis of abortion and acute disease caused by *S. tenella* and *S. arieticanis* has been limited by the low specificity and sensitivity of traditional diagnostic methods. The nested-PCRs based on the detection of the 18S rRNA genes of *S. tenella* or *S arieticanis* in blood samples of infected sheep are the first tests for species-specific diagnosis and differentiation of the two virulent *Sarcocystis* species infecting sheep.

Therefore, it will now be possible for the first time to collect accurate epidemiological data on the prevalence, incidence and transmission dynamics of infections with pathogenic *Sarcocystis* species, and on the importance of mixed infections with different cyst-forming coccidia in sheep.

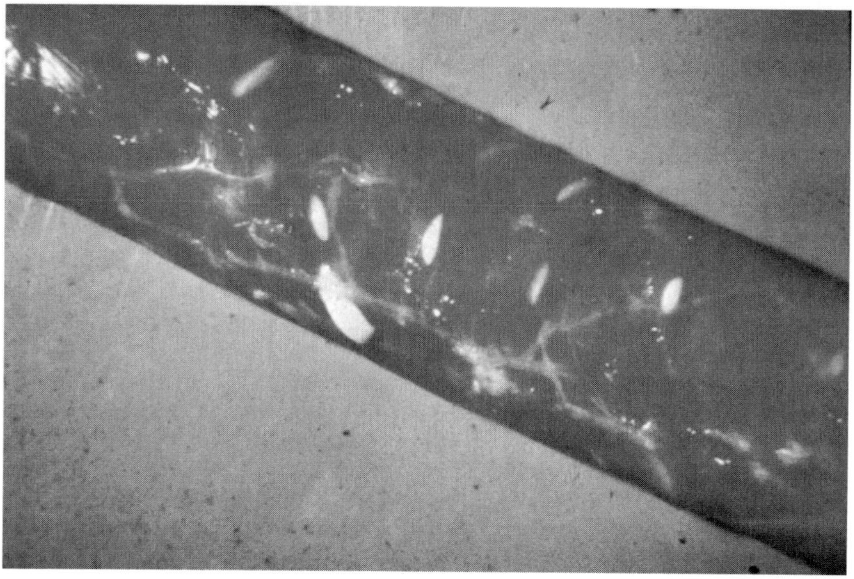

Fig. 3.3.14. Tissue cysts of *S. gigantea* in the oesophagus of a sheep. Copyright: Institut für Parasitologie, TiHo Hannover, Germany.

In future, it will be possible to identify geographic areas in which infections with *S. tenella* or *S. arieticanis* are endemic, as well as areas with endemic instability in which animals are at risk from acute sarcocystiosis or *Sarcocystis*-induced abortion.

Future Trends

Over the past decade, new techniques in immunology, protein chemistry and molecular biology have facilitated more advanced studies on the molecular composition and molecular biology of *Sarcocystis* species in various laboratories. The development of species-specific monoclonal antibodies and analyses of the molecular composition of some life-cycle stages of *Sarcocystis* species of cattle and sheep have shown that species-specific proteins and antigens exist in these species, although they are not highly abundant.

In addition, comparisons of rRNA genes of different *Sarcocystis* species have identified unique sequences in the rRNA of virulent *Sarcocystis* species that are suitable targets for species-specific identification. Thus, tools have become available that facilitate the development of methods for species-specific identification and differentiation of *Sarcocystis* species, as well as the identification and study of molecules that are associated with the pathogenicity of some of these parasites.

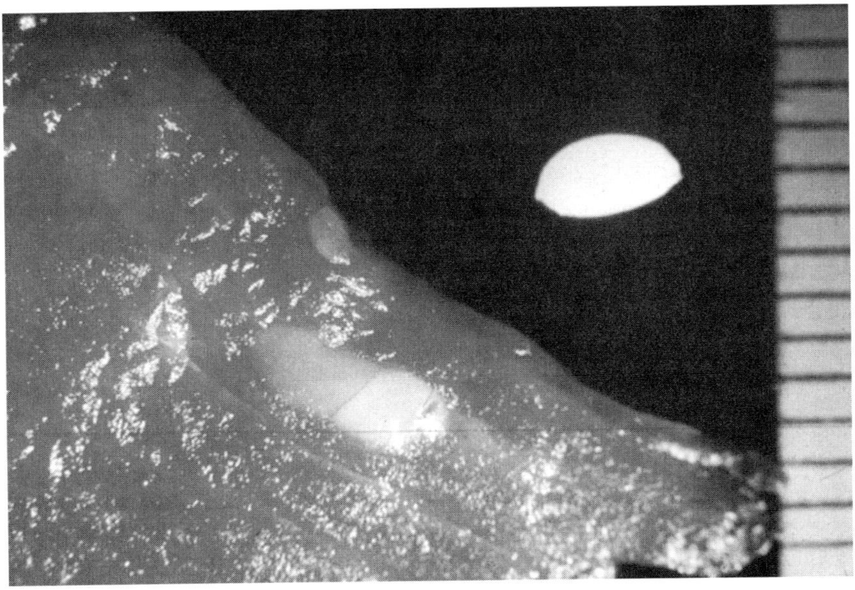

Fig. 3.3.15. Tissue cysts of *S. gigantea* in the oesophagus of a sheep. Copyright: Institut für Parasitologie, TiHo Hannover, Germany.

Serology

As described above, the usefulness of serological tests for the diagnosis of *Sarcocystis* infections has been limited by the high-antigenic cross-reactivity among different *Sarcocystis* spp. In addition, the *Sarcocystis* stages from which traditional antigens have been derived, i.e. cystozoites or merozoites, are obligately intracellular and, therefore, these antigen preparations lack standardization and are frequently contaminated with debris of host tissue (Rommel *et al.*, 1995).

A strategy for overcoming these limitations of immunological tests for the diagnosis of infections with *Sarcocystis* spp. may be the development of species-specific recombinant polypeptides that can be used as standardized antigens in diagnostic tests.

Thus far, five different gene fragments encoding antigenic polypeptides have been identified in a cDNA library constructed from cystozoite-derived mRNA of *S. tenella* (Mertens *et al.*, 1996). One of these gene fragments, termed *STC29*, encodes an antigenic portion of a native *S. tenella* polypeptide with an apparent molecular weight of 25,100. The recombinant STC29 polypeptide appears to be specific for *S. tenella*, and does not cross-react with sera of sheep infected with *S. arieticanis* or *T. gondii* when used as a diagnostic antigen in an ELISA. However, the ELISA based on the recombinant STC29 antigen shows only a low sensitivity and there is a need for other species-specific *S. tenella* polypeptides that can be used in combination with STC29 to increase the sensitivity of this test.

A comparison of *in vitro* translation products of *S. tenella* and *S. gigantea* has shown that cystozoite-derived mRNA of *S. tenella* translates at least another three polypeptides with apparent molecular weights of 80,000, 34,700 and 21,500, which are not translated by mRNA of *S. gigantea* (Tenter and Johnson, 1992). Therefore, these polypeptides are potential candidates for inclusion in an antigen mixture that may facilitate the development of a species-specific immunological test for the diagnosis of *S. tenella* infections in sheep.

While the development of species-specific immunological tests based on standardized recombinant antigens would be a great advance for the diagnosis of chronic sarcocystiosis, as well as epidemiological studies on infections with *Sarcocystis* spp. in intermediate hosts, it is important to realize that the development of such tests is time-consuming and expensive.

Several such tests would be required to accurately assess the significance of a single *Sarcocystis* species as the causative agent of disease, because domestic animals can be intermediate hosts for several different *Sarcocystis* species and it is possible that animals can be infected with two or more pathogenic species. In addition, even standardized immunological tests may be of little value for the diagnosis of acute sarcocystiosis and are not useful at all for epidemiological studies on infections with *Sarcocystis* species in definitive hosts.

Molecular biology

To date, only one diagnostic PCR assay for the detection of the virulent *S. tenella* and *S. arieticanis*, in sheep *intra vitam* is available (Heckeroth and Tenter, 1999a, b). This assay was shown to be highly sensitive and robust when used to examine material of different types, origin and condition. In principle, this method can be modified to an assay for the detection of *Sarcocystis* species in cattle or water buffaloes, as it has been shown that merozoites may be present in the bloodstream of calves after experimental infection (Ndiritu *et al.*, 1996).

The rapidly growing database of 18 rDNA sequences generated by phylogenetic studies of *Sarcocystis* species provides a rich source of sequences that can be exploited to develop similar species-specific PCRs. Alternatively, to develop species-specific oligonucleotides based on 18 rDNA sequences, it is not necessary to investigate the whole 18 rDNA of the respective *Sarcocystis* species.

Since the location of variable regions of interest for *Sarcocystis* species is known, it will be possible to design PCR primers that correspond to conserved flanking sequences of these regions and then only to sequence the 18 rDNA region of interest. This will generate information on putatively species-specific target sequences in a much shorter time than has been possible previously. Other target genes or sequences, e.g. derived by EST sequencing, have been described for other apicomplexan parasites such as *T. gondii*, *N. caninum* or *S. neurona*, but so far not for *Sarcocystis* species that infect ruminants (Li *et al.*, 2002).

PCR assays have a great potential for adaptation to the examination of parasite stages from canine or feline definitive hosts. The collection of epidemiological data on *Sarcocystis* infections in definitive hosts has been complicated by the fact that definitive hosts of *Sarcocystis* species do not develop parasite-specific antibodies.

Sarcocystis sporocysts can be found in faecal samples of canids of felids. However, the sporocysts of different *Sarcocystis* species are too similar morphologically to be differentiated by coproscopical methods. Therefore, the nested-PCRs described here can be employed in various ways in epidemiological studies that have not been possible with traditional diagnostic methods. In future, PCR assays will be invaluable for the detection and identification of *Sarcocystis* species in many clinical and epidemiological investigations.

References

Agarwal, M.C., Shah, H.L., Sharma, R.K. and Singh, K.P. (1987) A tissue mince method for detection of intact cysts of *Sarcocystis* in musculature. *Journal of Veterinary Parasitology* 1, 67–68.

Aldemir, O.S. and Dik, B. (2003) The diagnosis of *Sarcocystis* species in sheep using random amplified polymorphic DNA-PCR (RAPD-PCR). *Turkiye Parazitoloji Dergisi* 27, 255–259.

Anderson, M.L., Barr, B.C. and Conrad, P.A. (1994) Protozoal causes of reproductive failure in domestic ruminants. *The Veterinary Clinics of North America, Food Animal Practice* 10, 439–461.

Ashford, R.W. (1977) The fox, *Vulpes vulpes*, as a final host for *Sarcocystis* of sheep. *Annals of Tropical Medicine and Parasitology* 71, 29–34.

Barr, B.C., Anderson, M.L., Woods, L.W., Dubey, J.P. and Conrad, P.A. (1992) *Neospora*-like protozoal infections associated with abortion in goats. *Journal of Veterinary Diagnostic Investigation* 4, 365–367.

Barutzki, D. and Schaper, R. (2003) Endoparasites in dogs and cats in Germany, 1999–2002. *Parasitology Research* 90, Suppl. 3, S148–S150.

Bernard, P.N. and Bauche, J. (1912) Filariose et atherome aortique du buffle et du boeuf. *Bulletin de la Société de Pathologie Exotique* 5, 109–114.

Böttner, A., Charleston, W.A.G., Pomroy, W.E. and Rommel, M. (1987) The prevalence and identity of *Sarcocystis* in beef cattle in New Zealand. *Veterinary Parasitology* 24, 157–168.

Burgess, D.E., Speer, C.A. and Reduker, D.W. (1988) Identification of antigens of *Sarcocystis cruzi* sporozoites, merozoites, and bradyzoites with monoclonal antibodies. *The Journal of Parasitology* 74, 828–832.

Buxton, D. (1998) Protozoan infections (*Toxoplasma gondii, Neospora caninum* and *Sarcocystis* species) in sheep and goats: recent advances. *Veterinary Research* 29, 289–310.

Buxton, D. and Henderson, D. (1999) Infectious abortion in sheep. *In Practice* 21, 360–368.

Caldow, G.L., Gidlow, J.R. and Schock, A. (2000) Clinical, pathological and epidemiological findings in three outbreaks of ovine protozoan myeloencephalitis. *The Veterinary Record* 146, 7–10.

Carvalho, S.P. (1993) Prevalence and identity of *Sarcocystis* species cysts in cattle slaughtered at Lisbon. *Revista Portuguesa de Ciencias Veterinarias* 88, 36–41.

Cawthorn, R.J. and Speer, C.A. (1990) *Sarcocystis*: infection and disease of humans, livestock, wildlife and other hosts. In: Long, P.L. (ed.) *Coccidiosis of Man and Domestic Animals.* CRC Press, Boca Raton, Florida, pp. 91–120.

Claveria, F.G. and Cruz, M.J. (2000) *Sarcocystis levinei* infection in Philippine water buffaloes (*Bubalus bubalis*). *Parasitology International* 48, 243–247.

Claveria, F.G., Petersen, B., Macabagdal, M.R., Farolan, R.J., Farrol, M.A., Gonzalvo, F., Cadiz, R., Ajero, R., Roque, R. and Lozano, G. (1997) A survey of bovine, bubaline and swine sarcocystosis in the Philippines. *The Southeast Asian Journal of Tropical Medicine and Public Health* 28, Suppl. 1, 173–178.

Collins, G.H., Charleston, W.A.G. and Moriarty, K.M. (1976) *Sarcocystis* species in sheep. *New Zealand Veterinary Journal* 24, 123–124.

Collins, G.H., Atkinson, E. and Charleston, W.A.G. (1979) Studies on *Sarcocystis* species III: the macrocystic species of sheep. *New Zealand Veterinary Journal* 27, 204–206.

Da Silva, N.R.S., Rodrigues, R.J.D., de Araujo, F.A.P., Beck, C. and Olicheski, A.T. (2002) Detection of bovine *Sarcocystis cruzi* cysts in cardiac muscles: a new technique of concentration for diagnostics. *Acta Scientiae Veterinariae* 30, 127–129.

Dissanaike, A.S. and Kan, S.P. (1978) Studies on *Sarcocystis* in Malaysia. 1. *Sarcocystis levinei* n. species from the water buffalo *Bubalus bubalis. Zeitschrift für Parasitenkunde* 55, 127–138.

Dubey, J.P. (1976) A review of *Sarcocystis* of domestic animals and of other coccidia of cats and dogs. *Journal of the American Veterinary Medical Association* 169, 1061–1078.

Dubey, J.P. (1988) Lesions in sheep inoculated with *Sarcocystis tenella* sporocysts from canine feces. *Veterinary Parasitology* 26, 237–252.

Dubey, J.P. and Rommel, M. (1992) Durch protozoen bedingte aborte bei landwirtschaftlichen nutztieren. *Deutsche Tierärztliche Wochenschrift* 99, 355–362.

Dubey, J.P., Speer, C.A., Callis, G. and Blixt, J.A. (1982) Development of the sheep–canid cycle of *Sarcocystis tenella. Canadian Journal of Zooology* 60, 2464–2477.

Dubey, J.P., Fayer, R. and Speer, C.A. (1988) Experimental *Sarcocystis hominis* infection in cattle: lesions and ultrastructure of sarcocysts. *Journal of Parasitology* 74, 875–879.

Dubey, J.P., Speer, C.A. and Fayer, R. (1989a) *Sarcocystosis of Animals and Man.* CRC Press, Boca Raton, Florida.

Dubey, J.P., Speer, C.A., Munday, B.L. and Lipscomb, T.P. (1989b) Ovine sporozoan encephalomyelitis linked to *Sarcocystis* infection. *Veterinary Parasitology* 34, 159–163.

Egger, A. (1994) Sarkosporidien beim schlachtschaf in osttirol. *Wiener Tierärztliche Monatsschrift* 81, 247.

El-Rafaii, A.H., Abdel-Baki, G. and Selim, M.K. (1980) Sarcosporidia in goats in Egypt. *Journal of the Egyptian Society of Parasitology* 10, 470–472.

Engeland, I.V., Waldeland, H. and Andresen, O. (1994) Abort hos geit – en oversikt. *Norsk Veterinaertidsskrift* 106, 363–370.

Erber, M. and Burgkart, M. (1981) Wirtschaftliche verluste durch sarkosporidiose (*Sarcocystis ovicanis* und *S.* spec.) bei der mast von schaflämmern. *Der Praktische Tierarzt* 62, 422–427.

Fayer, R. and Dubey, J.P. (1986) Bovine sarcocystosis. *Compendium on Continuing Education for the Practicing Veterinarian* 8, 130–142.

Fayer, R. and Dubey, J.P. (1988) *Sarcocystis*-induced abortion and fetal death. *Progress in Clinical Biological Research* 281, 153–164.

Fayer, R. and Elsasser, T.H. (1991) Bovine sarcocystosis: how parasites negatively affect growth. *Parasitology Today* 7, 250–255.

Fischer, G. (1979) Die entwicklung von *Sarcocystis capracanis* n. spec. in der ziege. Thesis, Freie Universität Berlin, Germany.

Foreyt, W.J. (1989) *Sarcocystis* species in mountain goats (*Oreamnos americanus*) in Washington: prevalence and search for the definitive host. *Journal of Wildlife Diseases* 4, 619–622.

Fortier, G., Collobert, J.F., Viel, S. and Mariau, V. (1993) Prevalence de la sarcosporidiose musculaire bovine dans le Calvados. *Recueil de Medecine Vétérinaire* 169, 779–781.

Frenkel, J.K., Heydorn, A.O., Mehlhorn, H. and Rommel, M. (1979) Sarcocystinae: *nomina dubia* and available names. *Zeitschrift für Parasitenkunde* 58, 115–139.

Gajadhar, A.A., Marquardt, W.C. and Blair, C.D. (1992) Development of a model ribosomal hybridization assay for the detection of *Sarcocystis* and other coccidia. *Canadian Journal of Veterinary Research* 56, 208–213.

Gasbarre, L.C., Suter, P. and Fayer, R. (1984) Humoral and cellular immune responses in cattle and sheep inoculated with *Sarcocystis. American Journal of Veterinary Research* 45, 1592–1596.

Gestrich, R., Schmitt, M. and Heydorn, A.O. (1974) Pathogenität von *Sarcocystis tenella*-sporozysten aus den fäzes von hunden für lämmer. *Berliner und Münchener Tierärztliche Wochenschrift* 87, 362–363.

Ghaffar, F.A., Heydorn, A.O. and Mehlhorn, H. (1989) The fine structure of cysts of *Sarcocystis moulei* from goats. *Parasitology Research* 75, 416–418.

Gjerde, B. (1984a) A light microscopic comparison of the cysts of four species of *Sarcocystis* infecting the domestic reindeer (*Rangifer tarandus*) in northern Norway. *Acta Veterinaria Scandinavica* 25, 195–204.

Gjerde, B. (1984b) *Sarcocystis* infection in wild reindeer (*Rangifer tarandus*) from Hardangervidola in southern Norway: with a description of the cysts of *Sarcocystis hardangeri* n. species. *Acta Veterinaria Scandinavica* 25, 205–212.

Gjerde, B. (1984c) The fox as a definitive host for *Sarcocystis* species (Gjerde, 1984) from skeletal muscle of reindeer (*Rangifer tarandus*). With a proposal for *Sarcocystis tarandivulpes* n. species as replacement name. *Acta Veterinaria Scandinavica* 25, 403–410.

Gjerde, B. (1984d) *Sarcocystis hardangeri* and *Sarcocystis rangi* n. species from the domestic reindeer (*Rangifer tarandus*) in northern Norway. *Acta Veterinaria Scandinavica* 25, 411–418.

Gomes, A.G. and Lima, J.D. (1982) *Sarcocystis* (Lankester, 1882) em bovinos de minas gerais; ocorrência e métodos de diagnóstico. *Arquivos da Escola de Veterinaria Universidade Federal de Minas Gerais, Belo Horizonte* 34, 83–92.

Gothe, R. and Reichler, I. (1990) Zur befallshäufigkeit von kokzidien bei hundefamilien unterschiedlicher haltung und rassen in Süddeutschland. *Tierärztiche Praxis* 18, 407–413.

Gottstein, B. (1995) Zystenbildende kokzidien: *Toxoplasma, Neospora, Sarcocystis. Schweizerische Medizinische Wochenschrift* 125, 890–898.

Granstrom, D.E., Giles, R.C., Tuttle, P.A., Williams, N.M., Poonacha, K.B., Petrites-Murphy, M.B., Tramontin, R.R., Swerczek, T.W., Hong, C.B., Rezabek, G.B., Lyons, E.T. and Drudge, J.H. (1991) Immunohistochemical diagnosis of protozoan parasites in lesions of equine protozoal myeloencephalitis. *Journal of Veterinary Diagnostic Investigation* 3, 75–77.

Granstrom, D.E., Dubey, J.P., Davis, S.W., Fayer, R., Fox, J.C., Poonacha, K.B., Giles, R.C. and Comer, P.F. (1993) Equine protozoal myeloencephalitis: antigen analysis of cultured *Sarcocystis neurona* merozoites. *Journal of Veterinary Diagnostic Investigation* 5, 88–90.

Guay, J.M., Huot, A., Gagnon, S., Tremblay, A. and Levesque, R.C. (1992) Physical and genetic mapping of cloned ribosomal DNA from *Toxoplasma gondii*: primary and secondary structure of the 5S gene. *Gene* 114, 165–171.

Guclu, F., Aldem, O.S. and Guler, L. (2004) Differential identification of cattle *Sarcocystis* species by random amplified polymorphic DNA-polymerase chain reaction (RAPD-PCR). *Revue de Médecine Vétérinaire* 155, 440–444.

Gunning, R.F., Jones, J.R., Jeffrey, M., Higgins, R.J. and Williamson, A.G. (2000) *Sarcocystis* encephalomyelitis in cattle. *The Veterinary Record* 146, 328.

Gupta, R.S., Kushwah, H.S. and Kushwah, A. (1995) *Sarcocystis fusiformis*: some Krebs cycle enzymes in various fractions of sarcocysts of buffalo (*Bubalus bubalis*). *Veterinary Parasitology* 56, 1–5.

Hamir, A.N., Moser, G., Galligan, D.T., Davis, S.W., Granstrom, D.E. and Dubey, J.P. (1993) Immunohistochemical study to demonstrate *Sarcocystis* neurona in equine protozoal myeloencephalitis. *Journal of Veterinary Diagnostic Investigation* 5, 418–422.

Hasselmann, G. (1923) Parasitoses das carnes de consumo (4° nota prévia). *Brazil Medico* 2, 341.

Heckeroth, A.R. and Tenter, A.M. (1999a) Development and validation of species-specific Nested-PCRs for diagnosis of acute sarcocystiosis in sheep. *International Journal for Parasitology* 29, 1331–1349.

Heckeroth, A.R. and Tenter, A.M. (1999b) Diagnose von sarkozystiosen und *Sarcocystis*-bedingten aborten bei schafen. *Wiener Tierärztiche Monatsschrift* 86, 111–121.

Heydorn, A.O. (1985) Zur entwicklung von *Sarcocystis arieticanis* n. species. *Berliner und Münchener Tierärztliche Wochenschrift* 98, 231–241.

Heydorn, A.O. and Gestrich, R. (1976) Beiträge zum Lebenszyklus der sarkosporidien. VII. Entwicklungsstadien von *Sarcocystis ovicanis* im schaf. *Berliner und Münchener Tierärztliche Wochenschrift* 89, 1–5.

Heydorn, A.O. and Karaer, Z. (1986) Zur schizogonie von *Sarcocystis ovicanis*. *Berliner und Münchener Tierärztliche Wochenschrift* 99, 185–189.

Heydorn, A.O. and Unterholzner, J. (1983) Zur entwicklung von *Sarcocystis hircicanis* n. species. *Berliner und Münchener Tierärztliche Wochenschrift* 96, 275–282.

Heydorn, A.O., Gestrich, R., Mehlhorn, H. and Rommel, M. (1975) Proposal for a new nomenclature of the sarcosporidia. *Zeitschrift für Parasitenkunde* 48, 73–82.

Hinaidy, H.K. and Egger, A. (1994) Die sarkosporidien des schafes in Österreich. *Journal of Veterinary Medicine Series B* 41, 417–427.

Holmdahl, O.J.M., Mattson, J.G., Uggla, A. and Johansson, K.E. (1993) Oligonucleotide probes complementary to variable regions of 18S rRNA from *Sarcocystis* species. *Molecular and Cellular Probes* 7, 481–486.

Huong, L.T. and Uggla, A. (1999) *Sarcocystis dubeyi* n. species (Protozoa: Sarcocystidae) in the water buffalo (*Bubalus bubalis*). *The Journal of Parasitology* 85, 102–104.

Huong, L.T., Dubey, J.P., Nikkila, T. and Uggla, A. (1997a) *Sarcocystis buffalonis* n. species (Protozoa: Sarcocystidae) from the water buffalo (*Bubalus bubalis*) in Vietnam. *The Journal of Parasitology* 83, 471–474.

Huong, L.T., Dubey, J.P. and Uggla, A. (1997b) Redescription of *Sarcocystis levinei* Dissanaike and Kan, 1978 (Protozoa: Sarcocystidae) of the water buffalo (*Bubalus bubalis*). *The Journal of Parasitology* 83, 1148–1152.

Jeffrey, M. (1993) Sarcocystosis of sheep. *In Practice* 15, 2–8.

Jeffrey, M., O'Toole, D., Smith, T. and Bridges, A.W. (1988) Immunocytochemistry of ovine sporozoan encephalitis and encephalomyelitis. *Journal of Comparative Pathology* 98, 213–224.

Jeffrey, M., Low, J.C. and Uggla, A. (1989) A myopathy of sheep associated with sarcocystis infection and monensin administration. *Veterinary Record* 124, 422–426.

Joachim, A., Tenter, A.M., Jeffries, A.C. and Johnson, A.M. (1996) A RAPD-PCR derived marker can differentiate between pathogenic and non-pathogenic *Sarcocystis* species of sheep. *Molecular and Cellular Probes* 10, 165–172.

Kibenge, F.S.B., Cawthorn, R.J., Despres, D., McKenna, P.K. and Markham, R.J.F. (1991) Development of genomic probes to *Sarcocystis cruzi* (Apicomplexa). *Veterinary Parasitology* 40, 9–20.

Lankester, E.R. (1882) On *Drepanidium ranarum*, the cell-parasite of the frog's blood and spleen (Gaule's Würmschen). *Quarterly Journal of Microscopical Science* 22, 53–65.

Leek, R.G. (1986) Infection of sheep with frozen sporocysts of *Sarcocystis ovicanis*. *Proceedings of the Helminthological Society of Washington* 53, 297–298.

Leek, R.G. and Fayer, R. (1978) Sheep experimentally infected with *Sarcocystis* from dogs. II. Abortion and disease in ewes. *Cornell Vet* 68, 108–123.

Leek, R.G. and Fayer, R. (1979) Survival of sporocysts of *Sarcocystis* in various media. *Proceedings of the Helminthological Society of Washington* 46, 151–154.

Leek, R.G., Fayer, R. and Johnson, A.J. (1977) Sheep experimentally infected with *Sarcocystis* from dogs. I. Disease in young lambs. *The Journal of Parasitology* 63, 642–650.

Levine, N.D. (1970) Taxonomy of the sporozoa. *The Journal of Parasitology* 56, 208–209.

Levine, N.D. (1986) The taxonomy of *Sarcocystis* (Protozoa, Apicomplexa) species. *The Journal of Parasitology* 72, 372–382.

Li, Q.Q., Yang, Z.Q., Zuo, Y.X., Attwood, S.W., Chen, X.W. and Zhang, Y.P. (2002) A PCR-based RFLP analysis of *Sarcocystis cruzi* (Protozoa: Sarcocystidae) in Yunnan Province, PR China, reveals the water buffalo (*Bubalus bubalis*) as a natural intermediate host. *The Journal of Parasitology* 88, 1259–1261.

Lindsay, D.S., Dubey, J.P. and Blagburn, B.L. (1996) Finding the cause of parasite-induced abortions in cattle. *Veterinary Medicine* 91, 64–71.

Mackenstedt, U., Wagner, D., Heydorn, A.O. and Mehlhorn, H. (1990) DNA measurements and ploidy determination of different stages in the life cycle of *Sarcocystis muris*. *Parasitology Research* 76, 662–668.

Mackie, J.T., Rahaley, R.S. and Nugent, R. (1992) Suspected *Sarcocystis* encephalitis in a stillborn kid. *Australian Veterinary Journal* 69, 587–595.

Mackie, J.T. and Dubey, J.P. (1996) Congenital sarcocystosis in a Saanen goat. *The Journal of Parasitology* 82, 350–351.

MacPherson, J.M. and Gajadhar, A.A. (1993) Ribosomal RNA sequences for the specific detection of *Toxoplasma gondii* by hybridization assay. *Molecular and Cellular Probes* 7, 97–103.

MacPherson, J.M. and Gajadhar, A.A. (1994) Specific amplification of *Sarcocystis cruzi* DNA using a randomly primed polymerase chain reaction assay. *Veterinary Parasitology* 55, 267–277.

Mal'a, P. and Baranova, M. (1995) Zachytnost sarkocystozy u jatocnych zvierat pri veterinarnej prehliadke [Detection of sarcocystosis in slaughterhouse animals during a veterinary inspection]. *Veterinarni Medicina* 40, 97–100.

Markus, M.B. (1979) Technique for the separation of *Sarcocystis* from cardiac muscle. *The Journal of Parasitology* 65, 699.

Mason, F.E. (1910) *Sarcocystis* in the camel in Egypt. *Journal of Comparative Pathology and Therapeutics* 23, 168–176.

McEwen, B. and Carman, S. (2005) Animal health laboratory reports – cattle: bovine abortion update,1998–2004. *Canadian Veterinary Journal* 46, 46.

McKenna, P.B. and Charleston, W.A.G. (1988b) Recovery of *Sarcocystis gigantea* sporocysts from cat faeces. *Veterinary Parasitology* 26, 215–227.

McNally, J. (1993) New approaches to the control of infectious ovine abortion. *Veterinary Practice* 25, 4–5.

Melville, R.V. (1984) Reply to Frenkel, Mehlhorn, and Heydorn on protozoan *nomina dubia*. *The Journal of Parasitology* 70, 815.

Mertens, C.M., Tenter, A.M., Vietmeyer, C., Ellis, J.T. and Johnson, A.M. (1996) Production of a recombinant fusion protein of *Sarcocystis tenella* and evaluation of its diagnostic potential in an ELISA. *Veterinary Parasitology* 65, 185–197.

Mitchell, M.A. (1988) The prevalence of macroscopic sarcocysts in New Zealand cattle at slaughter. *New Zealand Veterinary Journal* 36, 35–38.

Morsy, T.A., Abdel-Mawla, M.M., Salama, M.M. and Hamdi, K.N. (1994) Assessment of intact *Sarcocystis* cystozoites as an ELISA antigen. *Journal of the Egyptian Society of Parasitology* 24, 85–91.

Moulé, L. (1886) Psorospermies du tissu musculaire du mouton. *Bulletin et Mémoires de la Société Centrale de Médecine Vétérinaire* 40, 125–129.

Moulé, L. (1888) Des sarcosporidies et leurs fréquence, principalement chez les animaux de boucherie. *Société des Sciences et Arts de Vitry-le-François* 14, 3–41.

Munday, B.L. (1981) Premature parturition in ewes inoculated with *Sarcocystis ovicanis*. *Veterinary Parasitology* 9, 17–26.

Munday, B.L. (1984) The effect of *Sarcocystis tenella* on wool growth in sheep. *Veterinary Parasitology* 15, 91–94.

Munday, B.L. (1986) Effects of different doses of dog-derived *Sarcocystis* sporocysts on growth rate and haematocrit in lambs. *Veterinary Parasitology* 21, 21–24.

Munday, B.L. and Obendorf, D.L. (1984) Development and growth of *Sarcocystis gigantea* in experimentally infected sheep. *Veterinary Parasitology* 15, 203–211.

Ndiritu, W., Cawthorn, R.J., Kibenge, F.S.B., Markham, R.J.F., Horney, B.S. and Chan, C.B. (1996) Use of genomic DNA probes for the diagnosis of acute sarcocystosis in experimentally infected cattle. *Veterinary Parasitology* 62, 9–25.

Neveu-Lemaire, M. (1912) II. Les sporozoaires. In: *Parasitologie des Animaux Domestiques. Maladies Parasitaires non-Bactériennes,* Part 2. J. Lamarre et Cie., Paris, pp. 198–317.

Obendorf, D.L. and Munday, B.L. (1987) Experimental infection with *Sarcocystis medusiformis* in sheep. *Veterinary Parasitology* 24, 59–65.

Odening, K. (1998) The present state of species-systematics in *Sarcocystis* (Lankester, 1882) (Protista, Sporozoa, Coccidia). *Systematic Parasitology* 41, 209–233.

O'Donoghue, P.J. and Ford, G.E. (1984) The asexual precyst development of *Sarcocystis tenella* in experimentally infected specific-pathogen-free lambs. *International Journal for Parasitology* 14, 345–355.

O'Donoghue, P.J. and Ford, G.E. (1986) The prevalence and intensity of *Sarcocystis* species infections in sheep. *Australian Veterinary Journal* 63, 273–278.

O'Donoghue, P.J. and Rommel, M. (1992) Australian–German collaborative studies on the immunology of *Sarcocystis* infections. *Angewandte Parasitologie* 33, 102–119.

O'Donoghue, P.J. and Weyreter, H. (1984) Examinations on the serodiagnosis of *Sarcocystis* infections. II. Class-specific immunoglobulin responses in mice, pigs and sheep. *Zentralblatt für Bakteriologie und Hygiene A* 257, 168–184.

O'Donoghue, P., Lumb, R., Smith, P., Brooker, J. and Mencke, N. (1990) Characterization of monoclonal antibodies against ovine *Sarcocystis* species antigens by immunoblotting and immuno-electron microscopy. *Veterinary Immunology and Immunopathology* 24, 11–25.

Orr, M.B., Collins, G.H. and Charleston, W.A.G. (1984) *Sarcocystis capracanis*: experimental infection of goats. II: Pathology. *International Goat and Sheep Research* 2, 202–211.

O'Toole, D., Duffell, S.J., Upcott, D.H. and Frewin, D. (1986) Experimental microcyst *Sarcocystis* infection in lambs: pathology. *The Veterinary Record* 119, 525–531.

O'Toole, D., Jeffrey, M., Challoner, D., Maybey, R. and Welch, V. (1993) Ovine myeloencephalitis-leukomalacia associated with a *Sarcocystis*-like protozoan. *Journal of Veterinary Diagnosis* 5, 212–225.

Railliet, A. (1886a) (*Miescheria tenella*). *Bulletin et Mémoires de la Société Centrale de Médecine Vétérinaire* 40, 130.

Railliet, A. (1886b) Psorospermies géantes dans l'oesophage et les muscles du mouton. *Bulletin et Mémoires de la Société Centrale de Médecine Vétérinaire* 40, 130–134.

Railliet, A. (1897) La douve pancréatique. *Bulletin et Mémoires de la Société Centrale de Médecine Vétérinaire* 51, 371–377.

Railliet, A. and Lucet, A. (1891) Note sur quelques espèces de coccidies encore peu étudiées. *Bulletin de la Société de Zoologie Paris* 16, 246–250.

RangaRao, G.S.C., Sharma, R.L. and Hemaprasanth (1994) Parasitic infections of Indian yak *Bos* (*poephagus*) *grunniens* – an overview. *Veterinary Parasitology* 53, 75–82.

Reiter, I., Weiland, G., Roscher, B., Meyer, J. and Frahm, K. (1981) Versuche zum serologischen nachweis der sarkosporidiose an experimentell mit sarkosporidien infizierten rindern und schafen. *Berliner und Münchener Tierärztliche Wochenschrift* 94, 425–430.

Rommel, M. (1989) Recent advances in the knowledge of the biology of the cyst-forming coccidia. *Angewandte Parasitologie* 30, 173–183.

Rommel, M., Tenter, A.M., Horn, K., Daugschies, A., Mertens, C., Heydorn, A.O., Entzeroth, R., Eschenbacher, K.H. and Klein, H. (1995) *Sarcocystis* species. In: Eckert, J., Braun, R., Shirley, M.W. and Coudert, P. (eds) *Guidelines on Techniques in Coccidiosis Research.* Office for Official Publications on the European Communities, Luxembourg, pp. 241–284.

Saito, M., Ohuchi, Y., Kobayashi, M., Haritani, M. and Itagaki, H. (1994) Preparation and applicability of *Sarcocystis cruzi* antigens and their anti-*S. cruzi* rabbit sera for serodiagnosis of bovine sarcocystosis. *The Journal of Veterinary Medical Science* 56, 589–591.

Saito, M., Shibata, Y., Kubo, M. and Itagaki, H. (1997) *Sarcocystis mihoensis* n. sp. from sheep in Japan. *The Journal of Veterinary Medical Science* 59, 103–106.

Sargison, N.D., Schock, A., Maclean, I.A., Rae, A., Heckeroth, A.R. and Clark, A.M. (2000) Unusual outbreak of sporozoan encephalomyelitis in Bluefaced Leicester ram lambs. *The Veterinary Record* 146, 225–226.

Savini, G., Dunsmore, J.D., Robertson, I.D. and Seneviratna, P. (1992) *Sarcocystis* species in Western Australian cattle. *Australian Veterinary Journal* 69, 201–202.

Savini, G., Dunsmore, J.D. and Robertson, I.D. (1993) A survey of Western Australian dogs for *Sarcocystis* species and other intestinal parasites. *Australian Veterinary Association* 70, 275–276.

Savini, G., Dunsmore, J.D. and Robertson, I.D. (1994a) Evaluation of a serological test system for the diagnosis of *Sarcocystis cruzi* infection in cattle using *S. cruzi* merozoite antigen. *Veterinary Parasitology* 51, 181–189.

Savini, G., Robertson, I.D. and Dunsmore, J.D. (1994b) Risk factors associated with the occurrence of sarcocystosis in Western Australia: results of a postal survey. *Preventive Veterinary Medicine* 19, 137–144.

Savini, G., Dunsmore, J.D. and Robertson, I.D. (1996) Studies on pathogenesis, tissue infection and congenital transmission in cows experimentally infected with *Sarcocystis cruzi* by various routes. *Veterinary Parasitology* 64, 319–327.

Schmidtova, D. and Breza, M. (1991) Species composition of ovine Sarcosporididae in Slovakia. *Folia Veterinaria* 35, 63–74.

Sharma, R.K. (1991) Studies on the life cycle, pathogenesis and treatment of *Sarcocystis capracanis* in goats. *Journal of Veterinary Parasitology* 5, 61–62.

Smith, T.S. and Herbert, I.V. (1986) Experimental microcyst *Sarcocystis* infection in lambs: serology and immunohistochemistry. *The Veterinary Record* 119, 547–550.

Taylor, M. (2000) Protozoal disease in cattle and sheep. *In Practice* 22, 604–617.

Tenter, A.M. (1987) Comparison of enzyme-linked immunosorbent assay and indirect fluorescent antibody test for the detection of IgG antibodies to *Sarcocystis muris. Zentralblatt für Bakteriologie und Hygiene A* 267, 259–271.

Tenter, A.M. (1988) Comparison of Dot-ELISA, ELISA and IFAT for the detection of IgG antibodies to *Sarcocystis muris* in experimentally infected and immunized mice. *Veterinary Parasitology* 29, 89–104.

Tenter, A.M. (1995) Current research on *Sarcocystis* species of domestic animals. *International Journal for Parasitology* 25, 1311–1330.

Tenter, A.M. and Johnson, A.M. (1992) Comparison of *in vitro* translation products of *Sarcocystis gigantea* and *Sarcocystis tenella*. *International Journal for Parasitology* 22, 153–164.

Tenter, A.M. and Johnson, A.M. (1997) Phylogeny of the tissue cyst-forming coccidia. *Advances in Parasitology* 39, 69–139.

Tenter, A.M., Luton, K. and Johnson, A.M. (1994) Species-specific identification of *Sarcocystis* and *Toxoplasma* by PCR amplification of small subunit ribosomal RNA gene fragments. *Applied Parasitology* 35, 173–188.

Tenter, A.M., Johnson, M.R. and Johnson, A.M. (1991a) Effect of tryptic or peptic digestion or mechanical isolation on the extraction of proteins, antigens and ribonucleic acids from *Sarcocystis muris* bradyzoites. *The Journal of Parasitology* 77, 194–199.

Tenter, A.M., Zimmerman, G.L. and Johnson, A.M. (1991b) Separation of antigens from *Sarcocystis* species using chromatofocusing. *The Journal of Parasitology* 77, 727–736.

Toparlak, M. (1987) Isolation of *Sarcocystis* cysts from bovine muscles by using a modified trypsin digestion technique. *Veteriner Fakültesi Dergisi, Ankara University* 34, 363–367.

Uggla, A. (2000) Sarcocystiosis. In: Martin, W.B. and Aitken, I.D. (eds) *Diseases of Sheep*, 3rd edn. Blackwell Science, Oxford, UK, pp. 389–392.

Uggla, A. and Buxton, D. (1990) Immune responses against *Toxoplasma* and *Sarcocystis* infections in ruminants: diagnosis and prospects for vaccination. *Revue Scientifique et Technique. Office International des Epizooties* 9, 441–462.

Vaughan, L. (1996) Abortion in sheep. *Irish Veterinary Journal* 49, 170–174.

Vercruysse, J., Fransen, J. and Van Goubergen, M. (1989) The prevalence and identity of *Sarcocystis* cysts in cattle in Belgium. *Journal of Veterinary Medicine B* 36, 148–153.

Vickers, M.C. and Brooks, H.V. (1983) Suspected *Sarcocystis* infection in an aborted bovine foetus. *New Zealand Veterinary Journal* 31, 166.

Waters, A.P. and McCutchan, T.F. (1990) Ribosomal RNA: nature's own polymerase-amplified target for diagnosis. *Parasitology Today* 6, 56–59.

Wei, T., Chang, P.Z., Doung, M.X., Wang, X.Y. and Xia, A.Q. (1985) Description of two new species of *Sarcocystis* from the yak (*Poephagus grunniens*) (in Chinese). *Sci. Agri. Sinica* 4, 80 (cf. Dubey *et al.*, 1989).

Weiland, G., Reiter, I. and Boch, J. (1982) Möglichkeiten und grenzen des serologischen nachweises von sarkosporidieninfektionen. *Berliner und Münchener Tierärztliche Wochenschrift* 95, 387–392.

Wenyon, C.M. (1926) Parasites of undetermined position. In: *Protozoology. A Manual for Medical Men, Veterinarians and Zoologists*, Vol. 1. Baillière, Tindall and Cox, London, pp. 760–778.

Xiao, B.N., Zeng, D.N., Zhang, C.G., Wang, M., Li, Y. and Gong, Z.F. (1993) Development of *Sarcocystis cruzi* in buffalo (*Bubalus bubalis*) and cattle (*Bos taurus*). *Acta Veterinaria et Zootechnica Sinica* 24, 185–192.

Yakimoff, W.L. and Sokoloff, I.I. (1934) Die sarkozysten des renntières und des maral (*Sarcocystis grüneri* n. species). *Berliner Tierärztliche Wochenschrift* 47, 772–774.

Yang, Z.Q., Zuo, Y.X., Ding, B., Chen, X.W., Luo, J. and Zhang, Y.P. (2001) Identification of *Sarcocystis hominis*-like (Protozoa: Sarcocystidae) cyst in water buffalo (*Bubalus bubalis*) based on 18S rRNA gene sequences. *The Journal of Parasitology* 87, 934–937.

Yang, Z.Q., Li, Q.Q., Zuo, Y.X., Chen, X.W., Chen, Y.J., Nie, L., Wei, C.G., Zen, J.S., Attwood, S.W., Zhang, X.Z. and Zhang, Y.P. (2002) Characterization of *Sarcocystis* species in domestic animals using a PCR-RFLP analysis of variation in the 18S rRNA gene: a cost-effective and simple technique for routine species identification. *Experimental Parasitology* 102, 212–221.

3.4 Tritrichomonosis

H. Sager, I. Ferre, K. Henning and L.M. Ortega-Mora

General Considerations

Despite increased awareness and proven control strategies, bovine tritrichomonosis continues to be a major source of economic loss in areas of the world where natural breeding is the norm (Martínez *et al.*, 1986; Behera *et al.*, 1987; Kvasnicka *et al.*, 1989; BonDurant *et al.*, 1990; Speer and White, 1991; Perez *et al.*, 1992; Copeland *et al.*, 1994; Owen-Rae *et al.*, 2004; Riley *et al.*, 1995).

By contrast, in Western Europe, due to the widespread use of artificial insemination (AI) and intensive testing of breeding animals, effective control of tritrichomonosis has been achieved in many areas.

However, in some areas of Spain, a prevalence of 2.9% has been reported in bulls from herds in which natural mating and AI are used as alternatives (Martín-Gómez *et al.*, 1996), and the personal experience of the authors suggests a re-emergence of the disease in extensive production systems in Southern Europe. A clear example is France, where the disease is reported sporadically. The World Organization for Animal Health (OIE) multi-annual animal disease status report shows the occurrence of bovine tritrichomonosis in France, Germany, Italy and Spain during the previous 5 years (http://www.oie.int).

The OIE has included bovine tritrichomonosis as a List B disease for many years, together with others transmissible diseases that are considered to be of socio-economic and/or public health importance within countries and that are significant in the international trade of animals and animal products. Reports about tritrichomonosis are normally submitted once a year from OIE member countries.

Bovine tritrichomonosis is also one of the diseases included in the European Union (EU) Directive that regulates conditions for the trade of semen. Particularly, collection of preputial washings – or, in the case of female animals, a vaginal mucus agglutination test for the diagnosis of tritrichomonosis – are recommended (European Union, 1988, 2004).

Most of the recommendations concerning standards for diagnostic tests described in the *Terrestrial Manual* from the OIE, or requirements from the EU directives concerning bovine semen trade, do not guarantee the exclusion of this disease in animals or their semen, and therefore we suggest that they are revised in light of present knowledge of diagnostic techniques for tritrichomonosis.

On the one hand, sample collection, transport and culture are key points in the diagnostic procedure and should be carried out following standard procedures to guarantee an adequate technique sensitivity. On the other hand, it is critical to use very specific diagnostic tests. This could be achieved using a PCR technique as a primary test, or as a second step with culture-positive samples. However, this would require a standardization and validation of this technique, with a large number of isolates over a wide geographical range.

Biology, Transmission and Clinical Signs

Tritrichomonas foetus is the causative agent of bovine tritrichomonosis, a venereal disease of cattle which causes reproductive failure. The biological, epidemiological and clinical aspects of the disease in cattle have been comprehensively reviewed by Skirrow and BonDurant (1988), Yule *et al.* (1989), and recently by BonDurant (2005). The natural hosts of *T. foetus* are the cattle *Bos taurus* and *Bos indicus*, and possibly the pig, horse, roe deer and cat.

The parasite is a flagellated pyriform eukaryote, included in Phylum Sarcomastigophora, Order Trichomonadida, measuring approximately 8–18 μm in length and 4–9 μm in width, characterized by three anterior and one posterior flagella, plus an undulating membrane running lengthwise along the side of the organism. The protozoa move rapidly, with a jerky, rolling motion when observed microscopically at 100–400 times magnification. The organism reproduces by longitudinal binary fission, and cysts have not been observed.

Three serotypes of *T. foetus*, based on agglutination reactions, are traditionally identified as *belfast*, *manley* and *brisbane*. *T. foetus* is confined to the reproductive tract of bulls and cows. It does not invade epithelium and parasites are found only on mucosal surfaces, in secretions and gland lumina.

In bulls, protozoa are found in the preputial cavity and sometimes in urethral orifice. The greatest concentrations of organisms are found on the penile mucosa and the adjacent posterior preputial mucosa. In cows, the uterus was thought to be the main site of infection, but several studies of naturally infected cows indicate that the *os cervix* is the preferred site.

Transmission of *T. foetus* is almost exclusively by coital contact between male and female. The parasite can survive in whole or diluted semen at 5°C, and can also withstand the conditions of semen cryopreservation. Mechanical transmission via insemination instruments, vaginal palpation and other contact of infected material with the female reproductive tract is possible. Infection through AI is rare.

The prevalence of infection in bulls seems to increase with age, presumably due to deepening of epithelial crypts in the prepuce of older bulls, which provide an ideal niche for the parasites. However, the number of matings during the lifetime of each bull must also play a part. The evidence suggests that, once infected, bulls tend to remain carriers for life. Breed predisposition to tritrichomonosis has not been unequivocally determined. Infection is self-limiting in cows, with clearance of parasites from the reproductive tract occurring after approximately 90–95 d. The phenomenon of carrier cows has been confirmed, but is rare.

Tritrichomonosis is unapparent in the bull. *T. foetus* infection produces little inflammatory response in the male reproductive tract. Grossly, there is usually no evidence of infection, and microscopically mild infiltrations of neutrophils, macrophages and lymphocytes are seen in the sub-epithelium of the penis and prepuce.

The pathogenesis of tritrichomonosis in the cow has not definitively been determined.

The progress of the infection appears to be rapid, and all regions of the female reproductive tract are colonized within 2 weeks of infection. The numbers of parasites present in the cervico-vaginal mucus fluctuate during the oestrus cycle, and the largest numbers are seen a few days before oestrus.

Following breeding and infection, signs of mild vaginitis, cervicitis and/or endometritis, such as mucopurulent vaginal discharge, may be observed, although generally there are no overt signs. Conception apparently proceeds normally, but almost all conceptuses are lost at some time early in gestation.

Most pregnancies are lost about 17 d after conception, when maternal recognition has occurred, but embryonic or fetal death can occur at any time up to the fifth month of gestation. These pregnancy losses are rarely noticed, and cows are therefore categorized as repeat breeders, infertile or non-pregnant. Infected beef herds manifest extended breeding (and subsequent calving) seasons, resulting in increased inter-calving intervals and reduced calving rates. In dairy herds, the disease tends to be more insidious, with an increase in repeat breeders, services per conception, days open and abortion rates.

Following fetal loss, there is a gradual process of repair, leading to restoration of a fertile uterus 2–6 months after infection, although retention of the *corpus luteum* or a macerated fetus can give rise to pyometra and chronic inflammation, which occasionally result in permanent infertility. Pyometra is seen in up to 10% of cows infected, presumably following fetal loss.

Diagnostic Techniques

Several techniques have been developed for the direct or indirect detection of *T. foetus*. However, only a few have been reliably validated for routine diagnosis and disease control.

The organisms may be diagnostically cultured *in vitro*, preferably in Diamond's medium (Diamond, 1983), on mammalian feeder cells (Balloh and Eisa, 1980) or in specific *Tritrichomonas* media that are commercially available (Ribeiro, 1990). A field culture test (InPouch™ TF) has been developed in the USA, which allows growth of the trichomonads and direct microscopic examination without further handling of the inoculum (Thomas *et al.*, 1990). Although the media are relatively selective, contamination with intestinal or coprophilic trichomonadid protozoa – which might be mistaken for *T. foetus* – cannot be excluded. Conversely, the PCR technique can be highly specific and sensitive (Felleisen, 1998). However, in many countries it is not accepted yet as an official method of investigation.

Indirect detection methods such as the intradermal (Kerr, 1944) or microagglutination test (Soto and Parma, 1989) are not used for routine diagnosis.

Parasite culture and identification

In this section, techniques for culture and identification of *T. foetus* are described, including information on isolation of the parasite, *in vitro* culture and identification by staining. The culture technique using mammalian feeder cells is not used for routine diagnosis and is only listed for the sake of completeness.

In vitro culture of T. foetus *(see Fig. 3.4.1)*

CULTURE IN MODIFIED DIAMOND'S MEDIUM (DIAMOND, 1957; FELLEISEN *ET AL.*, 1997)

Materials

- Peptone from pancreatically digested casein
- Yeast extract
- Maltose monohydrate
- L-Cysteine hydrochloride
- Ascorbic acid (vitamin C)
- di-Potassium hydrogen phosphate anhydrous
- Potassium dihydrogen phosphate
- Penicillin/streptomycin solution

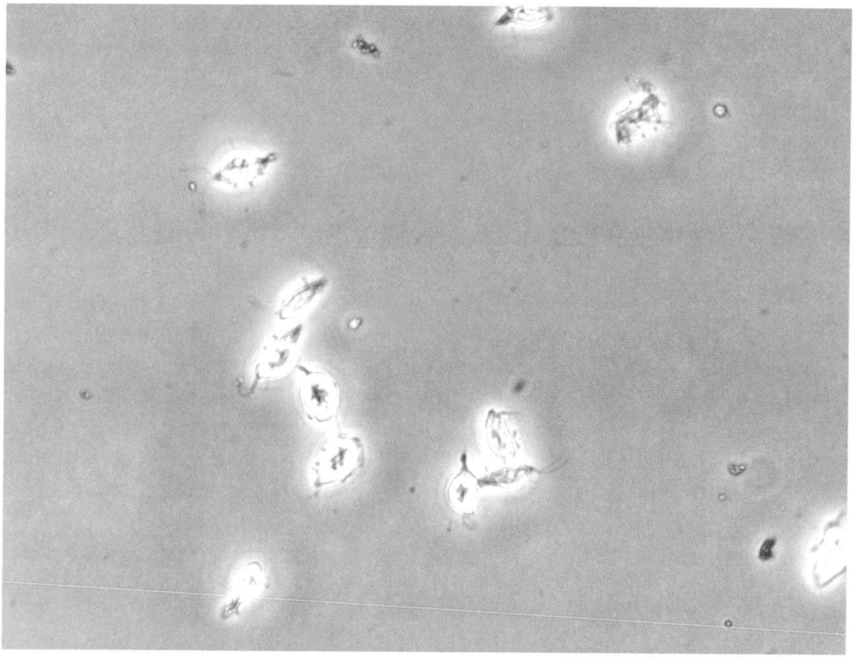

Fig. 3.4.1. *In vitro* culture of *Tritrichomonas foetus.*

- Amphotericin B solution
- Horse serum
- Sodium hydroxide solution 3 N
- Sterile distilled water
- 75 cm^2 tissue culture flasks
- Flat-bottomed plastic tubes
- Incubator

Method

NB: in order to avoid contamination, work should be carried out with full aseptic precautions.

MODIFIED DIAMOND'S MEDIUM

1. Dissolve in 80 ml sterile distilled water:
  - Peptone 2.0 g
  - Yeast extract 1.0 g
  - Maltose monohydrate 0.5 g
  - L-Cysteine hydrochloride 0.1 g
  - Ascorbic acid (vitamin C) 0.02 g
  - di-Potassium hydrogen phosphate 0.08 g

- Potassium dihydrogen phosphate 0.08 g.

2. Adjust to pH 6.9 using 3 N NaOH.
3. Adjust to 90 ml with sterile distilled water.
4. Autoclave.
5. Add 10% heat-inactivated horse serum and antibiotics (100 IU/ml penicillin, 100 μg/ml streptomycin, 2.5 μg/ml amphotericin B). Additionally, vancomycin (100 μg/ml) may be added.

Culture

Incubate *T. foetus* parasites at 37°C using tissue culture flasks or flat-bottomed plastic tubes. Observe the development of the parasites twice daily under phase contrast. Change medium once or twice per week.

CULTURE USING INPOUCH™ TF (SEE FIG. 3.4.2)

Materials

- InPouch™ TF test kit (Biomed Diagnostics, San Jose, California, USA)
- Incubator
- Microscope

Method

1. Add 0.5–1.0 ml of sample into the liquid of the pouch's upper chamber.
2. Express the contents of the upper pouch chamber into the lower.
3. Incubate the pouch vertically at 37°C.
4. Observe the development of the parasites twice daily (up to 3 d).

CULTURE USING MAMMALIAN FEEDER CELLS (BALLOH AND EISA, 1980)

Materials

- Membrane air pump (Pipet Boy acu)
- Inverted microscope
- Incubator
- Medium RPMI 1640
- L-glutamine
- Vitamins 100 ×
- Amino acids 100 ×
- Gentamicin
- Vancomycin
- Amphotericin B solution
- Bovine fetal serum
- Plastic flasks

Method

NB: in order to avoid contamination, work has to be carried out with full aseptic precautions.

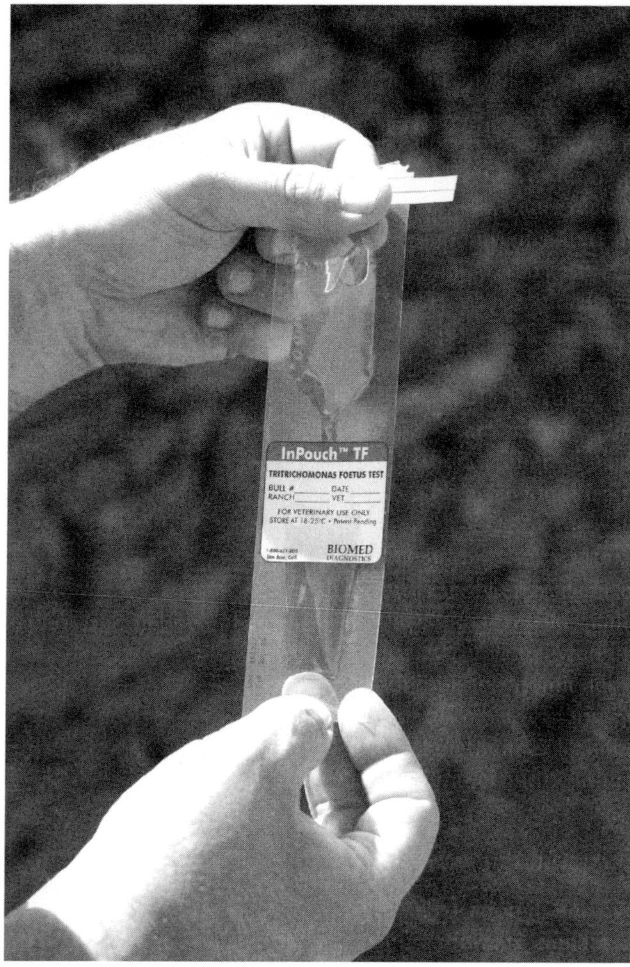

Fig. 3.4.2. Culture of *Tritrichomonas foetus* using InPouch™ TF.

1. Cell culture: buffalo-green-monkey (BGM) cells are grown in RPMI 1640 medium containing:

- 5% bovine fetal serum
- 2% amino acids
- 1% vitamins
- 1% glutamine
- 50 μg/ml gentamicin
- 2.5 μg/ml amphotericin B.

Additionally, vancomycin (100 μg/ml), penicillin (100 IU/ml) and streptomycin (100 μg/ml) may be added. The cells are cultured in plastic flasks.

2. Add *T. foetus* parasites.
3. Incubate at 37°C.
4. Observe the development of the parasites twice daily under phase contrast.
5. Change medium once or twice per week

STAINING OF *T. FOETUS* (CAMPERO *ET AL.*, 2003) (SEE FIG. 3.4.3)

Materials

- Methanol p.a.
- Giemsa solution for microscopy
- Distilled water
- Slides (76 × 26 mm)
- Absorbent paper

Method

1. Air-dried samples are fixed in methanol for 10 min.
2. Cover the slides with Giemsa's stain (working solution: 1 part Giemsa solution in 20 parts distilled water) for 20–30 min.
3. Wash with distilled water.
4. Slides are finally dried with absorbent paper.
5. Identification: trichomonadid protozoa with three anterior flagellae: *Tritrichomonas* sp. (Levy *et al.*, 2003).

Intradermal test

An intradermal test for the diagnosis of bovine tritrichomonosis has been reported (Kerr, 1944). A dose of 0.1 ml of the 'Tricin' antigen is injected

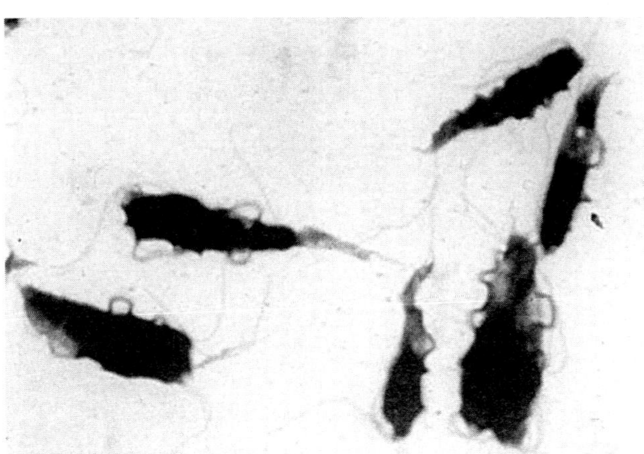

Fig. 3.4.3. Identification of *Tritrichomonas foetus* by staining (from Campero *et al.*, 2003).

intradermally and the reaction is measured 30–60 min later. The injection site is in the skin of the neck, similar to the site used for the tuberculin test. The reaction consists of a shallow plaque observed visually and showing an increase of > 2 mm in skin thickness. This test is not used routinely and is therefore not further described.

Polymerase chain reaction

Molecular-based techniques that use polymerase chain reaction (PCR) technology have been described for the identification of *T. foetus*. The PCR described in this chapter uses the primers TFR3 and TFR4, both flanking the 5.8S rRNA gene (Felleisen *et al.*, 1998). It results in a major amplification product of 347 bp. The PCR reaches a sensitivity of about 50 parasites per ml and amplifies DNA from *T. foetus*, *T. suis* and *T. mobilensis*.

The latter has been described as an intestinal parasite of the squirrel monkey (Culberson *et al.*, 1986) and is thus of lower importance in a differential diagnosis. *T. suis* is a parasite of swine, with predilection sites in the nasal cavity and the digestive tract. *T. foetus* and *T. suis* displayed identical sequences in their rRNA gene unit and were assumed to be of the same species (Felleisen *et al.*, 1997; Tachezky *et al.*, 2002), whereas the pathogenic potential of *T. suis* still remains unsolved (Cobo *et al.*, 2001).

Materials

- DNA isolation kit for cell-containing body fluids (e.g. Qiagen DNeasy™ tissue kit, Valencia, California, USA)
- H_2O demineralized
- PCR buffer 1x, supplemented with $MgCl_2$ to a final concentration of 2.5 mM (Perkin Elmer, N808-0152)
- dNTP
- Primers TFR3 and TFR4
- Positive control DNA

Method

PROCEDURE FOR DNA PREPARATION

1. Use 200 μl of sediment from preputial washings or vaginal swabs for DNA-isolation with the DNeasy™ Tissue Kit (Qiagen).
2. If the solution is very viscous due to mucus, a pre-treatment with N-acetylcysteine (NALC) has to be performed. Add an equal volume of NALC (500 mg/100 ml) to the viscous solution and incubate for 1 h at room temperature on a shaker.

3. Quantification of isolated DNA by spectrophotometry or gel-check is recommended.

PROCEDURE FOR POLYMERASE CHAIN REACTION (PCR) The following oligonucleotides are used for amplification of the 5.8S rRNA gene:

- TFR3 (5' to 3') CGGGTCTTCCTATATGAGACAGAACC
- TFR4 (5' to 3') CCTGCCGTTGGATCAGTTTCGTTAA

2. PCR is performed in 25 μl 1x PCR-buffer with 50 pmol of each primer, 200 μM dNTPs and two units of Taq DNA polymerase.
3. Samples are used directly (2.5 μl) and at a dilution of 1:10.
4. The following cycles are run:

- 20°C for 10 min
- 95°C for 2 min
- 40 cycles: denaturing at 94°C (30 s), annealing at 67°C (30 s) and extension at 72°C (30 s)
- 72°C for 15 min
- 4°C until removal from the thermocycler.

5. Samples are analysed by gel electrophoresis on 2% agarose gels and visualized by ethidium bromide staining.

Notes

- The use of uracil DNA glycosylase (UDG) helps to prevent carry-over contamination (Longo *et al.*, 1990). The nucleotide dTTP in the PCR reaction mix is therefore replaced by 400 μM dUTP. Isolated DNA is pre-incubated with UDG for 1 h at 37°C before use in the PCR (UDG will be inactivated at 95°C in the thermocycler).
- The use of an inhibition control (IC) detects PCR inhibition for each sample. The IC consists of approximately ten copies from a recombinant DNA fragment (size 528 bp), which contains the TFR3 and TFR4 primer target sites in flanking positions. Therefore, the reaction mix for each sample has to be prepared in duplicates. In the second sample, 2 μl (containing ten IC-copies) is added. The IC binds the same primers but gives an amplification product of larger size and can be differentiated from the diagnostic product.

Serology

The detection of humoral responses to *T. foetus* may be performed with samples of serum, vaginal mucus or preputial secretion and assayed by means of microagglutination of living protozoa (mucus agglutination test: Pierce, 1949; Soto and Parma, 1989). Systemic antibodies to *T. foetus* in bovine serum have been evaluated by ELISA (Skirrow and BonDurant, 1990; Cobo *et al.*, 2002) and haemolytic assay (BonDurant *et al.*, 1996), showing the ELISA to be more practical and more efficient (Cobo *et al.*, 2002). Currently, serological tests are not used for routine diagnosis of *T. foetus* infections in cattle.

Diagnostic Applications

The presence of *T. foetus* and tritrichomonosis within a bovine herd is initially indicated by lower than expected pregnancy rates and an extended calving season. Infection is inapparent in the bull, and mild vaginitis, cervicitis and/or endometritis, such as a mucopurulent vaginal discharge, is only occasionally found in some cows. Early abortion and occasional pyometra may occur, although the prevalence of these symptoms is often less than 5% in newly infected herds. As tritrichomonosis is so similar to *Campylobacter fetus venerealis* infection (vibriosis), the absence of this latter pathogen needs to be confirmed, as does the absence of other causes of early reproductive failure, such as lowered bull fertility and inadequate nutrition (Skirrow and BonDurant, 1988).

A definitive diagnosis is routinely made by direct observation of the active parasite in preputial or vaginal secretions and in placental and abomasal fluids from aborted fetuses, or by culturing it in one of several media. Recently, in an attempt to decrease the time spent on diagnostic culturing of *T. foetus* and to confirm parasite identification, efforts have been made to establish DNA-based detection methods (Ho *et al.*, 1994; Felleisen, 1997; Felleisen *et al.*, 1998; Nickel *et al.*, 2002).

Fetal diagnosis

T. foetus causes reproductive failure in cattle characterized by early embryonic death (Skirrow and BonDurant, 1988), and thus aborted fetuses are not frequent in diagnostic samples. However, tritrichomonosis can be diagnosed readily in an aborted fetus by examining the abomasal contents and placental fluids, both of which are frequently teeming with parasites. There are no pathognomonic macroscopic or microscopic lesions seen in the conceptus, therefore identification of the pathogen from either the conceptus or cow is necessary.

Detection of T. foetus *by direct observation or culture from placenta and fetal samples*

Laboratory diagnosis of *T. foetus*-induced abortion is routinely made by microscopic demonstration of trichomonadids in fresh placental fluids, uterine exudates, fetal abomasal contents or in cultures inoculated with these specimens.

Further studies are needed to standardize a PCR assay to demonstrate *T. foetus* in those samples or in cultures.

Detection of T. foetus *in formalin-fixed, paraffin wax-embedded histological sections of placenta and fetus*

T. foetus produces characteristic, but non-pathognomonic, placental and fetal lesions including mixed-cell placentitis and pyogranulomatous pneumonia, often with multinucleated giant cells (Rhyan *et al.*, 1988).

Trichomonads are often discernable in haematoxylin and eosin-stained sections of infected fetal lung and placenta, and special stains are helpful in differentiating the organisms from leucocytes (Parsonson *et al.*, 1976; Rhyan *et al.*, 1988). The organism cannot be reliably identified using these stains if present in low numbers or in exudates or necrotic tissue. For this reason, other techniques have been used to demonstrate *T. foetus* in histological sections.

T. foetus has been identified in bovine placenta using an indirect immunofluorescent antibody test (IFAT) employing specific monoclonal antibodies (Burgess and Knoblock, 1989), and an immunohistochemical (IHC) technique utilizing polyclonal antibodies, which has been used to demonstrate *T. foetus* in experimentally infected mice (Campero *et al.*, 1989).

An IHC technique using a monoclonal antibody has been evaluated as a diagnostic tool to specifically label *T. foetus* in histological sections of placenta and fetal lung from bovine abortions (Rhyan *et al.*, 1995). This technique provides a method for positive identification of *T. foetus* organisms in formalin-fixed tissues, and the labelled organisms are easily differentiated from surrounding leucocytes, stromal cells or necrotic debris. The technique is useful as an adjunct to culture in cases of bovine abortion or as an alternative when inadequate specimens are available for culture. In addition, its use as a research tool should provide valuable information in pathogenesis studies of this disease.

Recently, BonDurant *et al.* (2003) detected *T. foetus* DNA in formalin-fixed, paraffin wax-embedded sections of gut and lung tissues from two aborted bovine fetuses (naturally infected) using a PCR assay, originally developed for definitive laboratory identification of the pathogen from cultures of male reproductive tract fluids (Felleisen, 1997, 1998; Felleisen *et al.*, 1998).

Key points

- Samples from fresh placental fluids, uterine exudates and fetal abomasal contents.
- Detection of *T. foetus* by direct observation.
- Cultures inoculated with samples and direct observation afterwards.
- Detection of *T. foetus* in sections of formalin-fixed, paraffin wax-embedded samples of placenta and fetus by IFAT and IHC employing a suitable monoclonal antibody.

- Detection of *T. foetus* in sections of formalin-fixed, paraffin wax-embedded samples of placenta and fetus by PCR assay.

Post-natal detection of infection

Once a problem is recognized, herd diagnosis of tritrichomonosis is best accomplished by identification of the organisms in placental fluid, stomach contents of the aborted fetus (see above) or uterine washings, pyometra discharges or vaginal mucus. However, in infected herds, the most reliable material for diagnosis is either preputial or vaginal washings or preputial scrapings (Schönmann *et al.*, 1994).

As the bull is the most important maintenance host (Clark *et al.*, 1974; Parsonson *et al.*, 1974), and the organisms are found almost exclusively in the superficial layers of the preputial mucosa (Hammond and Bartlett, 1943; Parsonson *et al.*, 1974; Rhyan *et al.*, 1999), preputial specimens are the sample of choice.

The diagnosis is made on an individual animal basis by observing motile trophozoites in washes or scrapings from the prepuce of the bull or from short-term cultures of preputial washes or scrapings. Recently, a number of PCR assays have been described (Ho *et al.*, 1994; Felleisen *et al.*, 1998; Nickel *et al.*, 2002). Current diagnostic recommendations include triplicate samples at 2-week intervals from bulls, resulting in a 90–99% sensitivity using a Diamond's medium or a commercially available InPouch™ TF culture system (Todorovic and McNutts, 1967; Clark *et al.*, 1971; Kimsey *et al.*, 1980; Thomas *et al.*, 1990; Schönmann *et al.*, 1994; BonDurant, 1997; Parker *et al.*, 1999).

Sample collection methods

BULL A variety of techniques have been described for sample collection (Ball *et al.*, 1984; Irons *et al.*, 2002), and numerous papers have been written describing the various advantages and/or disadvantages of collection methods (Fitzgerald *et al.*, 1952; Tedesco *et al.*, 1979; Irons *et al.*, 2002). BonDurant (1985) and Kimsey (1986) described collection of smegma using an AI pipette, the method most commonly used today. Whether one uses a pipette method to scrape or otherwise collect cells and smegma from the glans penis of the bull or a douching system is usually dictated by circumstances. Field collection and handling of samples are usually more difficult than collection of samples taken at a clinic or laboratory.

Whatever method is used, care should be taken to keep the sample clean. For example, preputial hair should be trimmed from the preputial orifice, the orifice should be washed with soap and water and then thoroughly rinsed with clean water prior to collecting the sample (see Fig. 3.4.4). It is important to note that trichomonad recovery is increased if, before sampling, the prepuce is vigorously massaged to detach organisms from the penile surface. In addition, during sample collection the surface of the penis and prepuce should be scraped vigorously to dislodge trichomonads from the epithelial crypts.

If possible, bulls should be sexually rested for at least 1 week before sample collection to maximize the number of organisms in the preputial cavity (Bartlett, 1949).

Preputial material is generally collected by sheath washing or sheath scraping (see Fig. 3.4.5) combined with aspiration, which are both well described (BonDurant, 1997; Irons *et al.*, 2002) and have similar sensitivity (Shönmann *et al.*, 1994). Scraping the prepuce produces more positive samples than does aspiration (Tedesco *et al.*, 1979). However, the collection of preputial material by scraping with simultaneous aspiration has several practical advantages over preputial washing. Speed of collection, the ability to collect the sample without an assistant and the fact that contamination of the sample by urine is easily avoided are significant advantages.

Although *T. foetus* organisms do survive in urine (Parsonson *et al.*, 1974), dilution of the cellular content of the sample is undesirable. In addition, the use of disposable collection equipment eliminates the possibility of cross-contamination of samples or the transmission of organisms between successive animals. Mechanical transmission of *T. foetus* is a potential hazard whenever infected animals are examined (Goodger and Skirrow, 1986). A smaller

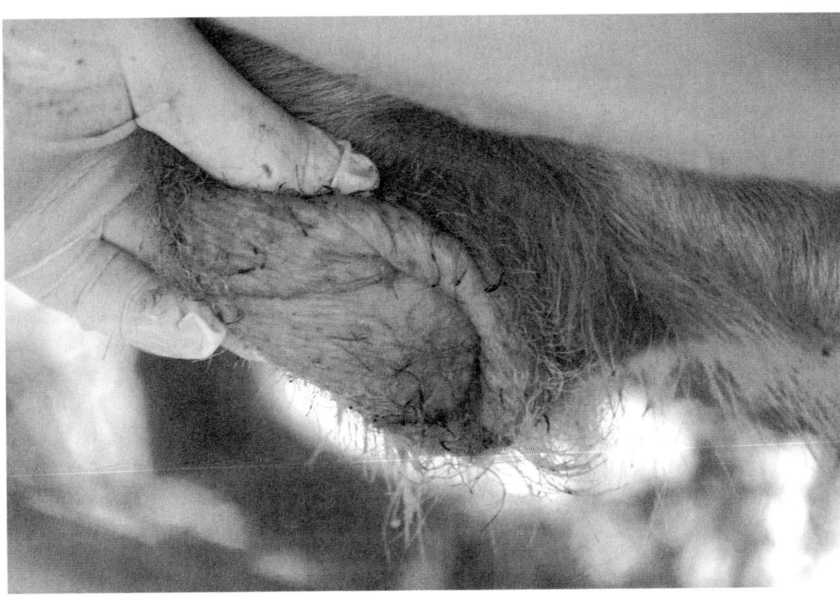

Fig. 3.4.4. In the bull, care should be taken in keeping the smegma sample clean; hair should be trimmed from the preputial orifice, which should be washed with soap and water.

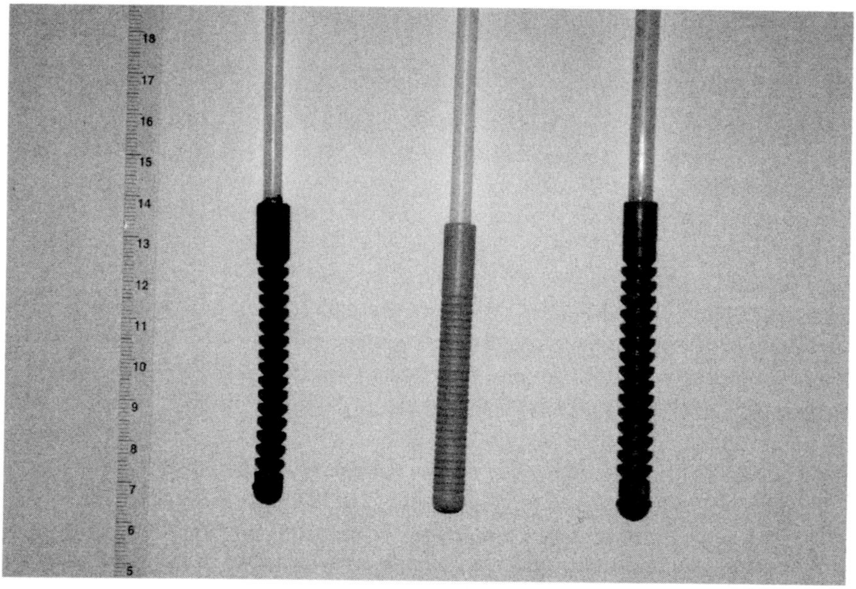

Fig. 3.4.5. Preputial smegma is generally collected by sheath washing or scraping, combined with aspiration.

volume of the sample obtained by scraping facilitates sample transport and laboratory processing.

Lastly, as the sample is collected primarily from the caudal area of the preputial cavity, there is less likelihood of contamination from the environment, particularly in bulls that have gross contamination of the anterior portion of the preputial cavity caused by habitual eversion of the *lamina interna*. This is in agreement with findings of other investigators sampling for *Campylobacter fetus* (Tedesco *et al.*, 1977).

Irons *et al.* (2002) evaluated the collection of preputial material by scraping and aspiration to diagnose *T. foetus* in bulls and found that preputial scraping was as effective as washing. Scraping was found to be quick, safe and offered other advantages (see above). However, it may be subject to greater operator variability than sheath washing.

Parker *et al.* (1999) compared the sensitivity of the diagnostic test when using two different sampling tools and a commercially available test kit (InPouch™ TF). They found that the sensitivity of the diagnostic test was the same whether the samples were collected with an AI pipette or with a metal brush. Moreover, they observed an increase in sensitivity when samples were collected on the right side of the bull, regardless of the sampling tool used. Recently, the same authors found that samples taken from the right side of the prepuce were four times as likely to be positive as samples taken from the left side (Parker *et al.*, 2003a).

Other factors such as sampling technique, culture media quality, sample handling and incubation and microscopic examination did not have a significant effect on the outcome of the diagnostic test. It may be preferable for right-handed samplers to collect samples from the right side of the bull, where possible. They concluded that it is important to make every effort to ensure that those factors that can be controlled are optimized, since other extraneous factors may be acting to decrease the sensitivity of the diagnostic test (Parker *et al.*, 2003b).

HEIFER AND COW Cervico-vaginal mucus (CVM) can be collected from trichomonad-infected cows using AI pipettes, with the samples being processed in the same way as for preputial smegma (Abbitt and Ball, 1978) (see Fig. 3.4.6).

Key points

- In bulls, smegma samples should be collected by sheath washing or sheath scraping, combined with aspiration.
- Steps should be taken to keep the samples clean such as: preputial hair should be trimmed from the preputial orifice and the orifice should be washed with soap and water.
- The prepuce should be vigorously massaged to detach trichomonads from the penile surface.

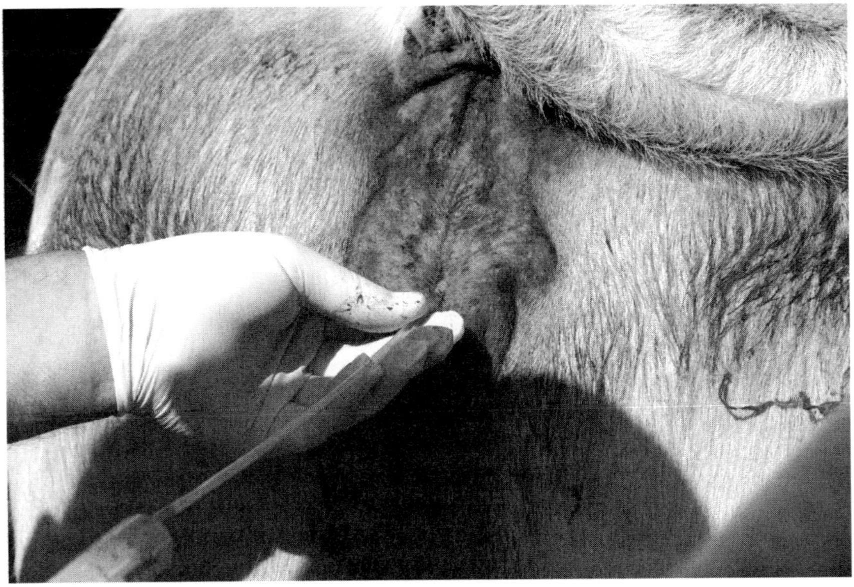

Fig. 3.4.6. Cervico-vaginal mucus can be collected from trichomonad-infected cows using an AI pipette.

- If possible, bulls should not have sexual activity for at least 1 week before sample collection.
- It is important to make every effort to ensure that factors that can be controlled are optimized, since other extraneous factors may be acting to decrease the sensitivity of the diagnostic test.
- In heifers and cows, CVM should be collected using AI pipettes.

Detection of T. foetus *by direct observation or culture*

FROM SMEGMA SAMPLES Once the smegma sample is collected, the material can be suspended in physiological saline and examined microscopically for trichomonads. The organism is relatively easy to discern. *T. foetus* is pyriform, 8–18 μm long and 4–9 μm wide, with three anterior and one posterior flagella and an undulating membrane. The organisms are recognized microscopically in clinical samples by their jerky and rolling progression, and are best detected by phase-contrast or dark-field microscopy. Recognition of a single active organism is sufficient to diagnose infection. However, other protozoa may occasionally be mistaken for *T. foetus* (see below).

Culture of the organisms is usually required because, in most cases, their number is not large enough to make a positive diagnosis by direct examination. The number of trichomonads in the preputial secretions is apparently relatively low, ranging from < 200 to > 80,000/ml (Skirrow and BonDurant, 1988). Ideally, once a smegma sample has been procured, it should be transferred on to culture medium as soon as possible. Direct examination is approximately 25% less sensitive than culture.

The decline in the number of viable organisms in samples prior to culture is well documented (Todorovic and McNutts, 1967; Tedesco *et al.*, 1979; Reece *et al.*, 1983; Skirrow *et al.*, 1985; Kittel *et al.*, 1998; Parker *et al.*, 1999). While immediate culture is the ideal, a delay of 24 h may result in a loss of approximately 10% in diagnostic sensitivity, with longer delays causing more drastic declines. This leads to the general recommendation by several countries, such as South Africa, requiring the use of overnight shipping of samples to the laboratory, and the utilization of enriched transport media where longer delays are unavoidable (Mukhufhi *et al.*, 2003).

TRANSPORT MEDIA Kimsey *et al.* (1980) compared a number of different transport media and concluded that refrigerated buffered saline (containing fetal bovine serum) or lactate Ringer's solution were highly effective if the sample was in transit for up 48 h. If transport time exceeded this, Kupferberg medium (Sprince and Kupferberg, 1947) was preferred. Lactate Ringer's solution is most frequently used due to its ready availability and relatively low cost. Milk-based transport media have also been used and found to be effective for up to 96 h if boiled first and used at 4°C (Reece *et al.*, 1983).

There is a negative correlation between the number of positive samples and the time elapsed from sample collection, hence transit time should be minimized and, if possible, should be < 12 h (Fitzgerald *et al.*, 1954; Kimsey *et al.*, 1980). From a practical point of view, culture medium may be used also as a transport medium. A commercially available transport and culture kit (InPouch™ TF) has been introduced and is now widely used (BonDurant, 1997).

CULTURE MEDIA There are different culture media available for isolation of *T. foetus*. Examples of such media include modified Plastridge (Fitzgerald *et al.*, 1954) and trypticase-yeast extract maltose (TYM) (Diamond, 1957). Other culture media such as: CPLM (cysteine/peptone/liver-infusion maltose) medium, BGPS (beef-extract/glucose/peptone serum) medium and Clausen's medium (neopeptone-lemco-liver extract glucose) have been used (Ribeiro, 1990; Eaglesome and Garcia, 1992).

Modified Plastridge is the medium of choice for initial culture of field samples, while TYM is used for most subsequent procedures including subculture, growth prior to experimental infection and freezing (Reece *et al.*, 1983; Skirrow and BonDurant, 1988). Long-term storage of *T. foetus* at room temperature can be accomplished in an eggshell-based medium (Schneider, 1942).

The diagnostic sensitivity of some media has been estimated (Skirrow *et al.*, 1985; Appell *et al.*, 1993; Schönmann *et al.*, 1994; Parker *et al.*, 1999). Diamond's medium is commonly used and its estimated diagnostic sensitivity ranged from 81.6 to 93.2% (Diamond, 1957; Skirrow *et al.*, 1985; Schönmann *et al.*, 1994). The estimated diagnostic sensitivity of the InPouch™ TF commercial culture kit ranges from 88.0 to 98.4% (Appell *et al.*, 1993; Schönmann *et al.*, 1994; Parker *et al.*, 1999).

Most reports (Thomas *et al.*, 1990; Borchardt *et al.*, 1992; Schönmann *et al.*, 1994) suggest that culture of preputial secretions, using a commercial selective medium in a pouch, is as efficacious or slightly more efficacious than using TYM medium or the closely related Diamond's medium system. Recently, Parker *et al.* (2003a) compared the diagnostic sensitivity of a commercially available culture kit (InPouch™ TF) and a culture test using Diamond's media and found that samples tested using the commercial kit were 6.95 times as likely to be positive as samples tested with a diagnostic test using Diamond's medium.

Culture of a single sample has been calculated to give only an 80–90% chance of obtaining a positive result from a bull known to be infected (Kimsey *et al.*, 1980; Skirrow *et al.*, 1985), and the cultures of at least three successive samples is required to minimize the risk of obtaining a false negative result. Culture of preputial material from bulls showing sensitivities of > 90% are attainable under optimal conditions (Todorovic and McNutts, 1967; Clark *et al.*, 1971; Kimsey *et al.*, 1980; Thomas *et al.*, 1990; Schönmann *et al.*, 1994; Parker *et al.*, 1999).

However, sensitivities of 70–90% are more representative of samples collected or processed under suboptimal conditions, as is often the case in the field (Todorovic and McNutts, 1967; Skirrow *et al.*, 1985; Schönmann *et al.*, 1994; Peter *et al.*, 1995; Gay *et al.*, 1996; Parker *et al.*, 1999; Rae *et al.*, 1999).

Specificity of diagnosis by culture methods may also be problematic, as the other motile protozoa found in preputial samples may also survive in culture media (Ribeiro, 1990; Taylor *et al.*, 1994; Felleisen *et al.*, 1997; BonDurant *et al.*, 1999). Unfortunately, culture sensitivity is even lower in cows (see below); the chance of obtaining a positive culture of a single sample from a cow known to be infected was estimated at only 59% (Clark *et al.*, 1983).

Specificity of diagnosis by culture methods may also be problematic, as false positives may occur through contamination of samples by intestinal (coprophilic) trichomonad protozoa, which may also survive in culture media and may be mistaken for *T. foetus* (Taylor *et al.*, 1994). Intestinal trichomonads rather than *T. foetus* have occasionally been found in virgin bull samples submitted for culture (BonDurant *et al.*, 1999; Campero *et al.*, 2003). Without additional microscopic or molecular confirmation for *T. foetus* (see below), some bulls are probably culled unnecessarily.

FROM CVM Except in the case of a trichomonad pyometra, the probability of detecting organisms in CVM by direct examination appears to be less than with culturing (Simmons and Laws, 1957; Skirrow and BonDurant, 1988). This is probably due to the fluctuations in trichomonad numbers throughout the oestrous cycle (Barlett and Hammond, 1945; Hammond and Barlett, 1945) and the viscous nature of CVM, which can affect both sample collection and movement of trichomonads.

Furthermore, the sensitivity of diagnosing tritrichomonosis in cows by culturing CVM samples from a naturally infected cow herd was 56% (Goodger and Skirrow, 1986) compared to experimental infection values of 85% (Skirrow and BonDurant, 1988) and 95% (Parsonson *et al.*, 1976). Skirrow and BonDurant (1988) found that, as the duration of infection increased, the number of positive samples obtained from heifers declined, which probably explains the discrepancies in the values reported by others.

Numerous studies (Kimsey *et al.*, 1980; Gregory *et al.*, 1990; Thomas *et al.*, 1990; Borchardt *et al.*, 1992; Appell *et al.*, 1993; Schönmann *et al.*, 1994) have compared different diagnostic techniques for *T. foetus* in bulls, but little information is available on diagnostic options in females. Kittel *et al.* (1998) showed that the commercially selective medium pouch system is considerably more sensitive than modified TYM medium for detecting infection in heifers, and that the difference in sensitivity is more evident than it is in bulls.

Key points

- *Tritrichomonas foetus* organisms are recognized microscopically in clinical samples by their jerky and rolling progression.
- Culture of the organisms is usually required because, in most of cases, the number of organisms is not large enough to make a positive diagnosis by direct examination.

- Transport media: refrigerated buffered saline containing bovine fetal serum or lactate Ringer's solution is highly effective if the sample is in transit for up 48 h, but if transport time exceeds this, Kupferberg medium should be used.
- Culture media: modified Plastridge is the medium of choice for initial culture of field samples, while TYM is used for most subsequent procedures including subculture, growth prior to experimental infection and freezing.
- From a practical point of view, culture media may also be used as transport media. A commercially available transport and culture kit (InPouch™ TF) has been introduced and is now widely used.
- Specificity of diagnosis by culture methods may also be problematic, as false positives may occur through contamination of samples by intestinal (coprophilic) trichomonad protozoa, which may also survive in culture media and may be mistaken for *T. foetus*.
- The commercially selective medium pouch system is considerably more sensitive than modified TYM medium for detecting infection in heifers.

Detection of T. foetus *DNA by PCR assay*

In an attempt to decrease the time spent on diagnostic culturing of *T. foetus*, and to confirm parasite identification, efforts have been made to establish DNA-based diagnostic methods. Amplification of DNA material by PCR is, potentially, a highly sensitive and specific diagnostic test for tritrichomonosis. PCR technology has also been shown to be effective in the diagnosis of the closely related *Tritrichomonas vaginalis* in humans, with a sensitivity of 95% and a specificity of 98% (Patel *et al.*, 2000).

PCR has been applied to the detection of *T. foetus* (Ho *et al.*, 1994; Felleisen *et al.*, 1998; Nickel *et al.*, 2002) and to the study of genetic relationships between trichomonad species and strains (Riley *et al.*, 1995; Felleisen, 1997; Felleisen *et al.*, 1998).

Several primer sets have been designed based on PCR amplification of a portion of the ribosomal RNA gene, specifically the 5.8 S ribosomal RNA and flanking internal transcribed spacer region ITS1 and ITS2 (Felleisen, 1997; Felleisen *et al.*, 1998; Nickel *et al.*, 2002).

SAMPLE COLLECTION While much emphasis has fallen on primer selection and the optimization of laboratory protocols to render satisfactory results, little attention has been given to sample collection and handling procedures (Mukhufhi *et al.*, 2003). Factors relating to sample collection and handling which might be expected to affect the test outcome include: (i) sample selection; (ii) collection method; (iii) transport medium; (iv) holding temperature; and (v) the time delay before processing can take place.

Of the two procedures for taking smegma samples, scraping is more traumatic than washing. Scraping samples often contain blood and it is known that components of blood inhibit the PCR. Scraping samples also contain other products of superficial epithelial damage in addition to thick and mucoid smegma. Smegma originating from sheath scrapings has been shown to reduce the sensitivity of PCR procedures (Ho *et al.*, 1994). On the other hand, sheath washes often contain substantial amounts of urine, which also has inhibitory properties. Scraping also provides a smaller and more concentrated sample than washing. The effects of these factors on the PCR diagnosis of *T. foetus* have not yet been reported (Mukhufhi *et al.*, 2003).

In addition, a time delay may lead to a decline in sensitivity of PCR testing due to degradation of the nucleic acids. *T. foetus* is known to secrete a range of hydrolytic enzymes (Thomford *et al.*, 1996). These enzymes cause rapid DNA breakdown following cell lysis (Turner and Muller, 1983; Wang and Wang, 1985). Other compounds, either from the organisms themselves, from other organisms which are present in the preputial cavity, from the preputial membrane or from the upper urogenital tract, may also affect the integrity of nucleic acids over time.

PCR TECHNIQUES The PCR assays described (Ho *et al.*, 1994; Felleisen *et al.*, 1998; Nickel *et al.*, 2002; Mukhufhi *et al.*, 2003) are therefore sensitive enough to detect the mean numbers of organisms that are typically found in smegma scrapings of experimentally (50 organisms/ml) and naturally (141 organisms/ml) infected bulls on the basis of studies by Hammond and Bartlett (1943). However, molecular analysis did not show an overall advantage in sensitivity when compared to parasite culture using the InPouch™ TF culture system for detection of *T. foetus* in bovine preputial scrapings (Hoevers *et al.*, 2003).

The two PCR assays described in the literature (Ho *et al.*, 1994; Felleisen *et al.*, 1998) showed a high specificity, with a sensitivity sufficient for the detection of DNA from just a single organism. However, the test sensitivity may be lower because of the inhibition of Taq DNA polymerase by inhibitors (Ho *et al.*, 1994). In order to improve the routine DNA extraction for the diagnosis of cattle tritrichomonosis, Chen and Li (2001) developed a method of extracting quality DNA from preputial smegma spiked with *T. foetus*. In addition, a single-tube nested PCR for the detection of *T. foetus* in feline faeces has been developed by Gookin *et al.* (2002).

Nickel *et al.* (2002) improved a PCR assay, previously described by Felleisen *et al.* (1998), to screen for the presence of *T. foetus* in bulls. The assay is simple and rapid to perform when compared with conventional culture methods followed by microscopic identification of the protozoan parasite. In addition, because there is only one specific PCR reaction product with the primer set described, the test can easily be taken to the next step for automation for analysis and detection of PCR product in large sample sets.

Campero *et al.* (2003) suggest that a two-step diagnostic strategy be developed for tritrichomonosis in bulls. The first step would involve culturing the preputial scrapings, a relatively inexpensive procedure that has been shown to be approximately 90% sensitive (Skirrow *et al.*, 1985; Borchardt *et al.*, 1992). The second step would be PCR confirmation of culture-positive samples. Before this two-step method can be widely applied, a quantitative estimate of the actual specificity of the PCR assay needs to be determined. This will require the acquisition and testing of a large number of isolates over a wide geographical range.

Recently, Mukhufhi *et al.* (2003) investigated the sensitivity of a PCR assay for diagnosis of *T. foetus* and the effects of sample collection method, storage and transport medium on the test. Their PCR assay showed a specificity of 98% and the detection limit of the assay was 100 organisms per sample (two organisms/ml of a 50 ml sheath-wash sample). The main feature of their findings was the decline in sensitivity of PCR testing of stored samples over time (from 90% at 6 h to 31% at 5 days). However, the sampling method (sheath washing or scraping) and the use of the DNA-preserving agent guanidinium thiocyanate did not prove advantageous.

BonDurant *et al.* (2003) found *T. foetus* DNA in formalin-fixed, paraffin-embedded endometrial samples from four of four experimentally infected heifers using the technique described by Felleisen *et al.* (1998).

Key points

- Amplification of DNA material by PCR is potentially a highly sensitive and specific diagnostic test for tritrichomonosis.
- The PCR assays described are therefore sensitive enough to detect the mean numbers of organisms that are typically found in smegma scrapings of experimentally and naturally infected bulls.
- Factors relating to sample collection and handling which may be expected to affect test outcome are sample selection, collection method, transport medium, holding temperature and time delay until processing can take place.
- A two-step diagnostic strategy has been developed for tritrichomonosis in bulls. The first step involves culture of preputial scrapings and the second step is the PCR confirmation of culture-positive samples.

Other diagnostic methods

Other tests for tritrichomonosis in cows are the vaginal mucus agglutination and intradermal tests. In addition, some serological methods have been described to detect serum-specific antibodies to *T. foetus.*

MUCUS AGGLUTINATION TEST Pierce (1947, 1949) developed a mucus agglutination test that detected 60% of naturally infected cows, but antibody levels varied according to the stage of oestrus. Mucus samples are collected with an AI pipette from the cervical region of the vagina, preferably a few days after oestrus. Any blood-containing mucus should not be used and the cow should be resampled because serum contains non-specific antibodies that will cause agglutination to occur. Antibodies appear in CVM about 6 weeks after infection, and persist for several months. The mucus agglutination test is most useful as a herd test, being capable of detecting latent or recently cleared infections. It is specific and does not cross-react with *C. fetus* or *Brucella abortus*, but lacks sensitivity.

INTRADERMAL TEST Kerr (1944) developed an intradermal test for tritrichomonosis using a trichloroacetic acid precipitate of *T. foetus* (Tricin). The injection site is the skin of the neck, similar to that used for the tuberculin test. The reaction consists of a shallow plaque observed visually and showing an increase of > 2 mm in skin thickness. The test was found to be valuable in herd diagnosis, but animals remained positive after recovery from the disease.

DETECTION OF SERUM-SPECIFIC ANTIBODIES TO *T. FOETUS* Systemic antibodies to *T. foetus* in bovine serum have been quantitated by ELISA (Skirrow and BonDurant, 1990) and haemolytic assay (BonDurant *et al.*, 1996). Recently, Cobo *et al.* (2002) employed these tests in heifers immunized with whole-cell and membrane vaccines against *T. foetus* and naturally challenged with an infected bull. Both serum tests described similar kinetics of serum antibodies to *T. foetus*; however, the ELISA test was more practical and more efficient (Cobo *et al.*, 2002).

DETECTION OF *T. FOETUS* ANTIGEN IN CVM The application of immunological methods to the detection of *T. foetus* infection has so far met with only limited success. The direct detection of antigen in clinical samples has proved far more difficult in the case of *T. foetus* than with its counterpart in humans, *T. vaginalis*. The development of a sensitive test has been complicated by cross-reactivity problems and by the low levels of antigen present during infection (Yule *et al.*, 1989b).

Conclusions

In conclusion, a definitive diagnosis is routinely made by direct observation of the active parasite in preputial or vaginal secretions collected by sheath washing or scraping and AI pipettes respectively, and in placental and abomasal fluids from aborted fetuses. Samples should be submitted to the laboratory as soon as possible. Culture of the samples is usually required because the number of trichomonads is not large enough to make a positive diagnosis by direct examination. Recently, in an attempt to decrease the time spent on diagnostic culturing of *T. foetus* and to confirm parasite identification, efforts have been made to establish PCR tests, and their routine applicability and advantage in terms of a rapid diagnosis and sample turnover have been demonstrated.

Future Trends

The traditional diagnostic test for *T. foetus* involves collection of preputial washings or scrapings or vaginal CVM samples, followed by culture in a growth medium and microscopic examination. The PCR described above has several advantages when compared with culture techniques:

- The high sensitivity allows pooling of the samples. Therefore, the number of reactions can be drastically reduced in countries with a low prevalence. Only pools yielding positive results have to be retested on an individual level.
- Contamination with coprophilic trichomonads can be excluded due to the high specificity of the PCR (Felleisen *et al.*, 1998; Hayes *et al.*, 2003).
- Environmental factors during transportation do not affect the sensitivity of the PCR, as is the case with culture techniques (Bryan *et al.*, 1999; Mukhufhi *et al.*, 2003; Parker *et al.*, 2003b).

The technique of sampling material from the preputial cavity or from cervico-vaginal mucosa and the stability and survival of trichomonads during transportation are key points and have a strong impact on the outcome of the results (Kimsey *et al.*, 1980; Ball *et al.*, 1984; BonDurant, 1997; Irons *et al.*, 2002). However, these factors can only partially be controlled by the diagnostic laboratories. It would therefore be advantageous if internationally accepted guidelines existed, describing in detail the sampling procedure and the optimal way of sending material. Furthermore, veterinarians should be trained in appropriate collection and handling of the material.

More recently, a rapid antigen detection system for *T. vaginalis* was described (Xenotype Diagnostics, Inc., San Antonio, Texas, USA). This test is based on the use of monoclonal antibodies targeting intracellular and surface secretory proteins of *T. vaginalis* (Kurth *et al.*, 2004). A comparison to the InPouch *in vitro* culture (regarded as gold standard) resulted in a test sensitivity of 90% (95% CI: 69.9–97.2%) and 76.7–79.4% (95% CI: 61.4–87.1%), respectively (Miller *et al.*, 2003; Kurth *et al.*, 2004). The same authors obtained a specificity of 92.5% (95% CI: 80.1–97.4%), and 97.1–99.8% (95% CI: 94.8–99.9%). These values turned out to be lower than those claimed by the manufacturer (sensitivity 100%, specificity 98.1%).

For diagnosis of *T. foetus* in cattle, no comparable antigen detection system exists. Various antigens of this parasite have been described previously, but they have been exclusively investigated in terms of their immunogenicity and potential as vaccine candidates. However, a rapid antigen detection system of high sensitivity and specificity would facilitate substantially the diagnosis of *T. foetus* in cattle. The results could be obtained immediately after sampling of the diagnostic material within a few minutes.

Serological techniques such as the ELISA are described for examination in female cattle (BonDurant *et al.*, 1993). These tests may be useful for potentially exposed herds. However, only a few bulls chronically infected with *T. foetus* show detectable antibodies (BonDurant *et al.*, 1996). It is therefore questionable whether serological tests are useful tools for the detection of potentially infected bulls. Thus, direct detection of the parasite is still the method of choice.

The definitive role of *T. foetus* and/or *T. suis* in cattle, pigs and cats, linked to the question of the potential pathogenicity or apathogenicity, needs to be addressed for epidemiological and control purposes (Gookin *et al.*, 2002). Molecular markers are needed to allow discrimination between different respective parasite populations. Furthermore, from the diagnostic point of view, there is a need for a quantitative PCR approach that would help to assess the infection intensity in a given sample, as this may also be related to a clinical course of infection.

References

Abbitt, B. and Ball, L. (1978) Diagnosis of trichomoniasis in pregnant cows by culture of cervical-vaginal mucus. *Theriogenology* 9, 267–270.

Appell, L.H., Mickelsen, W.D., Thomas, M.W. and Harmon, W.M. (1993) A comparison of techniques used for the diagnosis of *Tritrichomonas foetus* infections in beef bulls. *Agri-Practice* 14, 30–34.

Ball, L., Mortimer, R.G., Cheney, J.M. and Olson, J.D. (1984) Trichomoniasis: diagnosis, pathogenesis, treatment and control. *Bovine Practice* 16, 163–165.

Balloh, A. and Eisa, M. (1980) Isolation in tissue culture and tissue culture media of *Trichomonas gallinae* from pigeons in the Sudan. *Sudan Journal of Veterinary Research* 2, 63–67.

Bartlett, D.E. (1949) Procedures for diagnosing bovine venereal trichomoniasis and handling affected herds. *The Journal of the American Veterinary Medical Association* 114, 293–305.

Bartlett, D.E. and Hammond, D.M. (1945) Pattern of fluctuations in number of *Tritrichomonas foetus* occurring in the bovine vagina during initial infections. II. Application in diagnosis. *American Journal of Veterinary Research* 6, 91–95.

Behera, K.P., Mohanty, B.N., Mohanty, D.N. and Ray, S.K.H. (1987) Preliminary studies on the incidence of bovine trichomoniasis in Orissa. *Indian Veterinary Journal* 64, 112–114.

BonDurant, R.H. (1985) Diagnosis, treatment, and control of bovine trichomoniasis. *Compendium for Continuing Education* 7, S179–S188.

BonDurant, R.H. (1997) Pathogenesis, diagnosis and management of trichomoniasis in cattle. *Veterinary Clinics of North America* 13, 345–361.

BonDurant, R.H. (2005) Venereal diseases of cattle: natural history, diagnosis and the role of vaccines in their control. *Veterinary Clinics of North America Food Animal Practice* 21, 383–408.

BonDurant, R.H., Anderson, M.L., Blanchard, P., Hird, D., Elmi-Danaye, C., Palmer, C., Sischo, W.M., Suther, D., Utterback, W. and Weigler, B.J. (1990) Prevalence of trichomoniasis among beef herds in California. *The Journal of the American Veterinary Medical Association* 196, 1590–1593.

BonDurant, R.H., Corbeil, R.R. and Corbeil, L.B. (1993) Immunization of virgin cows with surface antigen TF1.17 of *Tritrichomonas foetus*. *Infection and Immunity* 61, 1385–1394.

BonDurant, R.H., van Hoosear, K.A., Corbeil, L.B. and Bernoco, D. (1996) Serological response to *in vitro*-shed antigen(s) of *Tritrichomonas foetus* in cattle. *Clinical and Diagnostic Laboratory Immunology* 3, 432–437.

BonDurant, R.H., Gajadhar, A., Campero, C.M., Johnson, E., Lun, Z.N., Nordhausen, R.W., Van-Hoosear, K.A., Villanueva, M.R. and Walker, R.L. (1999) Preliminary characterization of a *Tritrichomonas foetus*-like protozoan isolated from preputial smegma of virgin bulls. *Bovine Practice* 33, 124–127.

BonDurant, R.H., Campero, C.M., Anderson, M.L. and Van-Hoosear, K.A. (2003) Detection of *Tritrichomonas foetus* by polymerase chain reaction in cultured isolates, cervicovaginal mucus, and formalin-fixed tissues from infected heifers and fetuses. *Journal of Veterinary Diagnostic Investigations* 15, 579–584.

Borchardt, K.A., Norman, B.N., Thomas, M.W. and Harmon, W.M. (1992) Evaluation of a new culture method for diagnosing *Tritrichomonas foetus* infection. *Veterinary Medicine* 87, 104–112.

Bryan, L.A., Campbell, J.R. and Gajadhar, A.A. (1999) Effects of temperature on the survival of *Tritrichomonas foetus* in transport, Diamond's and InPouch TF media. *The Veterinary Record* 144, 227–232.

Burgess, D.E. and Knoblock, K.F. (1989) Identification of *Tritrichomonas foetus* in sections of bovine placental tissue with monoclonal antibodies. *Journal of Parasitology* 75, 977–980.

Campero, C.M., Ladds, P.W., Hirst, R.G., Vaughan, J.A. and Emery, D.E. (1989) Detection of *Tritrichomonas foetus* antigens in formalin-fixed, paraffin-embedded sections by the peroxidase antiperoxidase technique. *Australian Veterinary Journal* 66, 264–266.

Campero, C.M., Rodriguez-Dubra, C., Bolondi, A., Cacciato, C. Cobo, E., Perez, S., Odeon, A., Cipolla, A. and BonDurant, R.H. (2003) Two-step (culture and PCR) diagnostic approach for differentiation on non-*T. foetus* trichomonads from genitalia of virgin beef bulls in Argentina. *Veterinary Parasitology* 112, 167–175.

Chen, X.G. and Li, J. (2001) Increasing the sensitivity of PCR detection in bovine preputial smegma spiked with *Tritrichomonas foetus* by the addition of agar and resin. *Parasitology Research* 87, 556–558.

Clark, B.L., White, M.B. and Banfield, J.C. (1971) Diagnosis of *Tritrichomonas foetus* infection in bulls. *Australian Veterinary Journal* 47, 181–183.

Clark, B.L., Parsonson, I.M., White, M.B., Banfield, J.C. and Young, J.S. (1974) Control of trichomoniasis in a large herd of beef cattle. *Australian Veterinary Journal* 50, 424–426.

Clark, B.L., Dufty, J.H. and Parsonson, I.M. (1983) The effect of *Tritrichomonas foetus* infection on calving rates in beef cattle. *Australian Veterinary Journal* 60, 71–74.

Cobo, E.R., Cano, D. and Campero, C.M. (2001) Experimental infection with *Tritrichomonas suis* in heifers. *Veterinary Parasitology* 99, 73–78.

Cobo, E.R., Cano, D., Rossetti, O. and Campero, C.M. (2002) Heifers immunized with whole-cell and membrane vaccines against *Tritrichomonas foetus* and naturally challenged with an infected bull. *Veterinary Parasitology* 109, 169–184.

Copeland, S., Clarke, S., Krohn, G., Hurlburt, T., Walker, F. and Houston, L. (1994) Trichomoniasis in bulls from Saskatchewan and Manitoba. *Canadian Veterinary Journal* 35, 388–389.

Culberson, D.E., Pindak, F.F., Gardner, W.A. and Honigberg, B.M. (1986) *Tritrichomonas mobilensis* n. sp. (Zoomastigophorea: Trichomonadida) from the Bolivian squirrel monkey *Saimiri boliviensis boliviensis*. *Journal of Protozoology* 33, 301–304.

Diamond, L. (1957) The establishment of various trichomonads of animals and man in axenic cultures. *Journal of Parasitology* 43, 488–490.

Diamond, L.S. (1983) Lumen dwelling protozoa: entamoeba, trichomonads and giardia. In: Jensen, J.B. (ed.) In Vitro *Cultivation of Protozoan Parasites*. CRC Press, Boca Raton, Florida, pp. 65–109.

Eaglesome, M.D. and Garcia, M.M. (1992) Microbial agents associated with bovine genital tract infections and semen. Part 1. *Brucella, Leptospira, Campylobacter fetus* and *Tritrichomonas foetus*. *Veterinary Bulletin* 62, 743–775.

European Union (1988) Council Directive EEC 88/407 of 14 June 1988 laying down the animal health requirements applicable to intra-community trade in and imports of deep-frozen semen of domestic animals of the bovine species. *Official Journal of the European Communities Council* 194, 10–23.

European Union (2004) Modification of Annexe D of Council Directive EEC 88/407 of 14 June 1988 laying down the animal health requirements applicable to intra-community trade in and imports of deep-frozen semen of domestic animals of the bovine species. *Official Journal of the European Communities Council* 4.2.2004, 15–21.

Felleisen, R. (1997) Comparative sequence analysis of 5.8S rRNA genes and internal transcribed spacer (ITS) regions of trichomonadid protozoa. *Parasitology* 115, 111–119.

Felleisen, R.S.J. (1998) Comparative genetic analysis of tritrichomonadid protozoa by the random amplified polymorphic DNA technique. *Parasitology Research* 84, 153–156.

Felleisen, R.S.J., Schmid-Lambelet, N. and Gottstein, B. (1997) Comparative evaluation of methods for the diagnosis of bovine *Tritrichomonas foetus* infection. *Journal of Protozoology Research* 7, 90–101.

Felleisen, R.S.J., Lambelet, N., Bachmann, P., Nicolet, J., Müller, N. and Gottstein, B. (1998) Detection of *Tritrichomonas foetus* by PCR and DNA enzyme immunoassay based on rRNA gene unit sequences. *Journal of Clinical Microbiology* 36, 513–519.

Fitzgerald, P.R., Hammond, D.M., Miner, M.L. and Binns, W. (1952) Relative efficacy of various methods of obtaining preputial samples for diagnosis of trichomoniasis in bulls. *American Journal of Veterinary Research* 13, 452–457.

Fitzgerald, P.R., Hammond, D.M. and Shupe, J.L. (1954) The role of cultures in immediate and delayed examinations of preputial samples for *Tritrichomonas foetus*. *Veterinary Medicine* 49, 409–413.

Gay, J.M., Ebel, E.D. and Kearley, W.P. (1996) Commingled grazing as risk factor for trichomonosis in beef herds. *The Journal of the American Veterinary Medical Association* 209, 643–646.

Goodger, W.J. and Skirrow, S.Z. (1986) Epidemiologic and economic analyses of an usually long epizootic of trichomoniasis in a large California dairy herd. *The Journal of the American Veterinary Medical Association* 189, 772–776.

Gookin, J.L., Birkenheuer, A.J., Breitschwerdt, E.D. and Levy, M.G. (2002) Single-tube nested PCR for detection of *Tritrichomonas foetus* in feline feces. *Journal of Clinical Microbiology* 40, 4126–4130.

Gregory, M.W., Ellis, B. and Redwood, D.W. (1990) Comparison of sampling methods for the detection of *Tritrichomonas foetus* infection in bulls. *The Veterinary Record* 127, 16.

Hammond, D.M. and Bartlett, D.E. (1943) The distribution of *Tritrichomonas foetus* in the preputial cavity of infected bulls. *American Journal of Veterinary Research* 4, 143–149.

Hammond, D.M. and Bartlett, D.E. (1945) Pattern of fluctuations in numbers of *Tritrichomonas foetus* occurring in the bovine vagina during initial infections. I. Correlation with time of exposure and with subsequent estral cycles. *American Journal of Veterinary Research* 6, 84–90.

Hayes, D.C., Anderson, R.R. and Walker R.L. (2003) Identification of trichomonadid protozoa from the bovine preputial cavity by polymerase chain reaction and restriction fragment length polymorphism typing. *Journal of Veterinary Diagnostic Investigations* 15, 390–394.

Ho, M.S.Y., Conrad, P.A., Conrad, P.J., LeFebvre, R.B., Perez, E. and BonDurant, R.H. (1994) Detection of bovine trichomoniasis with a specific DNA probe and PCR amplification system. *Journal of Clinical Microbiology* 32, 98–104.

Hoevers, J., Snowden, K., Graham, S. and Barling, K. (2003) A comparison of PCR and *in vitro* culture for detection of *Tritrichomonas foetus* in bovine preputial scrapings. *Journal of Eukaryotic Microbiology* 50 (suppl.), 699–700.

Irons, P.C., Henton, M.M. and Bertschinger, H.J. (2002) Collection of preputial material by scraping and aspiration for the diagnosis of *Tritrichomonas foetus* in bulls. *Journal of the South African Veterinary Association* 73, 66–69.

Kerr, W.R. (1944) The intradermal test in bovine trichomoniasis. *The Veterinary Record* 56, 303-305.

Kimsey, P.B. (1986) Bovine trichomoniasis. In Morrow, D.A. (ed.) *Current Therapy in Theriogenology,* 2nd edn. W.B. Saunders, Philadelphia, pp. 275–279.

Kimsey, P.B., Darien, B.J., Kendrick, J.W. and Franti, C.E. (1980) Bovine trichomoniasis: diagnosis and treatment. *The Journal of the American Veterinary Medical Association* 177, 616–619.

Kittel, D.R., Campero, C., Van-Hoosear, K.A., Rhyan, J.C. and BonDurant, R.H. (1998) Comparison of diagnostic methods for detection of active infection with *Tritrichomonas foetus* in beef heifers. *The Journal of the American Veterinary Medical Association* 213, 519–522.

Kurth, A., Whittington, W.L., Golden, M.R., Thomas, K.K., Holmes, K.K. and Schwebke, J.R. (2004) Performance of a new, rapid assay for detection of *Trichomonas vaginalis. Journal of Clinical Microbiology* 42, 2940–2943.

Kvasnicka, W.G., Taylor, R.E.L., Huang, J.C., Hanks, D., Tronstad, R.J., Bosomworth, A. and May, M.R. (1989) Investigations on the incidence of bovine trichomoniasis in Nevada and of the efficacy of immunizing cattle with vaccines containing *Tritrichomonas foetus*. *Theriogenology* 31, 963–971.

Levy, G.M., Gookin, J.L., Poore, M., Birkenheuer, A.J., Dykstra, M.J. and Litaker, R.W. (2003) *Tritrichomonas foetus* and not *Pentatrichomonas hominis* is the etiologic agent of feline trichomonal diarrhea. *Journal of Parasitology* 89, 99–104.

Longo, M.C., Berninger, M.S. and Hartley, J.L. (1990) Use of uracil DNA glycosidase to control carry-over contamination in polymerase chain reaction. *Gene* 93, 125–128.

Martín-Gómez, S., González-Paniello, R., Pereira-Bueno, J. and Ortega-Mora, L.M. (1996) Prevalence of *Tritrichomonas foetus* infection in beef bulls in Northwestern Spain. *Veterinary Parasitology* 75, 265–268.

Martínez, A.H., Bardon, J.C., Nocedal, R.P., Cordeviola, J.M., Sarmiento, F. and Gau, J.A. (1986) Herd diagnosis of bovine trichomoniasis and campylobacteriosis by testing vaginal mucus from non-pregnant cows. *Veterinaria Argentina* 3, 962–966.

Miller, G.A., Klausner, J.D., Coates, T.J., Meza, R., Gaydos, C.A., Hardick, J., Leon, S. and Caceres, C.F. (2003) Assessment of a rapid antigen detection system for *Trichomonas vaginalis* infection. *Clinical and Diagnostic Laboratory Immunology* 10, 1157–1158.

Mukhufhi, N., Irons, P.C., Michel, A. and Peta, F. (2003) Evaluation of a PCR test for the diagnosis of *Tritrichomonas foetus* infection in bulls: effects of sample collection method, storage and transport medium on the test. *Theriogenology* 60, 1269–1278.

Nickel, D.D., Olson, M.E. and Schultz, G.A. (2002) An improved polymerase chain reaction assay for the detection of *Tritrichomonas foetus* in cattle. *Canadian Veterinary Journal* 43, 213–216.

Owen-Rae, D., Crews, J.E., Greiner, E.C. and Donovan, G.A. (2004) Epidemiology of *Tritrichomonas foetus* in beef bull populations in Florida. *Theriogenology* 61, 605–618.

Parker, S., Campbell, J., Ribble, C. and Gajadhar, A. (1999) Comparison of two sampling tools for the diagnosis of *Tritrichomonas foetus* in bulls and clinical interpretation of culture results. *The Journal of the American Veterinary Medical Association* 215, 231–235.

Parker, S., Campbell, J. and Gajadhar, A. (2003a) Comparison of the diagnostic sensitivity of a commercial available culture kit and a diagnostic culture test using Diamond's media for diagnosing *Tritrichomonas foetus* in bulls. *Journal of Veterinary Diagnostic Investigations* 15, 460–465.

Parker, S., Campbell, J., Ribble, C. and Gajadhar, A. (2003b) Sample collection factors affect the sensitivity of the diagnostic test for *Tritrichomonas foetus* in bulls. *The Canadian Journal of Veterinary Research* 67, 138–141.

Parsonson, I.M., Clark, B.L. and Dufty, J. (1974) The pathogenesis of *Tritrichomonas foetus* infection of the bull. *Australian Veterinary Journal* 50, 421–423.

Parsonson, I.M., Clark, B.L. and Dufty, J. (1976) Early pathogenesis and pathology of *Tritrichomonas foetus* infection in virgin heifers. *Journal of Comparative Pathology* 86, 59–66.

Patel, S.R., Wiese, W., Patel, S.C., Ohl, C., Byrd, J.C. and Estrada, C.A. (2000) Systematic review of diagnostic tests for vaginal trichomoniasis. *Infectious Diseases in Obstetrics and Gynecology* 8, 248–257.

Perez, E., Conrad, P.A., Hird, D., Ortuño, A., Chacon, J., BonDurant, R. and Noordhuizen, J. (1992) Prevalence and risk factors for *Tritrichomonas foetus* infection in cattle in northeastern Costa Rica. *Preventive Veterinary Medicine* 14, 155–165.

Peter, D.A., Fales, W.H., Miller, R.B., Youngquist, R.S., Randle, R.F., Ganjam, I.K. and Lybyer, J.L. (1995) *Tritrichomonas foetus* infection in a herd of Missouri cattle. *Journal of Veterinary Diagnostic Investigations* 7, 278–280.

Pierce, A.E. (1947) The demonstration of an agglutinin to *Tritrichomonas foetus* in the vaginal discharge of infected heifers. *Journal of Comparative Pathology* 57, 84–89.

Pierce, A.E. (1949) The mucus agglutination test for the diagnosis of bovine trichomoniasis. *The Veterinary Record* 61, 347–349.

Rae, D.O., Chenoweth, P.J., Genho, P.C., McIntosh, A.D., Crosby, E. and Moore, S.A. (1999) Prevalence of *Tritrichomonas foetus* in a bull population and effect on production in a large cow-calf enterprise. *The Journal of the American Veterinary Medical Association* 214, 1051–1055.

Reece, R.L., Dennett, D.P. and Johnson, R.H. (1983) Some observations on cultural and transport conditions for *Tritrichomonas foetus* var. Brisbane. *Australian Veterinary Journal* 60, 62–63.

Rhyan, J.C., Stackhouse, L.L. and Quinn, W.J. (1988) Fetal and placental lesions in bovine abortion due to *Tritrichomonas foetus*. *Veterinary Pathology* 25, 350–355.

Rhyan, J.C., Wilson, K.L., Burgess, D.E., Stackhouse, L.L. and Quinn, W.J. (1995) Immunohistochemical detection of *Tritrichomonas foetus* in formalin-fixed, paraffin-embedded sections of bovine placenta and fetal lung. *Journal of Veterinary Diagnostic Investigations* 7, 98-101.

Rhyan, J.C., Wilson, K.L., Wagner, B., Anderson, M.L., BonDurant, R.H., Burgess, D.E., Mutwiri, G.K. and Corbeil, L.B. (1999) Demonstration of *Tritrichomonas foetus* in the external genitalia and of specific antibodies in preputial secretions of naturally infected bulls. *Veterinary Pathology* 36, 406–411.

Ribeiro, L.M.M. (1990) An efficient medium for the isolation of *Tritrichomonas foetus*. *Onderstepoort Journal of Veterinary Research* 57, 209–210.

Riley, D.E., Wagner, B., Polley, L. and Krieger, J.N. (1995) PCR-based study of conserved and variable DNA sequences of *Tritrichomonas foetus* isolates from Saskatchewan, Canada. *Journal of Clinical Microbiology* 33, 1308–1313.

Schneider, M.D. (1942) A new thermostable medium for the prolonged bacteria-free cultivation of *Tritrichomonas foetus*. *Journal of Parasitology* 28, 428–429.

Schönmann, M.J., BonDurant, R.H., Gardner, I.A., Van-Hoosear, K., Baltzer, W. and Kachulis, C. (1994) Comparison of sampling and culture methods for the diagnosis of *Tritrichomonas foetus* infection in bulls. *The Veterinary Record* 134, 620–622.

Simmons, G.C. and Laws, L. (1957) Observations on bovine trichomoniasis. *Australian Veterinary Journal* 33, 249–253.

Skirrow, S.Z. and BonDurant, R.H. (1988) Bovine trichomoniasis. *Veterinary Bulletin* 58, 591–603.

Skirrow, S.Z. and BonDurant, R.H. (1990) Immunoglobulin isotype of specific antibodies in reproductive tract secretions and sera in *Tritrichomonas foetus*-induced heifers. *American Journal of Veterinary Research* 51, 645–653.

Skirrow, S.Z., BonDurant, R.H., Farley, J. and Correa, J. (1985) Efficacy of ipronidazole against trichomoniasis in beef bulls. *The Journal of the American Veterinary Medical Association* 187, 401–404.

Soto, P. and Parma, A.E. (1989) The immune response in cattle infected with *Tritrichomonas foetus. Veterinary Parasitology* 33, 343–348.

Speer, C.A. and White, M.W. (1991) Bovine trichomoniasis. Better diagnostics and control could save beef industry $650 million annually. *Large Animal Veterinary* 46, 18–20.

Sprince, H. and Kupferberg, A.B. (1947) The nutrition of protozoa. *Journal of Bacteriology* 53, 435–439.

Tachezy, J., Tachezy, R., Hampl, V., Sedinova, M., Vanacova, S., Vrlik, M., Van Ranst, M., Flegr, J. and Kuldaa, J. (2002) Cattle pathogen *Tritrichomonas foetus* (Riedmuller, 1928) and pig commensal *Tritrichomonas suis* (Gruby and Delafond, 1843) belong to the same species. *Journal of Eukaryotic Microbiology* 49, 154–163.

Taylor, M.A., Marshall, R.N. and Stack, M. (1994) Morphological differentiation of *Tritrichomonas foetus* from other protozoa of the bovine reproductive tract. *British Veterinary Journal* 150, 73–80.

Tedesco, L.F., Errico, F. and Del-Baglivi, L.P. (1977) Comparison of three sampling methods for the diagnosis of genital vibriosis in the bull. *Australian Veterinary Journal* 53, 470–472.

Tedesco, L.F., Errico, F. and Del-Baglivi, L.P. (1979) Diagnosis of *Tritrichomonas foetus* infection in bulls using two sampling methods and a transport medium. *Australian Veterinary Journal* 55, 322–324.

Thomas, M.W., Marmon, W.M. and White, C. (1990) An improved method for the detection of *Tritrichomonas foetus* infection by culture in bulls. *Agri-Practice* 11, 13–17.

Thomford, J.W., Talbot, J.A., Ikeda, J.S. and Corbeil, L.B. (1996) Characterization of extracellular proteinases of *Tritrichomonas foetus. Journal of Parasitology* 82, 112–117.

Todorovic, R. and McNutts, S.H. (1967) Diagnosis of *Tritrichomonas foetus* infection in bulls. *American Journal of Veterinary Research* 28, 1581–1590.

Turner, G. and Muller, M. (1983) Failure to detect extranuclear DNA in *Tritrichomonas foetus* and *Trichomonas vaginalis. Journal of Parasitology* 69, 234–236.

Wang, A.L. and Wang, C.C. (1985) Isolation and characterization of DNA from *Tritrichomonas foetus* and *Tritrichomonas vaginalis. Molecular and Biochemical Parasitology* 14, 323–335.

Yule, A., Skirrow, S.Z. and BonDurant, R.H. (1989) Bovine trichomoniasis. *Parasitology Today* 5, 373–376.

Yule, A., Skirrow, S.Z., Staats, J. and BonDurant, R.H. (1989) Development and preliminary assessment of a polyclonal antibody-based enzyme immunoassay for the detection of *Tritrichomonas foetus* antigen in breeding cattle. *Veterinary Parasitology* 31, 115–123.

Differential Diagnosis of Protozoal Abortion in Farm Ruminants

4

4.1 Non-infectious Causes of Abortion in Farm Ruminants

Michael Hässig

Since a large percentage of abortions are presumably of infectious origin, a thorough diagnosis of a non-infectious abortion can be achieved only by excluding infectious agents. From the pathological point of view, almost any substance can lead to abortion, depending on the concentration and the amount of the substance, as described from the time of Paracelsus from Hohenrain in the Middle Ages.

Abortion is such a general symptom that it can be linked to nearly every physical or chemical interaction. The non-infectious aetiology of abortion has to be divided into endogenous and exogenous origins. Of exogenous origin are innumerable quantities of xenobiotica, any of which can lead to toxicosis. Along with toxicosis go physical factors such as radiation (see Table 4.1.1). There are many other symptomatic abortifacients such as uterine torsion, ketosis and milk fever. For all non-infectious abortions, the interrelation of mother and fetus has to be recognized as one unit. This unit can be influenced by alteration of immunological, genetical and endocrinological aspects.

A lot of research has been carried out in recent times on the cellular and humoral interaction between fetus and mother. The importance of the fetal

© CAB International 2007. *Protozoal Abortion in Farm Ruminants*
(eds L.M. Ortega-Mora, B. Gottstein, F.J. Conraths and D. Buxton)

immune system is now recognized as playing a crucial role in the onset of abortion. Non-infectious modulators for abortion have to be kept in mind, such as feeding, vaccination, animal transport and stress.

Table 4.1.1. Selection of non-infectious exogenous abortifacients important in farming.

Chemical agents	Plants, plant substances or effects of plant ingestion	Physical agents
Herbicides	Phytoestrogene	Stray voltage
Chlornaphthalene	Acacia	Surgery
Lead	Fir needle	Trauma
Prostaglandins	Broom	Stress
Arsenic	Hypovitaminosis A	Radiation
Glucocorticoide	Mycotoxins	Extreme low temperature
Xylazine		Extreme high temperature
Dioxin		Transport
		Noise
		Crowding

4.2 Infectious Causes

Bovines

Franz J. Conraths

In this section, only infectious causes of abortion are listed: those that may have to be considered in differentiating neosporosis from other arbortifacients. It should be noted, however, that reproductive failure can be due to several non-infectious causes, including management deficiencies, environmental factors such as nutrition, exposure to toxins, physical damage and faulty genetics.

Viruses: Bovine Virus Diarrhoea Virus (BVDV), Bovine Herpesvirus 1 (BHV-1) and other bovine herpesviruses

BVDV

This pestivirus is transplacentally transmitted to the fetus (van Oirschot, 1983). A generalized fetal infection is often followed by cerebellar hypoplasia and different types of malformation, such as cataracts, retinal degeneration and hypoplasia, optic neuritis, skeletal malformations, hypotrichosis and general growth retardation. Mummification or abortion is a frequent consequence of congenital infection. A significant increase in abortions has been observed in mid-gestation (days 46–210; Rüfenacht *et al.*, 2001). During the last third of gestation, the infection is terminated by the ontogeny of the fetal immune system (Moennig and Liess, 1995). The reproductive consequences of infection with BVDV have been reviewed by Grooms (2004).

BHV-1

This virus can cause abortion storms in previously naïve herds, resulting in very substantial calf loss. For example, Stubbings and Cameron (1981) reported that 9.4% of the dams aborted in a 160-cow dairy herd during the 4 weeks of an outbreak. Testing paired serum samples may implicate BHV-1 if a more than fourfold increase in titres is observed. The virus can be isolated from the placenta and viral antigen detected in aborted fetuses. Focal necroses in liver, lung, spleen, thymus, lymph nodes and kidneys are usually detected in fetuses aborted due to infection with BHV-1.

Other viruses

The viruses BHV-4 and BHV-5 have also been implicated as causes of bovine abortion (Smith, 1997).

Diagnosis

CALVES AND HEIFERS There are antigen detection tests available for identification of persistently viraemic BVDV carriers that may spread the infection over long periods of time (Cornish *et al.*, 2005; Kuhne *et al.*, 2005; for a review, see Saliki and Dubovi, 2004). Serology can be used to identify BHV-1 infected animals (care: maternal antibodies in calves).

DAMS BVDV and BHV-1 infections can be diagnosed by antibody detection in the aborting dams. If paired serum samples are obtained immediately after the abortion and 3–4 weeks later, the comparison of titres can be used for diagnostic purposes (M. Beer, personal communication). If there is a significant increase in the titre, the respective virus is the likely cause of abortion. If the dam is already seropositive in the first serum sample, the result is inconclusive regarding the cause of abortion. If antibodies cannot be detected in any of the samples, BVDV and BHV-1 can be excluded as the cause of abortion.

FETAL DIAGNOSIS

Virus or genome detection in aborted fetuses

- BVDV: preferentially, virus isolation or detection of viral genome sequences by RT-PCR (Laamanen *et al.*, 1997; Hyndman *et al.*, 1998) from the blood or organs (preferably spleen; Martin Beer, personal communication) of the embryo or fetus should be performed. Due to the fact that the abortions take place with a time lag post-infection, the sensitivity of these techniques is limited, perhaps being better for the genome detection methods (Uruno *et al.*, 1998). Direct antigen detection in organs of the aborted fetus or embryo (e.g. immunofluorescence) is rather insensitive.

- BHV-1: virus isolation or detection of viral genome sequences by PCR (Rocha *et al.*, 1999; Takiuchi *et al.*, 2005) from the blood or organs of the embryo or fetus is the method of choice. Direct antigen detection in organs of the aborted fetus or embryo (e.g. immunofluorescence) is rather insensitive.

Fetal serology: detection of fetal antibodies is useful only in older fetuses if autolysis has not yet progressed. Blood or exudate may be obtained by cardial puncture.

Bacteria

Brucellosis, salmonellosis, leptospirosis, listeriosis, infection with *Campylobacter fetus* subspecies *fetus* or *venerealis, Coxiella burnetii, Ureaplasma diversum*, *Chlamydia*, *Mycoplasma*, other bacteria (e.g. *Actinomyces pyogenes*, *Streptococcus*), fungi (including *Aspergillus fumigatus*) and yeasts (e.g. *Candida* spp.).

Diagnosis

Diagnostic procedures rely on the isolation and identification of bacteria from the aborted fetus or the placenta and, for some, bacteria on serological testing of the dam. Any laboratory result must be interpreted against the background of the findings in the pathological – including histological – examination of the aborted fetus.

Parasites

Sarcocystis spp. have occasionally been described in association with abortion. It is possible, however, that in these cases abortion was an unspecific consequence of the infection and not the result of direct fetal damage due to infection with the parasite. Thus, the true importance of parasites of this genus as abortifacients in cattle remains unclear.

Tritrichomonas foetus has been implicated in causing bovine abortions, usually during the early stages of pregnancy, but is primarily regarded rather as a cause of infertility in cows. For more details, please refer to Chapter 3.3, this volume.

Diagnosis

Please refer to the Chapters 3.3 and 3.4, this volume.

Ovines and Caprines

Smaragda Sotiraki, Laura Kramer and David Buxton

Protozoa, bacteria and viruses can all cause abortion in sheep and goats. In reaching a diagnosis, an accurate case history (time of gestation when abortion occurs, prevalence, flock/herd status, etc.) and correct diagnostic procedures are essential. In this section, characteristic and differentiating features of the major causes of ovine and caprine abortion are reviewed. For the isolation of aetiological agents by culture, samples should be collected from the aborted fetus as soon as possible after expulsion from the dam, as the numbers of viable organisms are likely to decrease rapidly as autolysis proceeds. Histopathology will also yield more accurate results where autolysis is minimal.

Protozoa

Toxoplasmosis

CAUSAL AGENT *Toxoplasma gondii.*

CLINICAL EFFECTS Typically, *T. gondii* causes abortion late in pregnancy when ewes/does produce stillborn and/or weakly lambs/kids, often accompanied by a mummified fetus (see Chapter 3.2, this volume). Cotyledons in the accompanying placenta show macroscopically visible pale, necrotic foci, 2–3 mm in diameter. The intercotyledonary membranes are normal in appearance, but the mummified fetus is accompanied by a mummified, brown, leathery placenta. Histopathological lesions characteristic of a protozoal infection may be found in the liver and brain, while leukomalacia associated with anoxia may also be present in the latter.

DIAGNOSIS Clinical findings supported by serology (IFAT, ELISA) and histology, with immunohistochemistry and the PCR reserved for when a definitive diagnosis is required.

Neosporosis

CAUSAL AGENT *Neospora caninum.*

CLINICAL EFFECTS As outlined in Chapter 3.1, above, neosporosis is a major cause of fetal loss in cattle, and may infrequently cause losses in sheep and goats.

Clinical signs appear to be similar to those of toxoplasmosis, with abortion or birth of weak, sometimes ataxic, lambs. Pathological lesions in sheep and goats are also similar to those induced by *T. gondii*.

DIAGNOSIS Serology of mother (ELISA) and fetus (IFAT or ELISA) and histology. Immunohistochemistry and the PCR will probably be required for confirmation to distinguish from toxoplasmosis.

Sarcocystiosis

Sarcocystiosis is not generally considered to be a primary cause of abortion in farm ruminants, but it can cause unthriftiness, which may predispose to other problems, including fetal loss.

Bacteria

Chlamydial abortion

SYNONYMS Enzootic abortion of ewes (EAE); ovine enzootic abortion (OEA), ovine chlamydial abortion.

CAUSAL AGENT *Chlamydophila abortus* (formerly *Chlamydia psittaci* immunotype 1).

CLINICAL EFFECTS Chlamydial abortion in ewes and goats is a very significant cause of loss in the UK, Eire, France, Spain, Italy (Sardinia), Greece and some other countries. Typically, a ewe/doe aborts a well-preserved lamb/kid in the last 2 weeks of pregnancy. In sheep flocks, the appearance of a stillborn lamb at this time is often the first sign of trouble, although the affected ewe may have a vulval discharge and show behavioural changes for up to 48 h before this.

The stillborn lamb may look normal or show a degree of subcutaneous oedema. The placental membranes are usually thickened and reddened (sometimes yellowish), with adjacent cotyledons discoloured and necrotic. A dirty pink exudate may contaminate placental membranes and offspring.

Histopathology will show destruction of chorionic epithelial cells and inflammation in the hilus of placentomes, extending out into the intercotyledonary membranes where the resultant damage, oedema and inflammation gives rise to the characteristic macroscopic appearance of the placental membranes. Vasculitis and thrombosis of related vessels are common (necrotic placentitis is not a feature of abortion caused by *Listeria* or *Campylobacter* spp.) (see below).

DIAGNOSIS An impression smear of placental membranes stained with modified Ziehl-Neelsen's method (MZN) reveals cherry-red elementary bodies.

Histology of placental tissue and serology with the complement fixation test (CFT) and ELISA will confirm a diagnosis.

Campylobacter abortion

SYNONYMS Vibrionic abortion; vibriosis.

CAUSAL AGENTS *Campylobacter fetus fetus*; *C. jejuni.*

CLINICAL EFFECTS Infection in sheep is more common than in goats. In sheep, abortion usually occurs during the last third of pregnancy and may take the form of a severe abortion storm. The ewes generally appear well and develop a protective immunity, so that *Campylobacter* does not normally cause abortion in subsequent pregnancies. Infection may cause placentitis, with arteriolitis and thrombosis in the placentomes. Masses of cell free organisms may be found in the villous tips.

The fetal liver may be enlarged and contain numerous pale necrotic foci, although in some cases they may be visible only microscopically. In addition, macroscopically visible pleurisy and pericarditis may occur, but more often only microscopic clusters of neutrophil polymorphs may be seen in pulmonary alveoli.

DIAGNOSIS The flock history and clinical picture are important but, for a more certain diagnosis, culture of fetal or placental tissues or fetal stomach contents is necessary.

Salmonella abortion

CAUSAL AGENTS *Salmonella abortus ovis* and other *Salmonella* spp.

CLINICAL EFFECTS Salmonella abortion can also be caused by several other species of *Salmonella*, including *S. montivideo*, the most commonly recorded isolate, and other species such as *S. dublin*, *S. typhimurium*, *S. derby* and *S. hindmarsh*. Some salmonella infections, such as those caused by *S. montivideo*, produce abortion without causing illness in ewes, apart perhaps from a transient fever and mild diarrhoea. Other salmonella species however, such as *S. dublin* and *S. typhimurium*, may cause abortion following the establishment of severe septicaemic illness, which may end in death. Abortions from salmonella infections generally occur in the final third of pregnancy.

DIAGNOSIS Enteritis or septicaemia accompanied by abortion should raise the suspicion of salmonellosis (less commonly seen with *S. abortus ovis*). A definitive diagnosis may be achieved by culture of the organism from the aborted fetus/placenta, or from faeces, gut contents, internal organs or lymph nodes of the ewe. Serology (seroagglutination test, ELISA) and an allergic skin test may also be used for the diagnosis of *S. abortus ovis* infection.

Listeria abortion

CAUSAL AGENTS *Listeria monocytogenes* (and less frequently *L. ivanovii*)

CLINICAL EFFECTS Losses in sheep are usually sporadic and occur in the last third of pregnancy, with abortion storms being rare. Affected ewes often show a bloody or brown vaginal discharge prior to abortion. Other clinical signs – including enteritis, septicaemia (usually before abortion) and encephalitis (usually after abortion) – may occur in the flock. The finding of small abomasal erosions, a yellow-orange meconium and small yellowish, necrotic foci in the liver in aborted lambs may suggest listeriosis (although the latter may also be seen in cases of abortion due to *Campylobacter*).

DIAGNOSIS Isolation of the organism from the placenta and fetal tissues; histopathology of fetal tissues.

Leptospirosis

CAUSAL AGENT *Leptospira interrogans* (synonyms *L. pomona*, *grippotyphosa*, *australis*, *ballum*, *sejroe*).

CLINICAL EFFECTS Abortion occurs in late pregnancy and may be accompanied by full-term stillbirths and weak lambs. Agalactia may also occur. Characteristic histological lesions are not encountered.

DIAGNOSIS Isolation of the organism from the placenta and fetal tissues (culture, fluorescence antibody techniques) or serology (microscopic agglutination test).

Brucella abortion

CAUSAL AGENT *Brucella melitensis*.

CLINICAL EFFECTS This is a serious cause of abortion in sheep and goats, being most prevalent in the Mediterranean region and the Near East. Infection is usually asymptomatic until abortion occurs. *Brucella melitensis* causes a slowly developing necrotizing placentitis, so that abortion is usually seen in the later stages of pregnancy. It may be suspected on the basis of history and clinical examination, especially when several animals are involved.

DIAGNOSIS This can be confirmed by isolation of the organism from the products of abortion and by serology (Rose Bengal test, complement fixation test) and a hypersensitivity (allergen) test. Note that the latter cannot be used in flocks/herds vaccinated with the live Rev1 vaccine, as this induces hypersensitivity that can last for over 2 years.

Rickettsiae

Q fever

CAUSAL AGENT *Coxiella burnetii.*

CLINICAL EFFECTS This is a zoonotic infection that may cause a small number of sporadic cases of abortion in sheep and goats. A recent epidemiological study in Italy demonstrated that 38% of sheep herds and 47% of goat herds were seropositive for *C. burnetii*, and that 10 and 3% of ovine and caprine aborted fetuses, respectively, were PCR positive for the organism (Masala *et al.*, 2004), thus representing a considerable zoonotic risk to those helping with lambing or kidding, and to those ingesting untreated milk from such animals.

DIAGNOSIS Demonstration of the organism in the placenta and fetus with the modified Ziehl Neelsen's stain. Their appearance as short, cherry-red rods must be differentiated from the elementary bodies of *Chlamydophila abortus.* Histopathology and immunohistochemistry will reveal a suppurative inflammation, usually more superficial in nature than that of chlamydial abortion, and inclusions in chorionic epithelial cells fixed to, or sloughed from, the placental membrane.

Tick-borne fever

CAUSAL AGENTS *Anaplasma phagocytophilum* (synonyms *Ehrlichia phagoctophilia, Cytoecetes phagocytophilia*).

CLINICAL EFFECTS This organism is transmitted by ticks (*Ixodes ricinus* in Europe (UK, Eire, Norway, Finland) and *Rhipicephalus haemaphysaloides supino* in India). It causes an acute fever in susceptible sheep over a variable period of up to 3 weeks and, as a result, abortion may be induced in pregnant animals.

DIAGNOSIS Demonstration of the organism in neutrophil polymorphs in maternal blood smears stained with Leishman's and detection of seroconversion using the complement fixation test. Histopathology of aborted and stillborn lamb brains may reveal evidence of anoxic damage (Chianini *et al.*, 2004). While this lesion is arguably most commonly encountered in toxoplasmosis, it may also be seen in abortion caused by other agents as it reflects placental damage that may be the result of a number of causes.

Viruses

Border Disease

CAUSAL AGENT A pestivirus – border disease virus (BDV).

CLINICAL EFFECTS Border disease virus causes pathological changes and clinical signs in lambs/kids that become infected *in utero* either from persistently infected mothers or from ewes/does becoming infected for the first time while pregnant. Typically in sheep, disease is manifest as the birth of 'hairy shaker' lambs with a coarse, hairy birth coat and a tremble which may range from a mild uncontrollable tremor of the head to more violent shaking of the whole body, with ataxia so severe in some that they cannot stand. The head may appear domed and there may be skeletal defects of limbs and jaw.

However, the actual spectrum of disease may range from the birth of lambs of outwardly normal appearance to abortion and barren ewes. Fetuses exposed to infection in the first half of pregnancy are not able to recognize BDV as 'foreign', and so surviving fetuses are immunologically tolerant to the virus and remain persistently infected.

Thus, surviving lambs will be antibody negative, have circulating virus and will be a potent source of infection for naïve sheep. They may or may not show CNS and skeletal damage at birth. A proportion of ewes will abort or be barren following fetal death. In some cases, a severe placentitis may occur.

Border disease is less common in goats, but more pathogenic, and more likely to cause abortion than the birth of clinically affected kids (Loken, 1995).

DIAGNOSIS The case history, together with the clinical signs, is suggestive, but a tentative diagnosis may be confirmed by histopathology of the brain and spinal cord of affected fetuses and lambs, coupled with demonstration of viral antigen and/or antibody in maternal and fetal blood samples.

Arthropod-borne disease

CAUSAL AGENTS Bluetongue, Rift Valley Fever, Wesselsbron disease and Akabane disease are viral infections, spread by biting midges and mosquitoes, that can cause abortion, neonatal death and/or congenital malformations in sheep and goats.

Rift Valley Fever, a peracute/acute zoonotic disease in sheep and goats, mortality may be high and abortion storms may be seen in pregnant flocks/herds. The virus causes enlargement of the liver and icterus.

Wesselsbron disease presents with similar clinical signs, but may also induce congenital malformations. Both occur in Africa.

Bluetongue is present in subtropical and tropical areas of Africa, the Middle East, the Americas and Australia, and has recently spread into Europe (Belgium, France, Luxemburg, Germany, the Netherlands, Italy, Spain, Bulgaria, Turkey and the Balkans).

CLINICAL EFFECTS In adults, these include fever, depression, erosion and ulceration of the oral mucosa, and infection during pregnancy may lead to abortion, with central nervous system and skeletal abnormalities. In Akabane disease, which is endemic in Africa, Asia (not Russia) and Australia, infection in adult sheep may be inapparent but, in pregnant ewes, the fetal nervous system is affected. Abortion and stillbirth are also seen. The incidence of all of these conditions may be seasonal, in line with their insect vectors' activity.

DIAGNOSIS Location, timing and clinical signs. Virus isolation, serology and the PCR may all be appropriate.

DIFFERENTIAL DIAGNOSIS Differential diagnosis of the major causes of ovine and caprine abortion in Europe is shown in Table 4.1.2.

Table 4.1.2. Differential diagnosis of the major causes of infectious abortion in sheep and goats in Europe.

Agent	Time of abortion	Prevalence of abortion (%)[a]	Clinical signs/ macroscopic lesions	Definitive diagnosis
Toxoplasma gondii	Final 1–2 weeks	5–25	Necrotic foci in placenta	Histology and serology (IHC[b] + PCR)
Chlamydophila abortus	Final 2–3 weeks	5–30 (higher in goats)	Placentitis (thick red/yellowish membranes)	Smear, culture, (histology)
Campylobacter spp.	Late gestation	Sporadic–20	Abscessation in fetal liver (not always macroscopic)	Smear, culture of placenta, fetal stomach contents
Salmonella spp.	Final third	Sporadic–60	Inconsistent	Smear, culture, serology (allergic skin test)
Listeria spp	Final third	Sporadic	Brown vaginal discharge in ewes	Smear, culture (histology)

[a] This depends on herd status (naïve or immune).
[b] Immunohistochemistry.

References

Chianini, F., Adams, C. and Buxton, D. (2004) Tick-borne fever: neuropathological changes in ovine foetuses. *Veterinary Record* 155, 805–806.

Cornish, T.E., van Olphen, A.L., Cavender, J.L., Edwards, J.M., Jaeger, P.T., Vieyra, L.L., Woodard, L.F., Miller, D.R. and O'Toole, D. (2005) Comparison of ear notch immunohistochemistry, ear notch antigen-capture ELISA, and buffy coat virus isolation for detection of calves persistently infected with bovine viral diarrhoea virus. *Journal of Veterinarian Diagnostic Investigation* 17, 110–117.

Grooms, D.L. (2004) Reproductive consequences of infection with bovine viral diarrhoea virus. *Veterinary Clinics of North America Food Animal Practice* 20, 5–19.

Grunert, E. and Berchtold, M. (1982) In: Parey, P. (ed.) *Fertilitätsstörungen beim Weiblichen Rind.* Verlag, Berlin and Hamburg, Germany.

Hyndman, L., Vilcek, S., Conner, J. and Nettleton, P.F. (1998) A novel nested reverse transcription PCR detects bovine viral diarrhoea virus in fluids from aborted bovine fetuses. *Journal of Virology Methods* 71, 69–76.

Kovats, S., Main, E.K., Librach, C., Stubblebine, M., Fisher, S.J. and De Mars, R. (1990) A class I antigen, HLA-G, expressed in human trophoblasts. *Science* 248, 220–223.

Kramer, R. (1970) Sammelreferat: Lärmbedingte Schadensfälle unter besonderer Berücksichtigung von Aborten beim Rind. *Deutsche Tierärztliche Wochenschrift* 77, 553–576

Kuhne, S., Schroeder, C., Holmquist, G., Wolf, G., Horner, S., Brem, G. and Ballagi, A. (2005) Detection of bovine viral diarrhoea virus infected cattle–testing tissue samples derived from ear tagging using an Erns capture ELISA. *Journal of Veterinary Medicine. B: Infectious Diseases and Veterinary Public Health* 52, 272–277.

Laamanen, U.I., Neuvonen, E.P., Yliviuhkola, E.M. and Veijalainen, P.M. (1997) Comparison of RT-PCR assay and virus isolation in cell cultures for the detection of bovine viral diarrhoea virus (BVDV) in field samples. *Research in Veterinary Science* 63, 199–203.

Linklater, K.A. and Smith, M.C. (1993) *Color Atlas of Diseases and Disorders of the Goat and Sheep.* Wolfe, London.

Loken, T. (1995) Ruminant pestivirus infections in animals other than cattle and sheep. *Veterinary Clinics of North America Food Animal Practice* 11, 597–614.

Martin, W.B. and Aitken, I.D. (2000) In: Martin, W.B. and Aitken, I.D. (eds) *Diseases of Sheep.* Blackwell Science, Oxford, UK.

Masala, G., Porcu, R., Sanna, G., Chessa, G., Cillara, G., Chisu, V. and Tola, S. (2004) Occurrence, distribution, and role in abortion of *Coxiella burnetii* in sheep and goats in Sardinia, Italy. *Veterinary Microbiology* 99, 301–305.

Moennig, V. and Liess, B. (1995) Pathogenesis of intrauterine infections with bovine viral diarrhea virus. *Veterinary Clinics of North America Food Animal Practice* 11, 477–487.

Office International des Epizooties (OIE), World Organisation for Animal Health (2004) *Manual of Diagnostic Tests and Vaccines for Terrestrial Animals (Mammals, Birds and Bees)*, 5th edn. Vols 1 and 2. OIE, Paris.

Rocha, M.A., Barbosa, E.F., Guedes, R.M., Lage, A.P., Leite, R.C. and Gouveia, A.M. (1999) Detection of BHV-1 in a naturally infected bovine fetus by a nested PCR assay. *Veterinary Research Communitary* 23, 133–141.

Rodolakis, A., Nettleton, P. and Benkirane, A. (1998) *Manual for Laboratory Diagnosis of Infectious Abortions in Small Ruminants.* Food and Agriculture Organization of the United Nations, Rome.

Rüfenacht, J., Schaller, P., Audige, L., Knutti, B., Kupfer, U. and Peterhans, E. (2001) The effect of infection with bovine viral diarrhea virus on the fertility of Swiss dairy cattle. *Theriogenology* 56, 199–210.

Saliki, J.T. and Dubovi, E.J. (2004) Laboratory diagnosis of bovine viral diarrhea virus infections. *Veterinary Clinics of North America Food Animal Practice* 20, 69–83.

Smith, K.C. (1997) Herpesviral abortion in domestic animals. *Veterinary Journal* 153, 253–268.

Stubbings, D.P. and Cameron, I.R. (1981) Bovine abortion associated with infectious bovine rhinotracheitis virus infection. *Veterinary Record* 108, 101–102.

Takiuchi, E., Medici, K.C., Alfieri, A.F. and Alfieri, A.A. (2005) Bovine herpesvirus type 1 abortions detected by a semi-nested PCR in Brazilian cattle herds. *Research in Veterinary Science* 79, 85–88.

Tsakos, P., Siarkou, V., Kastanidou, C. and Papadopoulos, O. (2001) Epidemiological survey in sheep and goats for *Chlamydia* and *Brucella*. *Journal of the Hellenic Veterinary Society* 52, 410–421.

Uruno, K., Shibata, I. and Nakane, T. (1998) Detection of bovine viral diarrhea virus (BVDV) using reverse transcription polymerase chain reaction assay. *Journal of the Veterinary Medicine Science* 60, 867–870.

Van Oirschot, J.T. (1983) Congenital infections with non-arbo togaviruses. *Veterinary Microbiology* 8, 321–361.

Control Measures

5

5.1 Neosporosis

Franz J. Conraths, Gereon Schares, Luis M. Ortega-Mora and Bruno Gottstein

Introduction

Neospora caninum has been shown as a leading cause of abortion in cattle worldwide (Dubey and Lindsay, 1996; Hemphill and Gottstein, 2000; Dubey, 2003, 2005). It is considered as a disease of major importance in many countries. Serology has been shown as a valuable tool to examine herds for *N. caninum*-associated abortion problems (for a review, see Jenkins *et al.*, 2002). Serological techniques allow the adoption of some basic control measures, such as purchase and sale of seronegative animals for replacement.

Differences exist in various countries regarding the seroprevalence in individual animals and cattle herds, as well as in dairy and beef herds (Bartels *et al.*, 2006). Consequently, control options are also applied in a different fashion in various countries, ranging from no action taken over hygienic measures to vaccination, reproductive management and so-called 'test and cull' strategies. Any of these options should be chosen on the basis of a rational decision, including a cost–benefit analysis at the individual farm level.

© CAB International 2007. *Protozoal Abortion in Farm Ruminants* (eds L.M. Ortega-Mora, B. Gottstein, F.J. Conraths and D. Buxton)

Options for Control

Hygienic measures

Prevention of dogs and other potential definitive hosts from contaminating feedstuff and water with faeces

Dogs and coyotes (*Canis latrans*) are definitive hosts for *N. caninum* (McAllister *et al.*, 1998; Gondim *et al.*, 2004). Dogs can transiently shed oocysts with their faeces upon ingestion of *N. caninum*-infected body tissues from intermediate hosts (McAllister *et al.*, 1998; Basso *et al.*, 2001; Dijkstra *et al.*, 2001; Schares *et al.*, 2001; Gondim *et al.*, 2002).

Dogs kept together with a cattle herd ('farm dogs') have been identified as a risk factor for bovine neosporosis (Paré *et al.*, 1998; Mainar-Jaime *et al.*, 1999; Ould-Amrouche *et al.*, 1999; Schares *et al.*, 2003), and post-natal infection of cattle through ingestion of oocysts shed by dogs is likely (McAllister *et al.*, 2000; Gondim *et al.*, 2002; Schares *et al.*, 2002). It has also been reported that dogs newly introduced to a farm represent a risk, perhaps because they may arrive naïve with regard to *N. caninum* to the farm, become infected and shed oocysts (Dijkstra *et al.*, 2002).

An association between regional dog density and seropositivity in cattle has also been demonstrated, indicating that dogs kept outside farms might also be a risk factor for bovine neosporosis, e.g. through contamination of pastures with oocysts shed by pet dogs (Schares *et al.*, 2003). However, the risk of cattle becoming infected with *N. caninum* due to the contamination of pastures with oocysts has been estimated as low, in a risk assessment performed in Germany (Schares *et al.*, 2005); and the risk factor 'number of dogs kept on the farm' may have a greater impact than the factor 'dog density' within the local community (Schares *et al.*, 2004).

Contamination of feedstuff and water used for cattle and other intermediate hosts of *N. caninum* must therefore be prevented; appropriate hygiene regarding dog faeces on pastures is also recommended.

Safe disposal of aborted fetuses, placentae, fetal membranes, etc.

The idea of this measure is to prevent the access of dogs and other potential definitive hosts to infected tissues of intermediate hosts (bovine fetuses, placentae, bovine carcasses in the field, etc.).

Upon ingestion of *N. caninum*-infected body tissues from intermediate hosts, dogs and other definitive hosts can shed oocysts with their faeces (McAllister *et al.*, 1998; Basso *et al.*, 2001; Dijkstra *et al.*, 2001; Schares *et al.*, 2001; Gondim *et al.*, 2002, 2004; Rodrigues *et al.*, 2004). The infection risk for definitive hosts can be diminished if aborted fetuses, placentae, fetal membranes and other tissues of potentially infected intermediate hosts are disposed of safely so that dogs and other carnivores have no access.

Prevention of access of cattle to infected tissue

Horizontal transmission of neosporosis between cattle is unproven, but seems possible, since it has been shown that newborn calves can be infected with *N. caninum* tachyzoites added to colostra (Uggla *et al.*, 1998). It is unclear whether this potential route of transmission has any epidemiological relevance but, as a measure of precaution, the access of cattle to infected tissue should be prevented.

Control of rodents

Rats (*Rattus norvegicus*) from cattle farms located on the Island of Taiwan have been found seropositive to *N. caninum* (Huang *et al.*, 2004). A low frequency of infection with *N. caninum* was also detected in *Mus domesticus* (3%) and *R. norvegicus* (4.4%) populations investigated by a British group that used a nested PCR technique, based on primers from the Nc5 region of *N. caninum* (Hughes *et al.*, 2006).

Therefore, rats – and perhaps other rodents also – may not only be intermediate hosts but could also be a reservoir for the parasite, although transmission from rodents to dogs is as yet unproven. Regular rodent control should be implemented to reduce the potential risk of infection that may exist in a reservoir for *N. caninum* in rodents.

Reproduction management measures

Embryo transfer only to seronegative recipient cows

Transfer of embryos from infected dams into uninfected recipients can prevent endogenous transplacental transmission of *N. caninum* (Baillargeon *et al.*, 2001; Landmann *et al.*, 2002; Reichel and Ellis, 2002; Campero *et al.*, 2003). This technique may thus be used to recover uninfected calves from genetically valuable, but *N. caninum*-infected, dams.

By contrast, when embryos from seropositive or seronegative donors were transferred to seropositive recipients, five of six calves were seropositive at birth (Baillargeon *et al.*, 2001), although the abortion risk did not increase under this situation. As a consequence, pre-transfer testing of recipients for infection with *N. caninum* is highly recommended. Only uninfected cows should be used as recipients.

Artificial insemination of seropositive dairy dams with semen from beef bulls

The results of a study conducted in Spain, in two high-producing dairy farms with a mean seroprevalence of 28%, suggested that the use of beef bull semen could reduce the risk of abortion in dairy cows in those farms, and it was therefore proposed that this effect might be due to a favourable effect of cross-breed pregnancies on placental function (López-Gatius *et al.*, 2005). The effect observed in this study still awaits independent confirmation.

Test and cull

N. caninum-infected cows must be considered as a reservoir that may allow the parasite to spread slowly to other cattle in the herd by endogenous transplacental transmission or rapidly by horizontal spread, e.g. via ingestion of contaminated feedstuffs or water. As a consequence, farmers may decide to remove infected cows or their progeny from the (dairy) herd.

The 'test and cull' strategy includes the following options:

- Test and cull seropositive dams or seropositive aborting dams.
- Test and inseminate the progeny of seropositive dams with beef bull semen only.
- Test and exclude the progeny of seropositive dams from breeding.

To date, these options have also been successfully applied from an economic point of view in several countries (Hall *et al.*, 2005). However, the approaches can only be recommended for herds with predominantly endogenous transplacental ('vertical') transmission of the infection. Culled dams or dams excluded from breeding must be replaced only by seronegative animals. Before a test-and-cull strategy is adopted, the risk factors for infection (main route of transmission, i.e. endogenous transplacental transmission; presence of dogs; presence of other domestic or wildlife reservoirs) need to be analysed and taken into account (Haddad *et al.*, 2005).

A cost–benefit analysis for each farm should be performed before choosing any of these options. Computer programs are needed to facilitate these analyses.

Vaccination

It has been observed that the proportion of congenitally infected calves decreased with the increasing parity of the dams, suggesting that cows may develop a certain degree of immunity that may affect endogenous transplacental transmission (Romero *et al.*, 2002; Dijkstra *et al.*, 2003). It has thus been argued that vaccination may be an option in preventing endogenous transplacental transmission of *N. caninum* or abortions due to infection with this parasite.

At present it is uncertain, however, whether some specific vaccine formulations that have been tested exhibit effectivity and efficacy in that respect. Andrianarivo *et al.* (2000) reported that a POLYGEN-adjuvanted killed *Neospora caninum* tachyzoite preparation failed to prevent fetal infection in pregnant cattle following experimental intravenous or intramuscular experimental tachyzoite challenge.

A vaccine commercially available in a number of countries has yielded protection in a field study in two out of five herds in New Zealand, with an overall efficacy of 5.2–54% (Heuer *et al.*, 2003). The same vaccine had a 'reasonable effect on abortion' when tested in Costa Rica (Romero *et al.*, 2004), where protection was observed in 15 out of 25 herds in another field study. However, a slight negative effect was reported for six herds. The overall efficacy of the vaccine was calculated at 46%.

It must be emphasized that currently available vaccines do not allow the discrimination of vaccinated from infected cattle. As a consequence, after application of the vaccine, the infection status of an animal can no longer be reliably determined. All vaccinated cattle will have to be treated as infected animals, e.g. for trade purposes. Cattle vaccinated against *N. caninum* should therefore not be introduced into a *Neospora*-free herd. Since the analysis of the herd status regarding infection with *N. caninum* relies on serology, seroepidemiological approaches cannot be used in vaccinated herds. As a consequence, the diagnostic tools are restricted to analysing aborted fetuses and to testing pre-colostral samples of newborn calves in vaccinated herds.

So far, protection from transplacental transmission or from infection in general has not been shown for any vaccine in cattle. When pregnant heifers naturally infected with *N. caninum* were immunized with killed tachyzoites or left untreated, the immune responses against *N. caninum* during pregnancy and the immune responses of the offspring were both monitored. The results suggested that reactivation of a latent infection had occurred in the naturally infected heifers, regardless of their immunization status, and that immunization with the POLYGEN-adjuvanted killed *N. caninum* tachyzoite preparation had not been able to prevent vertical transmission in naturally infected heifers (Andrianarivo *et al.*, 2005).

At the WAAVP Conference in 2005, it was agreed in the workshop *Options for Control of Protozoal Abortion in Ruminants: Practical Experience* that a document should be prepared that described the scientific information required before a vaccine against bovine neosporosis could be licensed (Conraths and Ortega-Mora, 2005). This information should include:

- A statement on the objective of vaccination (i.e. protection against abortion, transplacental transmission or infection in general).
- Proof of efficacy in experimental studies performed in cattle.
- Proof of efficacy in field studies.
- Proof of safety.
- Proof of compatibility with diagnostic techniques, allowing one to distinguish vaccinated from infected cattle (e.g. addition of a marker to the vaccine; companion test).

In addition, instructions for the use of a vaccine (time, frequency of vaccination, mode of application, etc.) need to be verified by studies conducted according to scientific standards.

Chemotherapy

Currently, there is no chemotherapy for bovine neosporosis that has been shown to be safe and effective. Any efforts to treat cattle with existing drugs must therefore be discouraged at this stage.

However, interesting experimental studies have been conducted that may result in an option for chemotherapeutic control at a later stage. An effect of toltrazuril and its derivative ponazuril on tachyzoites of *N. caninum* has been shown *in vitro* (Darius *et al.*, 2004a, b) and *in vivo* in mice (Gottstein *et al.*, 2001, 2005; Ammann *et al.*, 2004; Darius *et al.*, 2004a, b) and calves (Kritzner *et al.*, 2002).

In calves treated with ponazuril, the parasite was no longer detectable in the brain and other organs (Kritzner *et al.*, 2002). In experimentally infected mice, evidence was obtained that treatment with toltrazuril may be able to block endogenous transplacental transmission of the infection (Gottstein *et al.*, 2005).

References

Ammann, P., Waldvogel, A., Breyer, I., Esposito, M., Müller, N. and Gottstein, B. (2004) The role of B- and T-cell immunity in toltrazuril-treated C57BL/6 WT, microMT and nude mice experimentally infected with *Neospora caninum*. *Parasitology Research* 93, 178–187.

Andrianarivo, A.G., Rowe, J.D., Barr, B.C., Anderson, M.L., Packham, A.E., Sverlow, K.W., Choromanski, L., Loui, C., Grace, A. and Conrad, P.A. (2000) A POLYGEN-adjuvanted killed *Neospora caninum* tachyzoite preparation failed to prevent fetal infection in pregnant cattle following i.v./i.m. experimental tachyzoite challenge. *International Journal for Parasitology* 30, 985–990.

Andrianarivo, A.G., Anderson, M.L., Rowe, J.D., Gardner, I.A., Reynolds, J.P., Choromanski, L. and Conrad, P.A. (2005) Immune responses during pregnancy in heifers naturally infected with *Neospora caninum* with and without immunization. *Parasitology Research* 96, 24–31.

Baillargeon, P., Fecteau, G., Paré, J., Lamothe, P. and Sauve, R. (2001) Evaluation of the embryo transfer procedure proposed by the International Embryo Transfer Society as a method of controlling vertical transmission of *Neospora caninum* in cattle. *Journal of the American Veterinary Medical Association* 218, 1803–1806.

Bartels, C.J.M., Arnaiz-Seco, J.I., Ruiz-Santa-Quitera, A., Björkman, C., Frössling, J., von Blumröder, D., Conraths, F.J., Schares, G., van Maanen, K., Wouda, W. and Ortega-Mora, L.M. (2006) Supranational comparison of *Neospora caninum* seroprevalences in cattle in Germany, The Netherlands, Spain and Sweden. *Veterinary Parasitology* 137, 17–27.

Basso, W., Venturini, L., Venturini, M.C., Hill, D.E., Kwok, O.C., Shen, S.K. and Dubey, J.P. (2001) First isolation of *Neospora caninum* from the feces of a naturally infected dog. *Journal of Parasitology* 87, 612–618.

Campero, C.M., Moore, D.P., Lagomarsino, H., Odeon, A.C., Castro, M. and Visca, H. (2003) Serological status and abortion rate in progeny obtained by natural service or embryo transfer from *Neospora caninum*-seropositive cows. *Journal of Veterinary Medicine B: Infectious Diseases and Veterinary Public Health* 50, 458–460.

Conraths, F.J. and Ortega-Mora, L.M. (2005) Options for control of protozoal abortion in ruminants: practical experience. Conclusions. *The 20th International Conference of the World Association for the Advancement of Veterinary Parasitology 2005: Workshop Session T.* Christchurch, New Zealand, 16–20 October 2005.

Darius, A.K., Mehlhorn, H. and Heydorn, A.O. (2004a) Effects of toltrazuril and ponazuril on the fine structure and multiplication of tachyzoites of the NC-1 strain of *Neospora caninum* (a synonym of *Hammondia heydorni*) in cell cultures. *Parasitology Research* 92, 453–458.

Darius, A.K., Mehlhorn, H. and Heydorn, A.O. (2004b) Effects of toltrazuril and ponazuril on *Hammondia heydorni* (syn. *Neospora caninum*) infections in mice. *Parasitology Research* 92, 520–522.

Dijkstra, T., Eysker, M., Schares, G., Conraths, F.J., Wouda, W. and Barkema, H.W. (2001) Dogs shed *Neospora caninum* oocysts after ingestion of naturally infected bovine placenta but not after ingestion of colostrum spiked with *Neospora caninum* tachyzoites. *International Journal for Parasitology* 31, 747–752.

Dijkstra, T., Barkema, H.W., Hesselink, J.W. and Wouda, W. (2002) Point source exposure of cattle to *Neospora caninum* consistent with periods of common housing and feeding and related to the introduction of a dog. *Veterinary Parasitology* 105, 89–98.

Dijkstra, T., Barkema, H.W., Eysker, M., Beiboer, M.L. and Wouda, W. (2003) Evaluation of a single serological screening of dairy herds for *Neospora caninum* antibodies. *Veterinary Parasitology* 110, 161–169

Dubey, J.P. (2003) Review of *Neospora caninum* and neosporosis in animals. *Korean Journal of Parasitology* 41, 1–16.

Dubey, J.P. (2005) Neosporosis in cattle. *The Veterinary Clinics of North America. Food Animal Practice* 21, 473–483.

Dubey, J.P. and Lindsay, D.S. (1996) A review of *Neospora caninum* and neosporosis. *Veterinary Parasitology* 67, 1–59.

Gondim, L.F.P., Gao, L. and McAllister, M.M. (2002) Improved production of *Neospora caninum* oocysts, cyclical oral transmission between dogs and cattle, and *in vitro* isolation from oocysts. *Journal of Parasitology* 88, 1159–1163.

Gondim, L.F.P., McAllister, M.M., Pitt, W.C. and Zemlicka, D.E. (2004) Coyotes (*Canis latrans*) are definitive hosts of *Neospora caninum*. *International Journal for Parasitology* 34, 159–161.

Gottstein, B., Eperon, S., Dai, W.J., Cannas, A., Hemphill, A. and Greif, G. (2001) Efficacy of toltrazuril and ponazuril against experimental *Neospora caninum* infection in mice. *Parasitology Research* 87, 43–48.

Gottstein, B., Razmi, G.R., Ammann, P., Sager, H. and Müller, N. (2005) Toltrazuril treatment to control diaplacental *Neospora caninum* transmission in experimentally infected pregnant mice. *Parasitology* 130, 41–48.

Haddad, J.P.A., Dohoo, I.R. and VanLeewen, J.A. (2005) A review of *Neospora caninum* in dairy and beef cattle – a Canadian perspective. *Canadian Veterinary Journal* 46, 230–243.

Hall, C.A., Reichel, M.P. and Ellis, J.T. (2005) *Neospora* abortions in dairy cattle: diagnosis, mode of transmission and control. *Veterinary Parasitology* 128, 231–241.

Hemphill, A. and Gottstein, B. (2000) A European perspective on *Neospora caninum. International Journal for Parasitology* 30, 877–924.

Heuer, C., Nicholson, C., Bielsa, J.M. and Weston, J.F. (2003) Efficacy of a vaccine against *Neospora caninum*-related abortions in New Zeland dairy herds. *Proceedings of the 10th Symposium of the International Society for Veterinary Epidemiology and Economy,* Vina del mar, Chile, 2003.

Huang, C.C., Yang, C.H., Watanabe, Y., Liao, Y.K. and Ooi, H.K. (2004) Finding of *Neospora caninum* in the wild brown rat (*Rattus norvegicus*). *Veterinary Research* 35, 283–290.

Hughes, J.M., Williams, R.H., Morley, E.K., Cook, D.A.N., Terry, R.S., Murphy, R.G., Smith, J.E. and Hide, G. (2006) The prevalence of *Neospora caninum* and co-infection with *Toxoplasma gondii* by PCR analysis in naturally occurring mammal populations. *Parasitology* 132, 29–36.

Jenkins, M., Baszler, T., Björkman, C., Schares, G. and Williams, D. (2002) Diagnosis and seroepidemiology of *Neospora caninum*-associated bovine abortion. *International Journal for Parasitology* 32, 631–636.

Kritzner, S., Sager, H., Blum, J., Krebber, R., Greif, G. and Gottstein, B. (2002) An explorative study to assess the efficacy of toltrazuril-sulfone (ponazuril) in calves experimentally infected with *Neospora caninum. Annals of Clinical Microbiology and Antimicrobials* 1, 4.

Landmann, J.K., Jillella, D., O'Donoghue, P.J., McGowan, M.R. (2002) Confirmation of the prevention of vertical transmission of *Neospora caninum* in cattle by the use of embryo transfer. *Australian Veterinary Journal* 80, 502–503.

Lopez-Gatius, F., Santolaria, P., Yaniz, J.L., Garbayo, J.M. and Almeria, S. (2005) The use of beef bull semen reduced the risk of abortion in *Neospora*-seropositive dairy cows. *Journal of Veterinary Medicine B: Infectious Diseases and Veterinary Public Health* 52, 88–92.

Mainar-Jaime, R.C., Thurmond, M.C., Berzal-Herranz, B. and Hietala, S.K. (1999) Seroprevalence of *Neospora caninum* and abortion in dairy cows in northern Spain. *Veterinary Record* 145, 72–75.

McAllister, M.M., Dubey, J.P., Lindsay, D.S., Jolley, W.R., Wills, R.A. and McGuire, A.M. (1998) Dogs are definitive hosts of *Neospora caninum. International Journal for Parasitology* 28, 1473–1478.

McAllister, M.M., Björkman, C., Anderson-Sprecher, R. and Rogers, D.G. (2000) Evidence of point-source exposure to *Neospora caninum* and protective immunity in a herd of beef cows. *Journal of the American Veterinary Medical Association* 217, 881–887.

Ould-Amrouche, A., Klein, F., Osdoit, C., Mohammed, H.O., Touratier, A., Sanaa, M. and Mialot, J.P. (1999) Estimation of *Neospora caninum* seroprevalence in dairy cattle from Normandy, France. *Veterinary Research* 30, 531–538.

Paré, J., Fecteau, G., Fortin, M. and Marsolais, G. (1998) Seroepidemiologic study of *Neospora caninum* in dairy herds. *Journal of the American Veterinary Medical Association* 213, 1595–1598.

Reichel, M.P. and Ellis, J.T. (2002) Control options for *Neospora caninum* infections in cattle – current state of knowledge. *New Zealand Veterinary Journal* 50, 86–92.

Rodrigues, A.A., Gennari, S.M., Aguiar, D.M., Sreekumar, C., Hill, D.E., Miska, K.B., Vianna, M.C. and Dubey, J.P. (2004) Shedding of *Neospora caninum* oocysts by dogs fed tissues from naturally infected water buffaloes (*Bubalus bubalis*) from Brazil. *Veterinary Parasitology* 124, 139–150.

Romero, J.J., Perez, E., Dolz, G. and Frankena, K. (2002) Factors associated with *Neospora caninum* serostatus in cattle of 20 specialised Costa Rican dairy herds. *Preventive Veterinary Medicine* 53, 263–273.

Romero, J.J., Perez, E. and Frankena, K. (2004) Effect of a killed whole *Neospora caninum* tachyzoite vaccine on the crude abortion rate of Costa Rican dairy cows under field conditions. *Veterinary Parasitology* 123, 149–159.

Schares, G., Heydorn, A.O., Cüppers, A., Conraths, F.J. and Mehlhorn, H. (2001) Cyclic transmission of *Neospora caninum*: serological findings in dogs shedding oocysts. *Parasitology Research* 87, 873–877.

Schares, G., Bärwald, A., Staubach, C., Söndgen, P., Rauser, M., Schröder, R., Peters, M., Wurm, R., Selhorst, T. and Conraths, F.J. (2002) p38-avidity-ELISA: examination of herds experiencing epidemic or endemic *Neospora caninum*-associated bovine abortion. *Veterinary Parasitology* 106, 293–305.

Schares, G., Bärwald, A., Staubach, C., Ziller, M., Klöß, D., Wurm, R., Rauser, M., Labohm, R., Dräger, K., Fasen, W., Hess, R.G. and Conraths, F.J. (2003) Regional distribution of bovine *Neospora caninum* infection in the German state of Rhineland-Palatinate modelled by Logistic regression. *International Journal for Parasitology* 33, 1631–1640.

Schares, G., Bärwald, A., Staubach, C., Ziller, M., Klöß, D., Schröder, R., Labohm, R., Dräger, K., Fasen, W., Hess, R.G., Conraths, F.J. (2004) Potential risk factors for bovine *Neospora caninum* infection in Germany are not under the control of the farmers. *Parasitology* 129, 301–309.

Schares, G., Tackmann, K., Ziller, M. and Conraths, F.J. (2005) Risikobewertung. Rinderaborte durch *Neospora caninum* – Welche Gefahren gehen von Hundekot auf Weiden aus? http://bfav.hnm.de/Risikobewertungen.1106.0.html (accessed 31 May 2006).

Uggla, A., Stenlund, S., Holmdahl, O.J., Jakubek, E.B., Thebo, P., Kindahl, H. and Björkman, C. (1998) Oral *Neospora caninum* inoculation of neonatal calves. *International Journal for Parasitology* 28, 1467–1472.

5.2 Toxoplasmosis

David Buxton

Cats as a Source of Infection

Clinical toxoplasmosis in sheep and goats follows a primary infection during pregnancy, initiated following the ingestion of food or water contaminated with sporulated *Toxoplasma gondii* oocysts. Epidemiological studies have led to the conclusion that pasture is perhaps the most common source of infection (Skjerve *et al.*, 1998), with fields treated with manure and bedding from farm buildings where cats live being one way in which infection may be disseminated (Faull *et al.*, 1986); cats defaecating in farm feeds, such as hay and stored grain, are another source of risk (Plant *et al.*, 1974). Water can also be a threat, not only to animals but also to people (Bowie *et al.*, 1997).

The potential risk of feline faecal contamination of foodstuffs can be illustrated by the hypothetical situation in which a cat, actively shedding oocysts in its faeces, defaecates in 10 t of grain stored on a farm. A single defaecation may contain as many as 10 million oocysts. If further processing of the feed dispersed these oocysts evenly throughout the grain, each kilogram would contain between five and 25 sheep-infective doses (McColgan *et al.*, 1988).

While in all probability contamination would be more variable than this, it serves to illustrate how cat faeces can create a large, potent, long-lasting source of infection for sheep and goats. Oocyst contamination of farm feeds and bedding, as well as pasture, is therefore closely related to the number and distribution of cats and, as discussed above (Chapter 3.2, this volume), with young cats presenting a particular risk.

On the other hand, it would seem likely that pelleted food poses much less of a threat to susceptible pregnant sheep and goats, as the manufacturing process raises the temperature of the pellet, resulting in the viability of any contaminating oocysts being significantly curtailed.

For some considerable time, it has been accepted that environmental contamination with oocysts is the only significant source of infection for sheep and goats. Also, exposure of ewes to infection later in gestation (see Chapter 3.2, this volume) can result in (vertical) transmission of *T. gondii* to the fetus (exogenous transplacental transmission; Trees and Williams, 2005). As the fetal immune system is sufficiently mature at this stage to control the parasite, clinically normal but infected lambs are born. Recent research has suggested that vertical transmission of *Toxoplasma* may also occur in persistently infected ewes (endogenous transplacental transmission) (Duncanson *et al.*, 2001; Morley *et al.*, 2005; Williams *et al.*, 2005).

In these studies, only the polymerase chain reaction was used to define 'infection', with no pathology or serology to support the conclusions reached and no testing to exclude other infectious causes of fetopathy. Another smaller, but more thorough, study supports the conclusion that infection does not commonly pass from persistently infected ewes to their offspring (Rodger *et al.*, 2006).

Thus, while in some, as yet undefined, circumstances other mechanisms may have a role to play, the weight of evidence from field and experimental investigations indicates that clinical toxoplasmosis is associated with primary infection of non-immune ewes during pregnancy. A long-lasting immunity develops following primary exposure, and animals are very unlikely to abort in subsequent years.

General management

During pregnancy, ewes and goats seronegative for *T. gondii* will be at risk if allowed access to an environment contaminated by cat faeces. Water and stored food should be kept free from soiling by cats as far as practically possible. Other measures to reduce environmental contamination by *Toxoplasma* oocysts should be aimed at reducing the number of cats capable of shedding. This is best achieved by maintaining a small, healthy population of mature cats (likely to be seropositive for *T. gondii*) that will help control rodents as well as minimize excretion of oocysts.

It is likely that a flock/herd, in which seroconversion to *T. gondii* can be demonstrated in a significant number of animals, is being maintained in an environment significantly contaminated with *T. gondii* oocysts. Young, seronegative, replacement stock will be at risk and so a case can be made for attempting to expose them to a contaminated environment before mating. While this will be very difficult to identify, the grazing around farm buildings where cats live, as well as pasture spread with manure from such buildings, is more likely to be infective. While this is good advice, it does not guarantee success and vaccination provides a more precise method of control.

Vaccination

Natural infection with *T. gondii* stimulates protective immunity in both sheep and goats (McColgan *et al.*, 1988), but attempts to create vaccines with killed toxoplasms, or certain of their antigens, have not been successful. A live vaccine for sheep, initially introduced in New Zealand (O'Connell *et al.*, 1988; Wilkins *et al.*, 1988) and later in Europe (Buxton and Innes, 1995) is currently marketed in the UK, Eire, France, Spain, Portugal and New Zealand (Toxovax® Intervet BV; AgVax, New Zealand).

Ewes in flocks that are 'at risk' should, in the first instance, all be vaccinated at least 3 weeks before mating. In subsequent years, it should only be necessary to vaccinate newcomers to the flock, usually young replacement stock. Although the manufacturers may recommend revaccination after 3 years, in many situations the animals would be expected to boost their immunity naturally on such farms. In this case, only one injection would be required in the life of the sheep. The vaccine has a shelf life of 7–10 days and is capable of infecting humans, so it must be handled with care, strictly according to the manufacturer's recommendations.

Other live vaccines for use in sheep, such as that against chlamydial abortion (Enzovax, Intervet BV; CEVAC Chlamydophila, CEVA Sante Animale) may be administered at the same time but at different sites, although it is arguably better that such injections are also separated by a couple of weeks.

As with sheep, the majority of goats previously exposed to infection with *T. gondii* develop a protective immunity to the parasite, so that they are protected against subsequent challenge during pregnancy (Obendorf *et al.*, 1990), although repeat abortions have been recorded (Dubey, 1982). Toxovax is not licensed for use in goats but would be expected to offer some protection against the parasite. Immunity induced in goats by experimental infection with the related coccidian parasite *Hammondia hammondi* has been shown to offer some cross-protection against challenge with *T. gondii*, both in non-pregnant (Dubey, 1981a) and pregnant animals (Dubey, 1981b; Munday and Dubey, 1988), but this avenue of research has not been pursued.

Chemoprophylaxis

Research has shown that a significant reduction in *Toxoplasma*-induced perinatal lamb mortality can be achieved by feeding decoquinate during pregnancy (Buxton *et al.*, 1996). It should be added to the feed to provide 2 mg/kg body weight/day from around mid-pregnancy. Decoquinate is most effective if it is already being fed to susceptible ewes at the time they encounter infection, rather than after infection has become established.

As it does not eliminate the parasite, challenged ewes become persistently infected and develop protective immunity to subsequent challenge. It is not suitable in management systems in which supplementary feed is not given. Monensin also has been shown to be effective when added to the feed (15 mg monensin/head/day) but, because of the low toxic:therapeutic ratio, it is no longer approved for use in sheep.

Treatment

A combination of pyrimethamine and sulfadimidine, well tried in human medicine, blocks folate synthesis and has been shown to be effective in the treatment of infected sheep (Buxton *et al.*, 1993). The drug combination, baquiloprim and sulfadimidine, also blocks folate synthesis and has given promising results in a controlled pilot study in non-pregnant sheep (Buxton, unpublished data).

References

Bowie, W.R., King, A.S., Werker, D.H., Isaac-Renton, J.L., Bell, A., Eng, S.B. and Marion, S.A. (1997) Outbreak of toxoplasmosis associated with municipal drinking water. *Lancet* 350, 173–177.

Buxton, D. and Innes, E.A. (1995) A commercial vaccine for ovine toxoplasmosis. *Parasitology* 10, S11–S16.

Buxton, D., Thomson, K.M. and Maley, S. (1993) Treatment of ovine toxoplasmosis with a combination of sulphamezathine and pyrimethamine. *Veterinary Record* 132, 409–411.

Buxton, D., Brebner, J., Wright, S., Maley, S.W., Thomson, K.M. and Millard, K. (1996) Decoquinate and the control of experimental ovine toxoplasmosis. *Veterinary Record* 138, 434–436.

Dubey, J.P. (1981a) Protective immunity against clinical toxoplasmosis in dairy goats vaccinated with *Hammondia hammondi* and *Hammondia heydorni*. *American Journal of Veterinary Research* 42, 2068–2070.

Dubey, J.P. (1981b) Prevention of abortion and neonatal death due to toxoplasmosis by vaccination of goats with the non-pathogenic coccidium *Hammondia hammondi*. *American Journal of Veterinary Research* 42, 2155–2157.

Dubey, J.P. (1982) Repeat transplacental transfer of *Toxoplasma gondii* in goats. *Journal of the American Veterinary Medical Association* 180, 1220–1221.

Duncanson, P., Terry, R.S., Smith, J.E. and Hide, G. (2001) High levels of congenital transmission of *Toxoplasma gondii* in a commercial sheep flock. *International Journal for Parasitology* 31, 1699–1703.

Faull, W.B., Clarkson, M.J. and Winter, A.C. (1986) Toxoplasmosis in a flock of sheep: some investigations into its source and control. *Veterinary Record* 119, 491–493.

McColgan, C., Buxton, D. and Blewett, D.A. (1988) Titration of *Toxoplasma gondii* oocysts in non-pregnant sheep and the effects of subsequent challenge during pregnancy. *Veterinary Record* 123, 467–470.

Morley, E.K., Williams, R.H., Hughes, J.M., Terry, R.S., Duncanson, P., Smith, J.E. and Hide, G. (2005) Significant familial differences in the frequency of abortion and *Toxoplasma gondii* infection within a flock of Charolais sheep. *Parasitology* 131, 181–185.

Munday, B.L. and Dubey, J.P. (1988) Prevention of *Toxoplasma gondii* abortion in goats by vaccination with oocysts of *Hammondia hammondi*. *Australian Veterinary Journal* 65, 150–153.

Obendorf, D.L., Statham, P. and Munday, B.L. (1990) Resistance to *Toxoplasma* abortion in female goats previously exposed to *Toxoplasma* infection. *Australian Veterinary Journal* 67, 233–234.

O'Connell, E., Wilkins, M.F. and Te Punga, W.A. (1988) Toxoplasmosis in sheep. II. The ability of a live vaccine to prevent lamb losses after an intravenous challenge with *Toxoplasma gondii*. *New Zealand Veterinary Journal* 36, 1–4.

Plant, J.W., Richardson, N. and Moyle, A.G. (1974) *Toxoplasma* infection and abortion in sheep associated with feeding of grain contaminated with cat faeces. *Australian Veterinary Journal* 50, 19–20.

Rodger, S.M., Maley, S.W., Wright, S.E., Mackellar, A., Wesley, F., Sales, J. and Buxton, D. (2006) The role of endogenous transmission in toxoplasmosis in sheep. *Veterinary Record* 159, 768–772.

Skjerve, E., Waldeland, H., Nesbakken, T. and Kapperud, G. (1998) Risk factors for the presence of antibodies to *Toxoplasma gondii* in Norwegian slaughter lambs. *Preventive Veterinary Medicine* 35, 219–227.

Trees, A.J. and Williams, D.J.L. (2005) Endogenous and exogenous transplacental infection in *Neospora caninum* and *Toxoplasma gondii*. *Trends in Parasitology* 21, 558–561.

Wilkins, M.F., O'Connell, E. and Te Punga, W.A. (1988) Toxoplasmosis in sheep. III. Further evaluation of the ability of a live *Toxoplasma gondii* vaccine to prevent

lamb losses and reduce congenital infection following experimental oral challenge. *New Zealand Veterinary Journal* 36, 86–89.
Williams, R.H., Morley, E.K., Hughes, J.M., Duncanson, P., Terry, R.S., Smith, J.E. and Hide, G. (2005) High levels of congenital transmission of *Toxoplasma gondii* in longitudinal and cross-sectional studies on sheep farms provides evidence of vertical transmission in ovine hosts. *Parasitology* 130, 301–307.

5.3 Sarcocystiosis

Franz J. Conraths

Introduction

Most *Sarcocystis* species are heteroxenous. Therefore, the continuation of the life cycle requires the presence of both intermediate and definitive hosts (Dubey *et al.*, 1989; Tenter and Johnson, 1997). Horizontal transmission occurs from intermediate to definitive host and from definitive to intermediate host, but not between different intermediate hosts or between different definitive hosts.

Ruminants are usually intermediate hosts of *Sarcocystis*. They become infected by ingesting sporocysts with contaminated food or water. Sporozoites are released from the sporocysts in the small intestine. Definitive hosts of *Sarcocystis* spp. infecting ruminants are canids, felids and primates, including humans. Most virulent *Sarcocystis* species cause disease only in their intermediate, but not in their definitive, hosts. In general, *Sarcocystis* species transmitted by canids seem to be more virulent than those transmitted by felids.

Hygienic Measures

1. Prevention of potential definitive hosts from contaminating pastures, sheds, feedstuffs and water with faeces. The definitive hosts of *Sarcocystis* pass oocysts or sporocysts with their faeces. Therefore, the contamination of pastures, sheds, feedstuff and water with human, canine and feline faeces must be avoided in order to prevent exposure of ruminants to infectious sporocysts (Caldow *et al.*, 2000).
2. Prevention of the spread of oocysts via sewage, sullage water and sludge. Since sewage, sullage water and sludge from sewage plants may contain infectious sporocysts, the contamination of pastures with potentially contaminated material must be avoided.
3. Prevention of infection of definitive hosts. The life cycle of *Sarcocystis* can be interrupted if the cystozoites contained in intermediate tissue are inactivated before being fed to the definitive host or being eaten by humans. Inactivation can be achieved by cooking (> 70°C) or freezing (4 d, –20°C).

References

Caldow, G.L., Gidlow, J.R. and Schock, A. (2000) Clinical, pathological and epidemiological findings in three outbreaks of ovine protozoan myeloencephalitis. *The Veterinary Record* 146, 7–10.
Dubey, J.P., Speer, C.A. and Fayer, R. (1989) *Sarcocystosis of Animals and Man.* CRC Press, Boca Raton, Florida.
Tenter, A.M. and Johnson, A.M. (1997) Phylogeny of the tissue cyst-forming coccidia. *Advances in Parasitology* 39, 69–139.

5.4 Tritrichomonosis

Carlos M. Campero and Bruno Gottstein

Introduction

Bovine venereal tritrichomonosis is caused by *Tritrichomonas foetus*, including cattle (*Bos taurus*, *B. indicus*) as the main host. Non-pathogenic species of trichomonads occur in the intestine of cattle, e.g. *Tetratrichomonas* sp. (Cobo *et al.*, 2004a). *Tritrichomonas suis* of pigs is indistinguishable from *T. foetus*, and there is a generalized agreement on considering both organisms as the same species.

T. foetus has also been isolated from young cats suffering from diarrhoea, and experimental infection of cats resulted in a transient colonization of the feline ileum, caecum and colon (Gookin *et al.*, 2001). Furthermore, an association between large bowel disease in cats and the presence of *T. foetus* (and not *Pentatrichomonas hominis*, as formerly assumed) has been documented (Levy *et al.*, 2003). Thus the cat could play a role in the epidemiology of the parasite as well.

Because of the widespread use of artificial insemination (AI), which has helped to control its distribution, bovine tritrichomonosis has disappeared in most of the Western European countries. However, the disease is still present, with pregnancy losses and low weaning rates in extensive areas of some European countries, North America, Canada, Australia, South Africa and South America, where beef cattle are raised on pasture and bred by natural service. As a consequence, this disease has a significant economic impact in affected areas (BonDurant, 1997).

Economic losses from tritrichomonosis in beef cattle are associated with culling and replacement of infected cattle, and the loss of calf crop caused by failure, delay in conception or impact on weaner calf production and revenue returns (BonDurant, 1997). Tritrichomonosis has been a List B disease (by OIE classification) for many years, and is therefore the subject of animal disease control in several countries. Since the introduction of artificial insemination, the economic importance of this disease has decreased, but there is evidence of a re-emergence in extensive husbandry in some European states.

As *T. foetus* may also have spread inapparently among other livestock animals such as pigs, a basic reconsideration of the epidemiological situation in Europe and the classification of virulence and pathogenicity among *Tritrichomonas* isolates is urgently required. Identifying risk factors associated with disease may improve prevention and control. Thus, *T. foetus* infection could be associated with several bull factors, including age, breed, herd and herd management practices, such as bull:cow ratio and bulls per breeding group (Rae *et al.*, 2004).

Once bovine tritrichomonosis has been diagnosed in the herd, the control programme and measures for implementation are different according to each situation. Control options vary in dependence on the prevalence found in different countries. They range from minimal monitoring action taken (e.g. in countries free of infection) over reproductive management (AI, culling positive bulls, etc.) to vaccination approaches (Corbeil, 1994; BonDurant, 1997).

Bulls become a permanent source of infection and a chronic carrier of *T. foetus* (Clark, 1971; Clark *et al.*, 1974a; Parsonson *et al.*, 1974; BonDurant, 1997), while in heifers and cows *T. foetus* persists in the genital secretions from 13 to 28 weeks (Skirrow and BonDurant, 1990a, b).

As a consequence, the disease control programmes focus on identifying and culling infected bulls, and also on culling non-pregnant cows. Prevention of transmission of the disease through culling practices relies on the ability to identify accurately infected animals. Control programmes at national or regional level have been implemented, based on detection of infected bulls and culling, in order to decrease the disease impact (Dillon *et al.*, 1995; Campero, 2000).

AI Centre Bull Testing

Although the parasite can survive in diluted semen and through the freezing process, the probability of transmitting infection through AI is not known. Therefore, bulls selected for entry into AI should preferably be from infection-free herds, and testing in quarantine is mandatory. In *T. foetus*-endemic countries, continued freedom from infection should then be confirmed by biannual testing.

Tritrichomonas foetus is unaffected by antibiotics in semen extenders, so if an AI bull is found to be infected, the entire stock of frozen semen – or at least, the semen collected from the date of the last negative test – should be destroyed.

Information of value in assessing the risk of *T. foetus* being present in imported semen includes the country of origin of the semen, the history and breed of the bull and the testing programme applied by the AI centre, with particular reference to the number of pre-entry and annual tests for the parasite.

Management and Measures in Beef Herds

Practical measures in extensive beef herds with large number of animals where bulls are year-breeding are not easy to implement. Some basic measures concerning replacement, as with purchase or sale, are relevant (Clark *et al.*, 1974b; BonDurant, 1997). The following measures are recommended for avoidance of the entrance of infected animals in free or infected herds.

Quarantine and testing of replacement bulls

One of the most relevant tools is to purchase only animals from disease-free herds or herds under a disease control programme, with records of excellency in reproductive performance. On the other hand, it is also crucial to test all replacement bulls. The primary goal is to prevent the introduction of the infection, through standard measures such as bull testing.

Reproductive management

Different reproductive management measures have been proposed to reduce the disease's impact (Kimsey, 1986; BonDurant, 1997):

1. AI in both dairy and beef cattle, with hygienic and good-quality semen, reduces the risk of disease transmission.
2. Pregnancy examination is mandatory for classified gestation, and also to be aware of reproductive performance and any infertility problem in the herd. Segregation of the females by reproductive status is another important aspect provided by this practice. Abortion risk is higher in females < 5 pregnancy months (Clark *et al.*, 1974b; BonDurant, 1997).
3. A limited breeding season: this first step should be carried out in order to limit the breeding season from 12 to 4–6 months initially, keeping bulls separated from cows in a well-fenced lot after the breeding season. The goal is to obtain a breeding season of 3–4 months' duration.
4. All open and aborted cows with an infertility problem should be culled because of the risk and the possibility of service as the origin of infection for new replacement *T. foetus*-negative bulls. The presence of carrier cows has been mentioned as another possible source of infection (Morgan, 1944; Skirrow, 1987; Mancebo *et al.*, 1995; BonDurant, 1997; Campero, 2000).
5. Immunization: in the presence of a significant infection risk, vaccination of all female herd members against tritrichomonosis and campylobacteriosis (two injections one month apart, the second in the pre-breeding season, followed by an annual booster) is a sound measure. The commercially available dual vaccine (Trich Guard V5L) offers some protection under these circumstances (BonDurant, 1997).

6. A monitoring system should be implemented to detect all cattle moving in or out of the herd. All replacement pregnant heifers or cows purchased should be kept separately until the end of the calving season. All females without calves should be culled.

Bull management

1. The best preventive management is never to allow untested bulls from an outside source to enter the herd without three negative cultures or PCR findings. Efforts to control *T. foetus* focus on tests with higher sensitivity, reduced cost and time efficacy, for diagnostic assays will be beneficial in the future. Culture of preputial samples from all breeding bulls should be done annually and every time new bulls are introduced into the herd (Campero *et al.*, 1993; BonDurant, 1997). Three negative culture results in bulls is a safe measure to apply in infected herds, as well as two negative results in bulls from a well-managed herd with good reproductive performance and zero presence of trichomonads.
2. In extensive farming systems, bulls should comprise no more than 3% of herd numbers and, in intensive systems, no more than 2.5%. An excessive number of bulls in service does not solve the problem (Acuña and Campero, 1999; Campero, 1999).
3. Replace all bulls after four breeding seasons. This measure reduces the risk of acquiring tritrichomonosis. Replacement with young bulls (1.5–2 years old) having tested negative reduces the prevalence (Clark *et al.*, 1971; Christensen *et al.*, 1977).
4. Immunization: protection with dual commercial vaccine (Trich Guard V5L) against tritrichomonosis in bulls does *not* work efficiently (BonDurant, 1997).
5. Infected bulls should be placed in isolated pasture with secure fencing until culling. Although bull to bull transmission of this disease is very unlikely, this hypothesis of a possible source of infection whas been mentioned (Kendrick, 1953) and must be considered (Parker *et al.*, 2003). Keeping infected and uninfected bulls together should be discouraged. Futhermore, mature bulls with genital campylobacteriosis housed together with uninfected young virgin bulls in a feedlot situation results in bull to bull transmission due to homosexual behaviour (C.M. Campero, unpublished information).

Young virgin bull replacement

Australian researchers have shown a lesser disease problem in infected herds when young bulls were used as a replacement for natural service (Clark *et al.*, 1971; Christensen *et al.*, 1977). However, these bulls must be cultured for trichomonads because of the possibility of infection (Campero, 2000). On the other hand, there is some concern regarding virgin bulls where other organisms, including non-*T. foetus* trichomonads, were observed. Recent reports in the USA, Canada and Argentina have shown the presence of trichomonads in preputial cultures from virgin bulls (BonDurant *et al.*, 1999; Campero *et al.*, 2003; Cobo *et al.*, 2003; Parker *et al.*, 2003).

Therefore, failure in culture specificity may be expected, since non-*T. foetus* trichomonads, morphologically similar and difficult to distinguish morphologically from *T. foetus* under light microscopy, can occasionally grow in cultures of genital secretions from the bull. The source of these trichomonads is speculated to be either contamination from the gastrointestinal tract (faecal contamination), from bulls sodomizing each other (Jezierski *et al.*, 1989) or during sample collection. These preputial trichomonads do not appear to be very long-lived (Cobo *et al.*, 2004a).

Finally, the stipulation of two consecutive negative cultures (if the bull comes from a negative commercial bull stud) should be established to certify the uninfected condition in all new replacement young bulls introduced to the herd.

Treatment

Different local and/or systemic treatments have been proposed for infected bulls, but most of these are cumbersome and difficult to apply, and drug-resistant strains have been found (Campero and Palladino, 1983; Campero *et al.*, 1987a, b). On the other hand, systemic treatment with drugs like metronidazole or dimetridazole produces adverse side-effects and resistant populations of trichomonads (Lauriente *et al.*, 1982; Palladino *et al.*, 1982a, b, 1983, 1985; Campero *et al.*, 1987a, b).

Ipronidazole in bulls and cows has been shown to be effective (Skirrow *et al.*, 1985), but the use of such drugs is not approved in either cows or bulls. The use of chemotherapeutic agents has been deemed to be illegal in many countries and is not permitted because of concerns regarding toxic residues in meat (BonDurant, 1997). As a consequence, infected bulls should be sacrificed.

Vaccination

Heifers/cows

The lack of an effective therapy against tritrichomonosis has encouraged research into the development of vaccines (Skirrow and BonDurant, 1988). Therefore, different experimental vaccines have been manufactured with inactivated whole cells, membranes or a purified surface antigen of *T. foetus*; these have been systemically applied to heifers and cows, reducing the number of infected females, the reproductive losses and the duration of the genital infection (Kvasnicka *et al.*, 1992; BonDurant *et al.*, 1993; Hudson *et al.*, 1993a, b; Gault *et al.*, 1995; Anderson *et al.*, 1996; Campero *et al.*, 1998, 1999; Corbeil *et al.*, 1998, 2001; Cobo *et al.*, 2002).

On the other hand, no advantages in the use of whole-cell vaccines were seen in cows in one publication (Herr *et al.*, 1991). Since the disease is of considerable economic importance, immunoprophylaxis could be very important, and this has resulted in a commercial vaccine becoming available in some countries (Kvasnicka *et al.*, 1989, 1992). This vaccine does not prevent infection, but allows clearance of the infection from vaccinated females before the peak time of abortion risk (70 gestation days).

A purified surface antigen (TF1.17) of *T. foetus* has been used for the challenge of immunized heifers: vaccinated animals had significantly greater clearance rate than control animals (BonDurant *et al.*, 1993). Systemic or local (intranasal or intravaginal boosters after systemic priming) immunization with surface TF1.17 antigen primarily induced IgA and IgG1 responses in genital secretions, and IgG1 antibodies in serum (BonDurant *et al.*, 1993; Corbeil *et al.*, 1998, 2001).

The use of membrane vaccines in bovine females produced different results. Campero *et al.* (1999) observed a shorter time of infection in the vaccinated females, whereas Hudson *et al.* (1993a) mentioned a better performance of a whole-cell vaccine compared with another vaccine based on *T. foetus* fractions.

In spite of all the advances achieved, there is still insufficient information about the performance of many of these vaccines tested under natural challenge conditions. Cobo *et al.* (2002) tested two types of vaccines against *T. foetus* in bovine females naturally challenged by mating with an infected bull: one formalized with whole cells and the other with cellular membranes of *T. foetus*. Fetal mortality for those vaccinated was significantly different between heifers vaccinated with membranes and control heifers.

Dual vaccine

Since both bovine venereal diseases, tritrichomonosis and campylobacteriosis, are similar in their pathogenesis, prevalence and economic effect, the development of effective dual prophylactic methods has considerable demand, particularly in areas with extensive cattle raising and natural breeding (Campero *et al.*, 1987a, 2005). Because identification and separation of infected animals is difficult and AI – commonly recommended to avoid venereal disease – is impractical under most range conditions, vaccines against both diseases are relevant. The only vaccine for *T. foetus* and *C. fetus* commercially available was tested against only *T. foetus* (Kvasnicka *et al.*, 1989, 1992).

The efficacy of dual vaccines (containing *T. foetus* and *C. foetus* antigens) in heifers (one experimental and the other commercial), tested under natural simultaneous infection conditions with both microorganisms, has recently been determined (Cobo *et al.*, 2004b). Vaccinated heifers resisted or quickly cleared both pathogens, had a higher pregnancy rate and a higher systemic immune response during and after the breeding period. The experimental vaccine was more efficient than the commercial vaccine.

Bulls

Tritrichomonas foetus in bulls is asymptomatic, colonizing for years and predominantly in the preputial cavity. Pathogen-specific antibodies have been documented in preputial fluids (Campero *et al.*, 1990; Campero and Ladds, 1991; Rhyan *et al.*, 1999). However, the local mucosal immune response in infected bulls, coupled to a lack of inflammatory reactions, appears to be ineffective in eliminating the infection.

Various research projects related to bull vaccination have been performed (Clark *et al.*, 1983, 1984; Soto and Parma, 1989; Campero *et al.*, 1990; Herr *et al.*, 1991). Assays using killed whole-cell vaccine were unsuccessful (Herr *et al.*, 1991); they also failed in bulls > 5 years of age (Clark *et al.*, 1984). Therefore, commercial Trich Guard V5L vaccine is not indicated for bulls (BonDurant, 1997). From a practical point of view, up to now, no commercial vaccine has been validated for prevention of infection in the bull.

General Considerations

Whenever possible, the use of AI is an excellent practice for disease control. However, practically speaking, changing from natural service to AI has proved to be difficult in many situations. Therefore, a list of measures is presented here that add support to control of the problem.

Endemic areas

1. Culture all bulls again after the breeding season, allow a mimimum of 1 month of sexual rest prior to sampling for culture and, afterwards, cull all infected bulls.
2. As a rule, when the breeding season starts, all the cows will have their calves by their side: do not keep together first-service heifers with mature cows. Replacement virgin heifers should be examined prior to the breeding season to discover any possible pregnancies (infection risk).
3. Do not give a second opportunity to open females: pasture separately until culling.
4. Fences: the nature of venereal diseases renders this point very important. Secure fences are an insurance against the possible transmission between infected and non-infected herds.

5. Service in community pastures and renting or leasing bulls without three (or better, four) negative culture results are risk factors that should be avoided.
6. All new replacement bulls introduced to the herd – as well as those bulls from commercial bull studs – should have had two consecutive negative cultures.
7. Immunization: annual vaccination programme of all females against tritrichomonosis and campylobacteriosis (two injections followed by an annual booster). This applies predominantly to highly endemic areas.
8. An estimated low culture test sensitivity suggests that one single negative finding is not sufficient to demonstrate freedom from infection. Repeated preputial samples from bulls decreases the probability of false negative results. These limitations of the diagnostic test sensitivity should be considered when designing control and eradication programmes.
9. The recommendation of two and four consecutive negative cultures, depending on whether the bull comes from a negative or endemic herd, respectively, is required to certify uninfected status (Parker *et al.*, 1999).
10. In bulls used for AI, numerous negative cultures are required before the semen is declared free of *T. foetus*. Therefore, a new diagnostic tool based on reliable PCR is needed. A recent assay with high sensitivity and reduced sampling time has shown excellent diagnostic performances (Grahn *et al.*, 2005).

Non-endemic areas

Bulls used for AI residing in non-emdemic countries still require negative testing during quarantine. Two temporally independent negative testings are required. It is suggested that replacement of one of these culture tests by a highly sensitive PCR allows the investigation of pooled samples. This combined approach yielded better sensitivity and offers the opportunity of specifying identified organisms by sequencing.

References

Acuña, C.M. and Campero, C.M. (1999) Examen clínico reproductivo en 22.994 toros de raza de carne en la pampa húmeda de Argentina. *Revista Therios* 28, 63–75.

Anderson, M.L., BonDurant, R.H., Corbeil, R.R. and Corbeil, L.B. (1996) Immune and inflammatory responses to reproductive tract infection with *Tritrichomonas foetus* in immunized and control heifers. *Journal of Parasitology* 82, 594–600.

BonDurant, R.H. (1997) Pathogenesis, diagnosis and management of trichomoniasis in cattle. *The Veterinary Clinics of North America. Food Animal Practice* 13, 345–361.

BonDurant, R.H., Corbeil, R. and Corbeil, L.B. (1993) Vaccination of virgin cows with surface antigen TF1.17 of *Tritrichomonas foetus. Infection Immunity* 61, 1385–1394.

BonDurant, R.H., Gajadhar, A., Campero, C.M., Johnson, E., Lun, Z.N., Nordhausen, R.W., Van Hoosear, K.A., Villanueva, M.R. and Walker, R.L. (1999) Preliminary characterization of a *Tritrichomonas foetus*-like protozoan isolated from preputial smegma of virgin bulls. *The Bovine Practitioner* 33, 124–127.

Campero, C.M. (1999) Selección y manejo de los toros en rodeos de cría. *Revista de Medicina Veterinaria* 80, 58–65.

Campero, C.M. (2000) Las enfermedades reproductivas de los bovinos: ayer y hoy. *Anales de la Academia Nacional de Agronomía y Veterinaria* 53, 88–112.

Campero, C.M. and Ladds, P.W. (1991) Cuantificación de inmunoglobulinas en fluidos genitales y células contenedoras de inmunoglobulinas en el tracto genital de toros vacunados y desafiados con *Tritrichomonas foetus*. *Revista de Medicina Veterinaria* 72, 36–39.

Campero, C.M. and Palladino, M.R. (1983) Presencia de cepas de *Tritrichomonas foetus* quimioresistentes en Argentina. *Gaceta Veterinaria* 45, 899–909.

Campero, C.M., Ballabene, N., Cipolla, A. and Zamora, A. (1987a) Dual infection of bulls with campylobacteriosis and trichomoniasis: treatment with dimetridazole chlorhydrate. *Australian Veterinary Journal* 64, 320–321.

Campero, C.M., Catena, M. and Demayo, M. (1987b) Tratamiento de toros infectados con *Trichomonas foetus* resistentes en rodeos de cría de la Provincia de Buenos Aires, Argentina. *Veterinaria Argentina* 4, 234–240.

Campero, C.M., Hirst, R.G., Ladds, P.W., Vaughan, J.A., Emery, D.L. and Watson, D.L. (1990) Measurement of antibody in serum and genital fluids of bull by ELISA after vaccination and challenge with *Tritrichomonas foetus*. *Australian Veterinary Journal* 67, 175–178.

Campero, C.M., Cano, D., Pinilla, G. and García, J.P. (1993) Infección a *Tritrichomonas foetus* en toros como secuela de la prueba de capacidad de servicio sobre vaca infectada. *Veterinaria Argentina* 10, 164–168.

Campero, C.M., Medina, D., Rossetti, O., Marcovecchio, F., Cosentino, B., Marcone, J. and Carracino, M. (1998) Vacunación subcutánea e intravaginal contra tricomoniasis en vaquillonas. *Revista de Medicina Veterinaria* 79, 347–353.

Campero, C.M., Rossetti, O., Medina, D., Bretschneider, G. and Roppel, M. (1999) Inmunización en vaquillonas mediante vacuna de membrana de *Tritrichomonas foetus*. *Veterinaria Argentina* 154, 250–262.

Campero, C.M., Rodriguez Dubra, C., Bolondi, A., Cacciato, C., Cobo, E., Perez, S., Odeon, A., Cipolla, A. and BonDurant, R.H. (2003) Two-step (culture and PCR) diagnostic approach for differentiation of non-*T. foetus* trichomonads from genitalia of virgin beef bulls in Argentina. *Veterinary Parasitology* 112, 167–175.

Campero, C.M., Anderson, M.L., Walker, R.L., Blanchard, P.C., Barbano, L., Chiu, P., Martínez, A., Combessies, G., Bardon, J.C. and Cordeviola, J. (2005) Identification of *Campylobacter foetus* in natural cases of bovine and ovine abortions. *Veterinary Medicine Series B* 52, 138–141.

Christensen, H.R., Clark, B.L. and Parsonson, I.M. (1977) Incidence of *Tritrichomonas foetus* in young replacement bulls following introduction into an infected herd. *Australian Veterinary Journal* 53, 132–134.

Clark, B.L. (1971) Venereal diseases of cattle. *Veterinary Review, University of Sydney, Australia* 11, 5–2.

Clark, B.L., White, M.B. and Banfield, J.C. (1971) Diagnosis of *Trichomonas foetus* infection in bulls. *Australian Veterinary Journal* 47, 181.

Clark, B.L., Parsonson, I.M. and Dufty, J.H. (1974a) Experimental infection of bulls with *Trichomonas foetus. Australian Veterinary Journal* 50, 189–191.

Clark, B.L., Parsonson, I.M., White, M.B., Banfield, J.C. and Young, J.S. (1974b) Control of trichomoniasis in a large herd of beef cattle. *Australian Veterinary Journal* 50, 424–426.

Clark, B.L., Dufty, J.H. and Parsonson, I.M. (1983) Immunisation of bulls against trichomoniasis. *Australian Veterinary Journal* 60, 178–179.

Clark, B.L., Emery, D.L. and Dufty, J.H. (1984) Therapeutic immunisation of bulls with the membranes and glycoproteins of *Tritrichomonas foetus* var. brisbane. *Australian Veterinary Journal* 61, 65–66.

Cobo, E.R., Cano, D., Rossetti, O. and Campero, C.M. (2002) Heifers immunized with whole-cell and membrane vaccines against *Tritrichomonas foetus* and naturally challenged with an infected bull. *Veterinary Parasitology* 109, 169–184.

Cobo, E.R., Campero, C.M., Mariante, R.M. and Benchimol, M. (2003) Ultrastructural study of a tetratrichomonad species isolated from prepucial smegma of virgin bulls. *Veterinary Parasitology* 117, 195–211.

Cobo, E.R., Cantón, G., Morrell, E., Cano, D. and Campero, C.M. (2004a) Failure to established infection with *Tetratrichomonas* sp. in the reproductive tracts of heifers and bulls. *Veterinary Parasitology* 12, 145–150.

Cobo, E.R., Morsella, C., Cano, D., Cipolla, A. and Campero, C.M. (2004b) Immunization in heifers with dual vaccines containing *Tritrichomonas foetus* and *Campylobacter fetus* antigens using systemic and mucosal routes. *Theriogenology* 62, 1367–1382.

Corbeil, L.B. (1994) Vaccination strategies against *Tritrichomonas foetus. Parasitology Today* 10, 103–106.

Corbeil, L.B., Anderson, M.L., Corbeil, R.R., Eddow, J.M. and BonDurant, R.H. (1998) Female reproductive tract immunity in bovine trichomoniasis. *American Journal of Reproduction and Immunology* 39, 189–198.

Corbeil, L.B., Munson, L., Campero, C. and BonDurant, R.H. (2001) Bovine trichomoniasis as a model for development of vaccines against sexually transmitted disease. *American Journal of Reproduction and Immunology* 45, 310–319.

Dillon, J.H., Spath, E.J., Casaro, A.P., Cipolla, A.L., Ibarra, O., Crenovich, H. and Bianchi, M.M. (1995) Programa regional de control de enfermedades venéreas de bovinos (Plan Toros). *Revista Argentina de Producción Animal* 15, 764–767.

Gault, R.A., Kvasnicka, W.G., Hanks, D., Hanks, M. and Hall, M.R. (1995) Specific antibodies in serum and vaginal mucus of heifers inoculated with a vaccine containing *Tritrichomonas foetus. American Journal of Veterinary Research* 56, 454–459.

Gookin, J.L., Breitschwerdt, E.B., Levy, M.G., Gager, R.B. and Benrud, J.G. (2001) Diarrhea associated with trichomonosis in cats. *Journal of the American Veterinary Medical Association* 218, 194–195.

Grahn, R.A., BonDurant, R.H., van Hoosear, K.A., Walker, R.L. and Lyons, L.A. (2005) An improved molecular assay for *Tritrichomonas foetus. Veterinary Parasitology* 127, 33–41.

Herr, S., Ribeiro, L.M., Claassen, E. and Myburgh, J.G. (1991) A reduction in the duration of infection with *Tritrichomonas foetus* following vaccination in heifers and the failure to demonstrate a curative effect in infected bulls. *Onderstepoort Journal of Veterinary Research* 58, 41–45.

Hudson, D.B., Ball, L., Cheney, J.M., Mortimer, R.G., Bowen, B.A., Marsh, D.J. and Peetz, R.H. (1993a) Development and testing of a bovine trichomoniasis vaccine. *Theriogenology* 39, 929–935.

Hudson, D., Ball, L., Cheney, J., Mortimer, R., Bowen, B., Marsh, D. and Peetz, R. (1993b) Testing of trichomoniasis vaccine in heifers mated to infected bulls. *Theriogenology* 39, 937–943.

Jezierski, T.A., Koziorowski, M., Goszczynski, J. and Sieradzka, I.S.O. (1989) Homosexual and social behaviours of young bulls of different geno- and phenotypes and plasma concentrations of some hormones. *Applied Animal Behaviour Sciences* 24, 101–113.

Kendrick, J.W. (1953) An outbreak of bovine trichomoniasis in a group of bulls used for artificial insemination. *Cornell Veterinary* 43, 231–240.

Kimsey, P.B. (1986) Bovine Trichomoniasis. In: Morrow, D. (ed.) *Current Therapy in Theriogenology*, 2nd edn. Saunders, Philadelphia, pp. 275–279.

Kvasnicka, W.G., Taylor, R.E., Huang, J.C., Hanks, D., Tronstad, R.J., Bosomworth, A. and Hall, M.R. (1989) Investigations of the incidence of bovine trichomoniasis in Nevada and of the efficacy of immunizing cattle with vaccines containing *Tritrichomonas foetus. Theriogenology* 31, 963–971.

Kvasnicka, W.G., Hanks, D., Huang, J.C., Hall, M.R., Sandblom, D., Chu, H.J., Chavez, L. and Acree, W.M. (1992) Clinical evaluation of the efficacy of inoculating cattle with a vaccine containing *Tritrichomonas foetus. American Journal of Veterinary Research* 53, 2023–2027.

Lauriente, P., Campero, C.M., Palladino, M.R., Silva, E. and Villar, J.A. (1982) Resultados de dos tratamientos en toros infectados con cepas de *T. foetus* quimioresistentes. *Gaceta Veterinaria* 44, 695–698.

Levy, M.G., Gookin, J.L., Poore, M., Birkenheuer, A.J., Dykstra, M.J. and Litaker, R.W. (2003) *Tritrichomonas foetus* and not *Pentatrichomonas hominis* is the etiologic agent of feline trichomonal diarrhea. *Journal of Parasitology* 89, 99–104.

Mancebo, O.A., Russo, A.M., Carabajal, L.L. and Monzon, C.M. (1995) Persistence of *Tritrichomonas foetus* in naturally infected cows and heifers in Argentina. *Veterinary Parasitology* 59, 7–11.

Morgan, B.B. (1944) Studies on the trichomonad carrier-cow problem. *Journal of Animal Sciences* 3, 437.

Palladino, M.R., Campero, C.M. and Villar, J.A. (1982a) Resistencia de *Trichomonas foetus* a tricomonicidas. Parte I. Acción terapéutica in vitro de cuatro drogas sobre cepas resistentes y sensibles. *Gaceta Veterinaria* 44, 165–174.

Palladino, M.R., Campero, C.M. and Villar, J.A. (1982b) Resistencia de *Trichomonas foetus* a tricomonicidas. Parte II. Prueba de hamster. *Gaceta Veterinaria* 44, 400–406.

Palladino, M.R., Campero, C.M. and Acuña, C. (1983) Utilización de diversas drogas tricomonicidas en toros. *Gaceta Veterinaria* 45, 1289–1295.

Palladino, M.R. and Campero, C.M. (1985) Utilización de un tratamiento local en toros infectados con cepas de *Tritrichomonas foetus* resistentes. *Veterinaria Argentina* 2, 73–74.

Parker, S., Campbell, J., Ribble, C. and Gajadhar, A. (1999) Comparison of two sampling tools for diagnosis of *Tritrichomonas foetus* in bulls and clinical interpretation of culture results. *Journal of American Veterinary Medical Association* 215, 231–235.

Parker, S., Campbell, J., McIntosh, K. and Gajadhar, A. (2003) Diagnosis of trichomoniasis in 'virgin' bulls by culture and polymerase chain reaction. *Canadian Veterinary Journal* 44, 732–734.

Parsonson, I.M., Clark, B.L. and Dufty, J. (1974) The pathogenesis of *Tritrichomonas foetus* infection in the bull. *Australian Veterinary Journal* 50, 421–423.

Rae, O.D., Crews, J.E., Greiner, E.C. and Donovan, G.A. (2004) Epidemiology of *Tritrichomonas foetus* in beef bull populations in Florida. *Theriogenology* 61, 605–618.

Rhyan, J.C., Wilson, K.L., Wagner, B., Anderson, M.L., BonDurant, R.H., Burgess, D.E., Mutwiri, G.K. and Corbeil, L.B. (1999) Demonstration of *Tritrichomonas foetus* in the external genitalia and of specific antibodies in preputial secretions of naturally infected bulls. *Veterinary Pathology* 36, 406–411.

Skirrow, S. (1987) Identification of trichomonad-carrier cows. *Journal of the American Veterinary Medical Association* 191, 553–554.

Skirrow, S.Z. and BonDurant, R.H. (1988) Bovine trichomoniasis. *Veterinary Bulletin* 58, 591–603.

Skirrow, S.Z. and BonDurant, R.H. (1990a) Induced *Tritrichomonas foetus* infection in beef heifers. *Journal of the American Veterinary Medical Association* 196, 885–889.

Skirrow, S.Z. and BonDurant, R.H. (1990b) Immunoglobulin isotype of specific antibodies in reproductive tract secretions and sera in *Tritrichomonas foetus*-infected heifers. *American Journal of Veterinary Research* 51, 645–653.

Skirrow, S.Z., BonDurant, R.H., Farley, J. and Correa, J. (1985) Efficacy of ipronidazole against trichomoniasis in beef bulls. *Journal of the American Veterinary Medical Association* 187, 401–404.

Soto, P. and Parma, A.E. (1989) The immune response in cattle infected with *Tritrichomonas foetus*. *Veterinary Parasitology* 33, 343–348.

Index